C++ Neural Networks

and Fuzzy Logic

Second Edition

C++ NEURAL NETWORKS
AND FUZZY LOGIC

SECOND EDITION

VALLURU RAO

&

HAYAGRIVA RAO

MIS:
PRESS

First Edition—1993

Printed in the United States of America.

Library of Congress Cataloging-in-Publication Data

Rao, Valluru
    C++ neural networks and fuzzy logic 2nd ed./ Valluru B. Rao and
       Hayagriva V. Rao.
          pm.   cm.
    ISBN 1-55851-552-6
    1. Neural networks (computer science).   2. Fuzzy systems.
    3. C++ (computer program language).   I. Title.   II. Rao, Hayagriva V.
QA76.87 .R36 1995
006.3--dc20   95-040347
                                                             CIP

10 9 8 7 6 5 4 3 2 1

Editor-in-Chief: Paul Farrell

Copy Edit Manager: Shari Chappell

Production Editor: Maya Riddick

Managing Editor: Cary Sullivan

Copy Editor: Suzanne Ingrao

Development Editors: Laura Lewin,
Andy Neusner

Assoc. Production Editor: Brian Oxman

DEDICATION

To the memory of

Vydehamma, Annapurnamma, Anandarao, Madanagopalarao, Govindarao, and Rajyalakshamma.

ACKNOWLEDGMENTS

We thank everyone at MIS:Press/Henry Holt and Co. who has been associated with this project for their diligence and support, namely, the Technical Editors of this edition and the first edition for their suggestions and feedback; Laura Lewin, the Editor, and all of the other people at MIS:Press for making the book a reality.

We would also like to thank Dr. Tarek Kaylani for his helpful suggestions, Professor R. Haskell, and our other readers who wrote to us, and Dr. V. Rao's students whose suggestions were helpful. Please E-mail us more feedback!

Finally, thanks to Sarada and Rekha for encouragement and support. Most of all, thanks to Rohini and Pranav for their patience and understanding through many lost evenings and weekends.

Contents

CHAPTER 5: A SURVEY OF NEURAL NETWORK MODELS   81

X

CHAPTER 6: LEARNING AND TRAINING 109

XII

XIII

CHAPTER 9: FAM: FUZZY ASSOCIATIVE MEMORY   217

CHAPTER 10: ADAPTIVE RESONANCE THEORY (ART)   243

CHAPTER 11: THE KOHONEN SELF-ORGANIZING MAP   271

CHAPTER 12: APPLICATION TO PATTERN RECOGNITION   305

CHAPTER 13: BACKPROPAGATION II   325

CHAPTER 14: APPLICATION TO FINANCIAL FORECASTING   377

CHAPTER 15: APPLICATION TO NONLINEAR OPTIMIZATION 417

Chapter 16: Applications of Fuzzy Logic   473

CHAPTER 17: FURTHER APPLICATIONS   511

xx

PREFACE

The number of models available in neural network literature is quite large. Very often the treatment is mathematical and complex. This book provides illustrative examples in C++ that the reader can use as a basis for further experimentation. A key to learning about neural networks to appreciate their inner workings is to experiment. Neural networks, in the end, are fun to learn about and discover. Although the language for description used is C++, you will not find extensive class libraries in this book. With the exception of the backpropagation simulator, you will find fairly simple example programs for many different neural network architectures and paradigms. Since backpropagation is widely used and also easy to tame, a simulator is provided with the capacity to handle large input data sets. You use the simulator in one of the chapters in this book to solve a financial forecasting problem. You will find ample room to expand and experiment with the code presented in this book.

There are many different angles to neural networks and fuzzy logic. The fields are expanding rapidly with ever-new results and applications. This book presents many of the different neural network topologies, including the BAM, the Perceptron, Hopfield memory, ART1, Kohonen's Self-Organizing map, Kosko's Fuzzy Associative memory, and, of course, the Feedforward Backpropagation network (aka Multilayer Perceptron). You should get a fairly broad picture of neural networks and fuzzy logic with this book. At the same time, you will have real code that shows you example usage of the models, to solidify your understanding. This is especially useful for the more complicated neural network architectures like the Adaptive Resonance Theory of Stephen Grossberg (ART).

The subjects are covered as follows:

- Chapter 1 gives you an overview of neural network terminology and nomenclature. You discover that neural nets are capable of solving complex problems with parallel computational architectures. The Hopfield network and feedforward network are introduced in this chapter.
- Chapter 2 introduces C++ and object orientation. You learn the benefits of object-oriented programming and its basic concepts.

- Chapter 3 introduces fuzzy logic, a technology that is fairly synergistic with neural network problem solving. You learn about math with fuzzy sets as well as how you can build a simple fuzzifier in C++.

- Chapter 4 introduces you to two of the simplest, yet very representative, models of: the Hopfield network, the Perceptron network, and their C++ implementations.

- Chapter 5 is a survey of neural network models. This chapter describes the features of several models, describes **threshold** functions, and develops concepts in neural networks.

- Chapter 6 focuses on learning and training paradigms. It introduces the concepts of supervised and unsupervised learning, self-organization and topics including backpropagation of errors, radial basis function networks, and conjugate gradient methods.

- Chapter 7 goes through the construction of a backpropagation simulator. You will find this simulator useful in later chapters also. C++ classes and features are detailed in this chapter.

- Chapter 8 covers the Bidirectional Associative memories for associating pairs of patterns.

- Chapter 9 introduces Fuzzy Associative memories for associating pairs of fuzzy sets.

- Chapter 10 covers the Adaptive Resonance Theory of Grossberg. You will have a chance to experiment with a program that illustrates the working of this theory.

- Chapters 11 and 12 discuss the Self-Organizing map of Teuvo Kohonen and its application to pattern recognition.

- Chapter 13 continues the discussion of the backpropagation simulator, with enhancements made to the simulator to include momentum and noise during training.

- Chapter 14 applies backpropagation to the problem of financial forecasting, discusses setting up a backpropagation network with 15 input variables and 200 test cases to run a simulation. The problem is approached via a systematic 12-step approach for preprocessing data and setting up the problem. You will find a number of examples of financial forecasting highlighted from the literature. A resource guide for neural networks in finance is included for people who would like more information about this area.

- Chapter 15 deals with nonlinear optimization with a thorough discussion of the Traveling Salesperson problem. You learn the formulation by Hopfield and the approach of Kohonen.

- Chapter 16 treats two application areas of fuzzy logic: fuzzy control systems and fuzzy databases. This chapter also expands on fuzzy relations and fuzzy set theory with several examples.
- Chapter 17 discusses some of the latest applications using neural networks and fuzzy logic.

In this second edition, we have followed readers' suggestions and included more explanations and material, as well as updated the material with the latest information and research. We have also corrected errors and omissions from the first edition.

Neural networks are now a subject of interest to professionals in many fields, and also a tool for many areas of problem solving. The applications are widespread in recent years, and the fruits of these applications are being reaped by many from diverse fields. This methodology has become an alternative to modeling of some physical and nonphysical systems with scientific or mathematical basis, and also to expert systems methodology. One of the reasons for it is that absence of full information is not as big a problem in neural networks as it is in the other methodologies mentioned earlier. The results are sometimes astounding, even phenomenal, with neural networks, and the effort is at times relatively modest to achieve such results. Image processing, vision, financial market analysis, and optimization are among the many areas of application of neural networks. To think that the modeling of neural networks is one of modeling a system that attempts to mimic human learning is somewhat exciting. Neural networks can learn in an unsupervised learning mode. Just as human brains can be trained to master some situations, neural networks can be trained to recognize patterns and to do optimization and other tasks.

In the early days of interest in neural networks, the researchers were mainly biologists and psychologists. Serious research now is done by not only biologists and psychologists, but by professionals from computer science, electrical engineering, computer engineering, mathematics, and physics as well. The latter have either joined forces, or are doing independent research parallel with the former, who opened up a new and promising field for everyone.

In this book, we aim to introduce the subject of neural networks as directly and simply as possible for an easy understanding of the methodology. Most of the important neural network architectures are covered, and we earnestly hope that our efforts have succeeded in presenting this subject matter in a clear and useful fashion.

We welcome your comments and suggestions for this book, from errors and oversights, to suggestions for improvements to future printings at the following E-mail addresses:

V. Rao   rao@cse.bridgeport.edu

H. Rao   ViaSW@aol.com

CHAPTER 1

INTRODUCTION TO NEURAL NETWORKS

NEURAL PROCESSING

How do you recognize a face in a crowd? How does an economist predict the direction of interest rates? Faced with problems like these, the human brain uses a web of interconnected processing elements called *neurons* to process information. Each neuron is autonomous and independent; it does its work *asynchronously*, that is, without any synchronization to other events taking place. The two problems posed, namely recognizing a face and forecasting interest rates, have two important characteristics that distinguish them from other problems: First, the problems are complex, that is, you can't devise a *simple* step-by-step *algorithm* or precise formula to give you an answer; and second, the data provided to resolve the problems is equally complex and may be noisy or incomplete. You could have forgotten your glasses when you're trying to recognize that face. The economist may have at his or her disposal thousands of pieces of data that may or may not be relevant to his or her forecast on the economy and on interest rates.

The vast processing power inherent in biological neural structures has inspired the study of the structure itself for hints on organizing human-made computing structures. *Artificial neural networks*, the subject of this book, covers the way to organize synthetic neurons to solve the same kind of difficult, complex problems in a similar manner as we think the human brain may. This chapter will give you a sampling of the terms and nomenclature used to talk about neural networks. These terms will be covered in more depth in the chapters to follow.

NEURAL NETWORK

A *neural network* is a computational structure inspired by the study of biological neural processing. There are many different types of neural networks, from relatively simple to very complex, just as there are many theories on how biological neural processing works. We will begin with a discussion of a *layered feed-forward* type of neural network and branch out to other paradigms later in this chapter and in other chapters.

A layered feed-forward neural network has *layers,* or subgroups of processing elements. A layer of processing elements makes independent computations on data that it receives and passes the results to another layer. The next layer may in turn make its independent computations and pass on the results to yet another layer. Finally, a subgroup of one or more processing elements determines the output from the network. Each processing element makes its computation based upon a weighted sum of its inputs. The first layer is the *input layer* and the last the *output layer.* The layers that are placed between the first and the last layers are the *hidden layers.* The processing elements are seen as units that are similar to the neurons in a human brain, and hence, they are referred to as *cells, neuromimes,* or *artificial neurons.* A *threshold function* is sometimes used to qualify the output of a neuron in the output layer. Even though our subject matter deals with artificial neurons, we will simply refer to them as neurons. Synapses between neurons are referred to as *connections,* which are represented by edges of a directed graph in which the nodes are the artificial neurons.

Figure 1.1 is a layered feed-forward neural network. The circular nodes represent neurons. Here there are three layers, an input layer, a hidden layer, and an output layer. The directed graph mentioned shows the connections from nodes from a given layer to other nodes in other layers. Throughout this book you will see many variations on the number and types of layers.

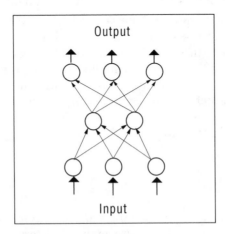

FIGURE 1.1 A TYPICAL NEURAL NETWORK.

OUTPUT OF A NEURON

Basically, the *internal activation* or raw output of a neuron in a neural network is a weighted sum of its inputs, but a threshold function is also used to determine the final value, or the output. When the output is 1, the neuron is said to *fire*, and when it is 0, the neuron is considered not to have fired. When a threshold function is used, different results of activations, all in the same interval of values, can cause the same final output value. This situation helps in the sense that, if precise input causes an activation of 9 and noisy input causes an activation of 10, then the output works out the same as if noise is filtered out.

To put the description of a neural network in a simple and familiar setting, let us describe an example about a daytime game show on television, *The Price is Right*.

CASH REGISTER GAME

A contestant in *The Price is Right* is sometimes asked to play the *Cash Register Game*. A few products are described, their prices are unknown to the contestant, and the contestant has to declare how many units of each item he or she would like to (pretend to) buy. If the total purchase does not exceed the amount specified, the contestant wins a special prize. After the contestant announces how many items of a particular product he or she wants, the price of that product is revealed, and it is rung up on the cash register. The contestant must be careful, in this case, that the total does not exceed some nominal value, to earn the associated prize. We can now cast the whole operation of this game, in terms of a neural network, called a *Perceptron*, as follows.

Consider each product on the shelf to be a neuron in the input layer, with its input being the unit price of that product. The cash register is the single neuron in the output layer. The only connections in the network are between each of the neurons (products displayed on the shelf) in the input layer and the output neuron (the cash register). This arrangement is usually referred to as a neuron, the cash register in this case, being an *instar* in neural network terminology. The contestant actually determines these connections, because when the contestant says he or she wants, say five, of a specific product, the contestant is thereby assigning a weight of 5 to the connection between that product and the cash register. The total bill for the purchases by the contestant is nothing but the weighted sum of the unit prices of the different products offered. For those items the contestant does not choose to purchase, the implicit weight assigned is 0. The application of the dollar limit to the bill is just the application of a threshold, except that the threshold value should not be exceeded for the outcome from this network to favor the contestant, winning him or her a good prize. In a Perceptron, the way the threshold works is that an output neuron is supposed to fire if its activation value *exceeds* the threshold value.

WEIGHTS

The weights used on the connections between different layers have much significance in the working of the neural network and the characterization of a network. The following actions are possible in a neural network:

1. Start with one set of weights and run the network. (NO TRAINING)
2. Start with one set of weights, run the network, and modify some or all the weights, and run the network again with the new set of weights. Repeat this process until some predetermined goal is met. (TRAINING)

TRAINING

Since the output(s) may not be what is expected, the weights may need to be altered. Some rule then needs to be used to determine how to alter the weights. There should also be a criterion to specify when the process of successive modification of weights ceases. This process of changing the weights, or rather, updating the weights, is called *training*. A network in which learning is employed is said to be subjected to *training*. Training is an external process or regimen. Learning is the desired process that takes place internal to the network.

FEEDBACK

If you wish to train a network so it can recognize or identify some predetermined patterns, or evaluate some function values for given arguments, it would be important to have information fed back from the output neurons to neurons in some layer before that, to enable further processing and adjustment of weights on the connections. Such feedback can be to the input layer or a layer between the input layer and the output layer, sometimes labeled the *hidden layer*. What is fed back is usually the error in the output, modified appropriately according to some useful paradigm. The process of feedback continues through the subsequent cycles of operation of the neural network and ceases when the training is completed.

SUPERVISED OR UNSUPERVISED LEARNING

A network can be subject to *supervised* or *unsupervised* learning. The learning would be supervised if external criteria are used and matched by the network output, and if not, the learning is unsupervised. This is one broad way to divide different neural network approaches. Unsupervised approaches are also termed *self-organizing*. There is more interaction between neurons, typically with feedback and intralayer connections between neurons promoting self-organization.

Supervised networks are a little more straightforward to conceptualize than unsupervised networks. You apply the inputs to the supervised network along with an expected response, much like the Pavlovian conditioned stimulus and response regimen. You mold the network with stimulus-response pairs. A stock market forecaster may present economic data (the *stimulus*) along with metrics of stock market performance (the *response*) to the neural network to the present and attempt to predict the future once training is complete.

You provide unsupervised networks with only stimulus. You may, for example, want an unsupervised network to correctly classify parts from a conveyor belt into part numbers, providing an image of each part to do the classification (the stimulus). The unsupervised network in this case would act like a look-up memory that is indexed by its contents, or a *Content-Addressable-Memory (CAM)*.

NOISE

Noise is perturbation, or a deviation from the actual. A data set used to train a neural network may have inherent noise in it, or an image may have random speckles in it, for example. The response of the neural network to noise is an important factor in determining its suitability to a given application. In the process of training, you may apply a metric to your neural network to see how well the network has learned your training data. In cases where the metric stabilizes to some meaningful value, whether the value is acceptable to you or not, you say that the network *converges*. You may wish to introduce noise intentionally in training to find out if the network can learn in the presence of noise, and if the network can converge on noisy data.

6

MEMORY

Once you train a network on a set of data, suppose you continue training the network with new data. Will the network forget the intended training on the original set or will it remember? This is another angle that is approached by some researchers who are interested in preserving a network's *long-term memory (LTM)* as well as its *short-term memory (STM)*. Long-term memory is memory associated with learning that persists for the long term. Short-term memory is memory associated with a neural network that decays in some time interval.

CAPSULE OF HISTORY

You marvel at the capabilities of the human brain and find its ways of processing information unknown to a large extent. You find it awesome that very complex situations are discerned at a far greater speed than what a computer can do.

Warren McCulloch and Walter Pitts formulated in 1943 a model for a nerve cell, a neuron, during their attempt to build a theory of self-organizing systems. Later, Frank Rosenblatt constructed a Perceptron, an arrangement of processing elements representing the nerve cells into a network. His network could recognize simple shapes. It was the advent of different models for different applications.

Those working in the field of artificial intelligence (AI) tried to hypothesize that you can model thought processes using some symbols and some rules with which you can transform the symbols.

A limitation to the symbolic approach is related to how knowledge is representable. A piece of information is localized, that is, available at one location, perhaps. It is not distributed over many locations. You can easily see that distributed knowledge leads to a faster and greater inferential process. Information is less prone to be damaged or lost when it is distributed than when it is localized. Distributed information processing can be fault tolerant to some degree, because there are multiple sources of knowledge to apply to a given problem. Even if one source is cut off or destroyed, other sources may still permit solution to a problem. Further, with subsequent learning, a solution may be remapped into a new organization of distributed processing elements that exclude a faulty processing element.

In neural networks, information may impact the activity of more than one neuron. Knowledge is distributed and lends itself easily to parallel computation. Indeed there are many research activities in the field of hardware design of neural network processing engines that exploit the parallelism of the neural network paradigm. Carver Mead, a pioneer in the field, has suggested analog VLSI (very large scale integration) circuit implementations of neural networks.

NEURAL NETWORK CONSTRUCTION

There are three aspects to the construction of a neural network:

1. **Structure**—the architecture and topology of the neural network
2. **Encoding**—the method of changing weights
3. **Recall**—the method and capacity to retrieve information

Let's cover the first one—*structure*. This relates to how many layers the network should contain, and what their functions are, such as for input, for output, or for feature extraction. Structure also encompasses how interconnections are made between neurons in the network, and what their functions are.

The second aspect is *encoding*. Encoding refers to the paradigm used for the determination of and changing of weights on the connections between neurons. In the case of the multilayer feed-forward neural network, you initially can define weights by randomization. Subsequently, in the process of training, you can use the *backpropagation algorithm*, which is a means of updating weights starting from the output backwards. When you have finished training the multilayer feed-forward neural network, you are finished with encoding since weights do not change after training is completed.

Finally, *recall* is also an important aspect of a neural network. Recall refers to getting an expected output for a given input. If the same input as before is presented to the network, the same corresponding output as before should result. The type of recall can characterize the network as being *autoassociative* or *heteroassociative*. Autoassociation is the phenomenon of associating an input vector with itself as the output, whereas heteroassociation is that of recalling a related vector given an input vector. You have a fuzzy remembrance of a phone number. Luckily, you stored it in an autoassociative neural network. When you apply the fuzzy remembrance, you retrieve the actual phone number. This is a use of autoassociation. Now if you want the individual's name associated with a given phone number, that would require heteroassociation. Recall is closely related to the concepts of STM and LTM introduced earlier.

The three aspects to the construction of a neural network mentioned above essentially distinguish between different neural networks and are part of their design process.

Sample Applications

One application for a neural network is *pattern classification,* or *pattern matching*. The patterns can be represented by binary digits in the discrete cases, or real numbers representing analog signals in continuous cases. Pattern classification is a form of establishing an autoassociation or heteroassociation. Recall that associating different patterns is building the type of association called heteroassociation. If you input a corrupted or modified pattern A to the neural network, and receive the true pattern A, this is termed autoassociation. What use does this provide? Remember the example given at the beginning of this chapter. In the human brain example, say you want to recall a face in a crowd and you have a hazy remembrance (input). What you want is the actual image. Autoassociation, then, is useful in recognizing or retrieving patterns with possibly incomplete information as input. What about heteroassociation? Here you associate A with B. Given A, you get B and sometimes vice versa. You could store the face of a person and retrieve it with the person's name, for example. It's quite common in real circumstances to do the opposite, and sometimes not so well. You recall the face of a person, but can't place the name.

Qualifying for a Mortgage

Another sample application, which is in fact in the works by a U.S. government agency, is to devise a neural network to produce a quick response credit rating of an individual trying to qualify for a mortgage. The problem to date with the application process for a mortgage has been the staggering amount of paperwork and filing details required for each application. Once information is gathered, the response time for knowing whether or not your mortgage is approved has typically taken several weeks. All of this will change. The proposed neural network system will allow the complete application and approval process to take three hours, with approval coming in five minutes of entering all of the information required. You enter in the applicant's employment history, salary information, credit information, and other factors and apply these to a trained neural network. The neural network, based on prior training on thousands of case histories, looks for patterns in the applicant's profile and then produces a yes or no rating of worthiness to carry a particular mortgage. Let's now continue our discussion of factors that distinguish neural network models from each other.

COOPERATION AND COMPETITION

We will now discuss *cooperation* and *competition*. Again we start with an example feed forward neural network. If the network consists of a single input layer and an output layer consisting of a single neuron, then the set of weights for the connections between the input layer neurons and the output neuron are given in a *weight vector*. For three inputs and one output, this could be $\mathbf{W} = \{w_1, w_2, w_3\}$. When the output layer has more than one neuron, the output is not just one value but is also a vector. In such a situation each neuron in one layer is connected to each neuron in the next layer, with weights assigned to these interconnections. Then the weights can all be given together in a two-dimensional *weight matrix*, which is also sometimes called a *correlation matrix*. When there are in-between layers such as a hidden layer or a so-called Kohonen layer or a Grossberg layer, the interconnections are made between each neuron in one layer and every neuron in the next layer, and there will be a corresponding correlation matrix. *Cooperation* or *competition* or both can be imparted between network neurons in the same layer, through the choice of the right sign of weights for the connections. Cooperation is the attempt between neurons in one neuron aiding the prospect of firing by another. Competition is the attempt between neurons to individually excel with higher output. *Inhibition*, a mechanism used in competition, is the attempt between neurons in one neuron decreasing the prospect of another neuron's firing. As already stated, the vehicle for these phenomena is the connection weight. For example, a positive weight is assigned for a connection between one node and a cooperating node in that layer, while a negative weight is assigned to inhibit a competitor.

To take this idea to the connections between neurons in consecutive layers, we would assign a positive weight to the connection between one node in one layer and its nearest neighbor node in the next layer, whereas the connections with distant nodes in the other layer will get negative weights. The negative weights would indicate competition in some cases and inhibition in others. To make at least some of the discussion and the concepts a bit clearer, we preview two example neural networks (there will be more discussion of these networks in the chapters that follow): the feed-forward network and the Hopfield network.

EXAMPLE—A FEED-FORWARD NETWORK

A sample feed-forward network, as shown in Figure 1.2, has five neurons arranged in three layers: two neurons (labeled x_1 and x_2) in layer 1, two neurons (labeled x_3 and x_4) in layer 2, and one neuron (labeled x_5) in layer 3. There are arrows connecting the neurons together. This is the direction of information flow. A feed-forward network has information flowing forward only. Each arrow that connects neurons has a *weight* associated with it (like, w_{31} for example). You calculate the *state,* **x,** of each neuron by summing the weighted values that flow into a neuron. The state of the neuron is the output value of the neuron and remains the same until the neuron receives new information on its inputs.

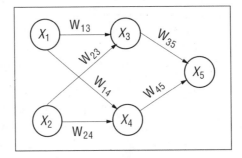

FIGURE 1.2   A FEED-FORWARD NEURAL NETWORK WITH TOPOLOGY 2-2-1.

For example, for $\mathbf{x_3}$ and $\mathbf{x_5}$:

$$\mathbf{X_3} = W_{23}\,\mathbf{X_2} + W_{13}\,\mathbf{X_1}$$
$$\mathbf{X_5} = W_{35}\,\mathbf{X_3} + W_{45}\,\mathbf{X_4}$$

We will formalize the equations in Chapter 7, which details one of the training algorithms for the feed-forward network called *Backpropagation.*

Note that you present information to this network at the leftmost nodes (layer 1) called the *input layer.* You can take information from any other layer in the network, but in most cases do so from the rightmost node(s), which make up the *output layer.* Weights are usually determined by a supervised training algorithm, where you present examples to the network and adjust weights appropriately to achieve a desired response. Once you have completed training, you can use the network without changing weights, and note the response for inputs that you apply. Note that a detail not yet shown is a nonlinear scaling function that limits the range of the weighted sum. This scaling function has the effect of clipping very large values in positive and negative directions for each neuron so that the cumulative summing that occurs across the network stays within reasonable bounds. Typical real number ranges for neuron inputs and outputs are –1 to +1 or 0 to +1. You will see more about this network and applications for it in Chapter 7. Now let us contrast this neural network with a completely different type of neural network, the Hopfield network, and present some simple applications for the *Hopfield network.*

EXAMPLE—A HOPFIELD NETWORK

The neural network we present is a Hopfield network, with a single layer. We place, in this layer, four neurons, each connected to the rest, as shown in Figure 1.3. Some of the connections have a positive weight, and the rest have a negative weight. The network will be presented with two input patterns, one at a time, and it is supposed to recall them. The inputs would be binary patterns having in each component a 0 or 1. If two patterns of equal length are given and are treated as vectors, their *dot product* is obtained by first multiplying corresponding components together and then adding these products. Two vectors are said to be *orthogonal,* if their dot product is 0. The mathematics involved in computations done for neural networks include matrix multiplication, transpose of a matrix, and transpose of a vector. Also see Appendix B. The inputs (which are stable, stored patterns) to be given should be orthogonal to one another.

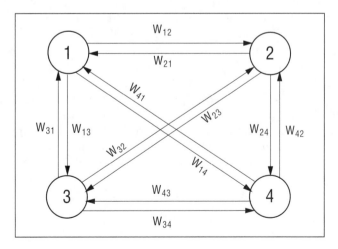

FIGURE 1.3 LAYOUT OF A HOPFIELD NETWORK.

The two patterns we want the network to recall are $\mathbf{A} = (1, 0, 1, 0)$ and $\mathbf{B} = (0, 1, 0, 1)$, which you can verify to be orthogonal. Recall that two vectors $\mathbf{A}$ and $\mathbf{B}$ are orthogonal if their dot product is equal to zero. This is true in this case since

$$A_1B_1 + A_2 B_2 + A_3B_3 + A_4B_4 = (1\text{x}0 + 0\text{x}1 + 1\text{x}0 + 0\text{x}1) = 0$$

The following matrix W gives the weights on the connections in the network.

$$W = \begin{matrix} 0 & -3 & 3 & -3 \\ -3 & 0 & -3 & 3 \\ 3 & -3 & 0 & -3 \\ -3 & 3 & -3 & 0 \end{matrix}$$

We need a threshold function also, and we define it as follows. The threshold value θ is 0.

$$f(t) = \begin{cases} 1 & \text{if } t >= \theta \\ 0 & \text{if } t < \theta \end{cases}$$

We have four neurons in the only layer in this network. We need to compute the activation of each neuron as the weighted sum of its inputs. The activation at the first node is the dot product of the input vector and the first column of the weight matrix (0 -3 3 -3). We get the activation at the other nodes similarly. The output of a neuron is then calculated by evaluating the threshold function at the activation of the neuron. So if we present the input vector $\mathbf{A}$, the dot product works out to 3 and f(3) = 1. Similarly, we get the dot products of the second, third, and fourth nodes to be –6, 3, and –6, respectively. The corresponding outputs therefore are 0, 1, and 0. This means that the output of the network is the vector (1, 0, 1, 0), same as the input pattern. The network has recalled the pattern as presented, or we can say that pattern $\mathbf{A}$ is stable, since the output is equal to the input. When $\mathbf{B}$ is presented, the dot product obtained at the first node is –6 and the output is 0. The outputs for the rest of the nodes taken together with the output of the first node gives (0, 1, 0, 1), which means that the network has stable recall for $\mathbf{B}$ also.

In Chapter 4, a method of determining the weight matrix for the Hopfield network given a set of input vectors is presented.

NOTE

So far we have presented easy cases to the network—vectors that the Hopfield network was specifically designed (through the choice of the weight matrix) to recall. What will the network give as output if we present a pattern different from both **A** and **B**? Let **C** = (0, 1, 0, 0) be presented to the network. The activations would be –3, 0, –3, 3, making the outputs 0, 1, 0, 1, which means that **B** achieves stable recall. This is quite interesting. Suppose we did intend to input **B** and we made a slight error and ended up presenting **C**, instead. The network did what we wanted and recalled **B**. But why not **A**? To answer this, let us ask is **C** closer to **A** or **B**? How do we compare? We use the distance formula for two four-dimensional points. If (a, b, c, d) and (e, f, g, h) are two four-dimensional points, the distance between them is:

$$\sqrt{[(a-e)^2 + (b-f)^2 + (c-g)^2 + (d-h)^2]}$$

The distance between **A** and **C** is $\sqrt{3}$, whereas the distance between **B** and **C** is just 1. So since **B** is closer in this sense, **B** was recalled rather than **A**. You may verify that if we do the same exercise with **D** = (0, 0, 1, 0), we will see that the network recalls **A**, which is closer than **B** to **D**.

Hamming Distance

When we talk about closeness of a bit pattern to another bit pattern, the *Euclidean distance* need not be considered. Instead, the *Hamming distance* can be used, which is much easier to determine, since it is the number of bit positions in which the two patterns being compared differ. Patterns being strings, the Hamming distance is more appropriate than the Euclidean distance.

NOTE   The weight matrix W we gave in this example is not the only weight matrix that would enable the network to recall the patterns A and B correctly. You can see that if we replace each of 3 and –3 in the matrix by say, 2 and –2, respectively, the resulting matrix would also facilitate the same performance from the network. For more details, consult Chapter 4.

Asynchronous Update

The Hopfield network is a *recurrent* network. This means that outputs from the network are fed back as inputs. This is not apparent from Figure 1.3, but is clearly seen from Figure 1.4.

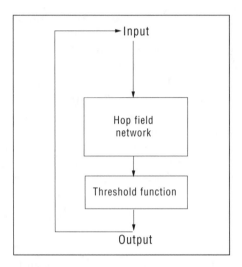

FIGURE 1.4 FEEDBACK IN THE HOPFIELD NETWORK.

The Hopfield network always stabilizes to a fixed point. There is a very important detail regarding the Hopfield network to achieve this stability. In the examples thus far, we have not had a problem getting a stable output from the network, so we have not presented this detail of network operation. This detail is the need to update the network *asynchronously*. This means that changes do not occur simultaneously to outputs that are fed back as inputs, but rather occur for one vector component at a time. The true operation of the Hopfield network follows the procedure below for input vector **Invec** and output vector **Outvec:**

1. Apply an input, Invec, to the network, and initialize **Outvec = Invec**

2. Start with i = 1

3. Calculate $Value_i$ = DotProduct ($Invec_i$, $Column_i$ of Weight matrix)

4. Calculate **Outvec$_i$** = **f(Value$_i$)** where **f** is the threshold function discussed previously

5. Update the input to the network with component **Outvec$_i$**

6. Increment i, and repeat steps 3, 4, 5, and 6 until **Invec = Outvec** (note that when i reaches its maximum value, it is then next reset to 1 for the cycle to continue)

Now let's see how to apply this procedure. Building on the last example, we now input **E** = (1, 0, 0, 1), which is at an equal distance from **A** and **B**.

Without applying the asynchronous procedure above, but instead using the shortcut procedure we've been using so far, you would get an output **F** = (0, 1, 1, 0). This vector, **F**, as subsequent input would result in **E** as the output. This is incorrect since the network oscillates between two states. We have updated the entire input vector synchronously.

Now let's apply asynchronous update. For input **E**, (1,0,0,1) we arrive at the following results detailed for each update step, in Table 1.1.

TABLE 1.1 EXAMPLE OF ASYNCHRONOUS UPDATE FOR THE HOPFIELD NETWORK

Step	i	Invec	Column of Weight vector	Value	Outvec	notes
0		1001			1001	initialization : set Outvec = Invec = Input pattern
1	1	1001	0 -3 3 -3	-3	0001	column 1 of Outvec changed to 0
2	2	0001	-3 0 -3 3	3	0101	column 2 of Outvec changed to 1
3	3	0101	3 -3 0 -3	-6	0101	column 3 of Outvec stays as 0
4	4	0101	-3 3 -3 0	3	0101	column 4 of Outvec stays as 1
5	1	0101	0 -3 3 -3	-6	0101	column 1 stable as 0
6	2	0101	-3 0 -3 3	3	0101	column 2 stable as 1
7	3	0101	3 -3 0 -3	-6	0101	column 3 stable as 0
8	4	0101	-3 3 -3 0	3	0101	column 4 stable as 1; stable recalled pattern = 0101

Binary and Bipolar Inputs

Two types of inputs that are used in neural networks are *binary* and *bipolar* inputs. We have already seen examples of binary input. Bipolar inputs have one of two values, 1 and −1. There is clearly a one-to-one mapping or correspondence between them, namely having -1 of bipolar correspond to a 0 of binary. In determining the weight matrix in some situations where binary strings are the inputs, this mapping is used, and when the output appears in bipolar values, the inverse transformation is applied to get the corresponding binary string. A simple example would be that the binary string 1 0 0 1 is mapped onto the bipolar string 1 −1 −1 1; while using the inverse transformation on the bipolar string −1 1 −1 −1, we get the binary string 0 1 0 0.

Bias

The use of threshold value can take two forms. One we showed in the example. The activation is compared to the threshold value, and the neuron fires if the threshold value is attained or exceeded. The other way is to add a value to the activation itself, in which case it is called the *bias*, and then determining the output of the neuron. We will encounter bias and *gain* later.

ANOTHER EXAMPLE FOR THE HOPFIELD NETWORK

You will see in Chapter 12 an application of *Kohonen's feature map* for pattern recognition. Here we give an example of pattern association using a Hopfield network. The patterns are some characters. A pattern representing a character becomes an input to a Hopfield network through a bipolar vector. This bipolar vector is generated from the pixel (picture element) grid for the character, with an assignment of a 1 to a black pixel and a -1 to a pixel that is white. A grid size such as 5x7 or higher is usually employed in these approaches. The number of pixels involved will then be 35 or more, which determines the dimension of a bipolar vector for the character pattern.

We will use, for simplicity, a 3x3 grid for character patterns in our example. This means the Hopfield network has 9 neurons in the only layer in the network. Again for simplicity, we use two *exemplar patterns*, or reference patterns, which are given in Figure 1.5. Consider the pattern on the left as a representation of the character "plus", +, and the one on the right that of "minus", - .

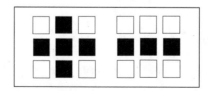

FIGURE 1.5 THE "PLUS" PATTERN AND "MINUS" PATTERN.

The bipolar vectors that represent the characters in the figure, reading the character pixel patterns row by row, left to right, and top to bottom, with a 1 for black and -1 for white pixels, are C+ = (-1, 1, -1, 1, 1, 1, -1, 1, -1), and C- = (-1, -1, -1, 1, 1, 1, -1, -1, -1). The weight matrix W is:

$$W = \begin{matrix}
0 & 0 & 2 & -2 & -2 & -2 & 2 & 0 & 2 \\
0 & 0 & 0 & 0 & 0 & 0 & 0 & 2 & 0 \\
2 & 0 & 0 & -2 & -2 & -2 & 2 & 0 & 2 \\
2 & 0 & -2 & 0 & 2 & 2 & -2 & 0 & -2 \\
2 & 0 & -2 & 2 & 0 & 2 & -2 & 0 & -2 \\
2 & 0 & -2 & 2 & 2 & 0 & -2 & 0 & -2 \\
2 & 0 & 2 & -2 & -2 & -2 & 0 & 0 & 2 \\
0 & 2 & 0 & 0 & 0 & 0 & 0 & 0 & 0 \\
2 & 0 & 2 & -2 & -2 & -2 & 2 & 0 & 0
\end{matrix}$$

The activations with input C+ are given by the vector (-12, 2, -12, 12, 12, 12, -12, 2, -12). With input C-, the activations vector is (-12, -2, -12, 12, 12, 12, -12, -2, -12).

When this Hopfield network uses the threshold function

$$f(x) = \begin{cases} 1 & \text{if } x > 0 \\ -1 & \text{if } x \leq 0 \end{cases}$$

the corresponding outputs will be C+ and C-, respectively, showing the stable recall of the exemplar vectors, and establishing an autoassociation for them. When the output vectors are used to construct the corresponding characters, you get the original character patterns.

Let us now input the character pattern in Figure 1.6.

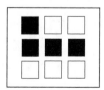

FIGURE 1.6 CORRUPTED "MINUS" PATTERN.

We will call the corresponding bipolar vector **A** = (1, -1, -1, 1, 1, 1, -1, -1, -1). You get the activation vector (-12, -2, -8, 4, 4, 4, -8, -2, -8) giving the output vector, **C-** = (-1, -1, -1, 1, 1, 1, -1, -1, -1). In other words, the character -, corrupted slightly, is recalled as the character - by the Hopfield network. The intended pattern is recognized.

We now input a bipolar vector that is different from the vectors corresponding to the exemplars, and see whether the network can store the corresponding pattern. The vector we choose is **B** = (1, -1, 1, -1, -1, -1, 1, -1, 1). The corresponding neuron activations are given by the vector (12, -2, 12, -4, -4, -4, 12, -2, 12) which causes the output to be the vector (1, -1, 1, -1, -1, -1, 1, -1, 1), same as B. An additional pattern, which is a 3x3 grid with only the corner pixels black, as shown in Figure 1.7, is also recalled since it is autoassociated, by this Hopfield network.

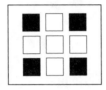

FIGURE 1.7   PATTERN RESULT.

If we omit part of the pattern in Figure 1.7, leaving only the top corners black, as in Figure 1.8, we get the bipolar vector **D** = (1, -1, 1, -1, -1, -1, -1, -1, -1). You can consider this also as an incomplete or corrupted version of the pattern in Figure 1.7. The network activations turn out to be (4, -2, 4, -4, -4, -4, 8, -2, 8) and give the output (1, -1, 1, -1, -1, -1, 1, -1, 1), which is B.

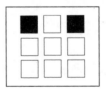

FIGURE 1.8   A PARTLY LOST PATTERN OF FIGURE 1.7.

SUMMARY

In this chapter we introduced a neural network as a collection of processing elements distributed over a finite number of layers and interconnected with positive or negative weights, depending on whether cooperation or competition (or inhibition) is intended. The activation of a neuron is basically a weighted sum of its inputs. A threshold function determines the output of the network. There may be layers of neurons in between the input layer and the output layer, and some such middle layers are referred to as hidden layers, others by names such as Grossberg or Kohonen layers, named after the researchers Stephen Grossberg and Teuvo Kohonen, who proposed them and their function. Modification of the weights is the process of training the network, and a network subject to this process is said to be learning during that phase of the operation of the network. In some network operations, a feedback operation is used in which the current output is treated as modified input to the same network.

You have seen a couple of examples of a Hopfield network, one of them for pattern recognition.

Neural networks can be used for problems that can't be solved with a known formula and for problems with incomplete or noisy data. Neural networks seem to have the capacity to recognize patterns in the data presented to it, and are thus useful in many types of pattern recognition problems.

CHAPTER 2

C++ AND OBJECT ORIENTATION

INTRODUCTION TO C++

C++ is an *object-oriented* programming language built on the base of the C language. This chapter gives you a very brief introduction to C++, touching on many important aspects of C++, so you would be able to follow our presentations of the C++ implementations of neural network models and write your own C++ programs.

The C++ language is a superset of the C language. You could write C++ programs like C programs (a few of the programs in this book are like that), or you could take advantage of the *object-oriented* features of C++ to write object-oriented programs (like the backpropagation simulator of Chapter 7). What makes a programming language or programming methodology object oriented? Well, there are several indisputable pillars of object orientation. These features stand out more than any other as far as object orientation goes. They are *encapsulation, data hiding, overloading, polymorphism,* and the grand-daddy of them all: *inheritance.* Each of the pillars of object-orientation will be discussed in the coming sections, but before we tackle these, we need to answer the question, What does all this object-oriented stuff buy me ? By using the object-oriented features of C++, in conjunction with *Object-Oriented Analysis and Design(OOAD),* which is a methodology that fully utilizes object orientation, you can have well-packaged, reusable, extensible, and reliable programs and program segments. It's beyond the scope of this book to discuss OOAD, but it's recommended you read Booch or Rumbaugh to get more details on OOAD and how and why to change your programming style forever! See the reference section in the back of this book for more information on these readings. Now let's get back to discussing the great object-oriented features of C++.

Encapsulation

In C++ you have the facility to encapsulate data and the operations that manipulate that data, in an appropriate object. This enables the use of these collections of data and function, called *objects* , in programs other than the program for which they were originally created. With objects, just as with the traditional concept of subroutines, you make functional blocks of code. You still have language-supported abstractions such as scope and separate compilation available. This is a rudimentary form of encapsulation. Objects carry encapsulation a step further. With objects, you define not only the way a function operates, or its *implementation*, but also the way an object can be accessed, or its *interface.* You can specify access differently for different entities. For example, you could make function **do_operation()** contained inside **Object A** accessible to **Object B** but not to **Object C.** This access qualification can also be used for data members inside an object. The encapsulation of data and the intended operations on them prevents the data from being subjected to operations not meant for them. This is what really makes objects reusable and portable! The operations are usually given in the form of functions operating upon the data items. Such functions are also called *methods* in some object-oriented programming languages. The data items and the functions that manipulate them are combined into a structure called a *class.* A class is an abstract data type. When you make an instance of a class, you have an object. This is no different than when you instantiate an integer type to create variables *i* and *j.* For example, you design a class called **ElectronicBook**, with a data element called **ArrayofPages**. When you instantiate your class you make objects of *type* ElectronicBook. Suppose that you create two of these called **EB_Geography** and **EB_History**. Every object that is instantiated has its own data member inside it, referred to by **ArrayOfPages.**

Data Hiding

Related to the idea of encapsulation is the concept of *data hiding.* Encapsulation hides the data from other classes and functions in other classes. Going back to the **ElectronicBook** class, you could define functions like **GetNextPage, GetPreviousPage,** and **GetCurrentPage** as the only means of accessing information in the **ArrayofPages** data member, by functions that access the **ElectronicBook** object. Although there may be a hundred and one other attributes and data elements in the class **ElectronicBook,** these are all hidden from view. This makes programs more reliable, since publishing a specific interface to an object prevents inadvertent access to data in ways that were not designed or accounted for. In C++, the access to an object, and its encapsulated data and functions is treated very carefully, by the use of keywords *private*, *protected*, and *public*. One has the opportunity to make access specifications for data objects and functions as being private, or protected, or public while defining a class. Only when the declaration is made as public do other functions and objects have access to the object and its components without question. On the other hand, if the declaration happens to be as private, there is no possibility of such access. When the declaration given is as protected, then the access to data and functions in a class by others is not as free as when it is public, nor as restricted as when it is private. You can declare one class as derived from another class, which will be discussed shortly. So-called derived classes and the declaring class do get the access to the components of the object that are declared protected. One class that is not a derived class of a second class can get access to data items and functions of the second class if it is declared as a *friend* class in the second. The three types of declarations of access specification can be different for different components of an object. For example, some of the data items could be declared public, some private, and the others protected. The same situation can occur with the functions in an object. When no explicit declaration is made, the default specification is as private.

CONSTRUCTORS AND DESTRUCTORS AS SPECIAL FUNCTIONS OF C++

Constructors and *destructors* are special functions in C++. They define how an object is created and destroyed. You cannot have a class defined in a C++ program without declaring and defining at least one constructor for it. You may omit declaring and then defining a destructor only because the compiler you use will create a default destructor. More than one constructor, but only one destructor, can be declared for a class.

Constructors are for the creation of an object of a class and for initializing it. C++ requires that every function has a return type. The only exceptions are constructors and destructors. A constructor is given the same name as the class for which it is a constructor. It may take arguments or it may not need them. Different constructors for the same class differ in the number and types of arguments they take. It is a good idea to provide for each class at least a default constructor that does not take any arguments and does not do anything except create an object of that class type. A constructor is called at the time an object of its class is needed to be created.

A destructor also is given the same name as the class for which it is a destructor, but with the tilde (~) preceding the name. Typically, what is done in a destructor is to have statements that ask the system to delete the various data structures created for the class. This helps to free-up allocated memory for those data structures. A destructor is called when the object created earlier is no longer needed in the program.

Dynamic Memory Allocation

C++ has keywords *new* and *delete*, which are used as a pair in that order, though separated by other statements of the program. They are for making dynamic allocation of memory at the time of creation of a class object and for freeing-up such allocated memory when it is no longer needed. You create space on the heap with the use of new. This obviates the need in C++ for malloc, which is the function for dynamic memory allocation used in C.

Overloading

Encapsulation of data and functions would also allow you to use the same function name in two different objects. The use of a name for a function more than once does not have to be only in different object declarations. Within the same object one can use the same name for functions with different functionality, if they can be distinguished in terms of either their return type or in terms of their argument types and number. This feature is called *overloading*. For example, if two different types of variables are data items in an object, a commonly named function can be the addition, one for each of the two types of variables—thus taking advantage of overloading. Then the function addition is said to be overloaded. But remember that the function main is just about the only function that cannot be overloaded.

Polymorphism and Polymorphic Functions

A *polymorphic* function is a function whose name is used in different ways in a program. It can be also declared virtual, if the intention is late binding. This enables it to be bound at run time. Late binding is also referred to as dynamic binding. An advantage in declaring a function in an object as virtual is that, if the program that uses this object calls that function only conditionally, there is no need to bind the function early, during the compilation of the program. It will be bound only if the condition is met and the call of the function takes place. For example, you could have a polymorphic function called **draw()** that is associated with different graphical objects, like for rectangle, circle, and sphere. The details or methods of the functions are different, but the name **draw()** is common. If you now have a collection of these objects and pick up an arbitrary object without knowing exactly what it is (via a pointer, for example), you can still invoke the draw function for the object and be assured that the right draw function will be bound to the object and called.

OVERLOADING OPERATORS

You can overload operators in addition to overloading functions. As a matter of fact, the system defined left shift operator << is also overloaded in C++ when used with **cout,** the C++ variation of the C language **printf** function. There is a similar situation with the right shift operator >> in C++ when used with **cin,** the C++ variation of the C language **scanf** function. You can take any operator and overload it. But you want to be cautious and not overdo it, and also you do not create confusion when you overload an operator. The guiding principle in this regard is the creation of a code-saving and time-saving facility while maintaining simplicity and clarity. Operator overloading is especially useful for doing normal arithmetic on nonstandard data types. You could overload the multiplication symbol to work with complex numbers, for example.

INHERITANCE

The primary distinction for C++ from C is that C++ has classes. Objects are defined in classes. Classes themselves can be data items in other classes, in which case one class would be an element of another class. Of course, then one class is a member, which brings with it its own data and functions, in the second class. This type of relationship is referred to as a "has-a" relationship: Object A has an Object B inside it.

A relationship between classes can be established not only by making one class a member of another but also by the process of deriving one class from another. One class can be derived from another class, which becomes its base class. Then a hierarchy of classes is established, and a sort of parent–child relationship between classes is established. The derived class inherits, from the base class, some of the data members and functions. This type of relationship is referred to as an "is-a" relationship. You could have class **Rectangle** be derived from class **Shape**, since **Rectangle** is a **Shape**. Naturally, if a class A is derived from a class B, and if B itself is derived from a class C, then A inherits from both B and C. A class can be derived from more than one class. This is how *multiple inheritance* occurs. Inheritance is a powerful mechanism for creating base functionality that is passed onto next generations for further enhancement or modification.

Derived Classes

When one class has some members declared in it as protected, then such members would be hidden from other classes, but not from the derived classes. In other words, deriving one class from another is a way of accessing the protected members of the parent class by the derived class. We then say that the derived class is inheriting from the parent class those members in the parent class that are declared as protected or public.

In declaring a derived class from another class, access or visibility specification can be made, meaning that such derivation can be public or the default case, private. Table 2.1 shows the consequences of such specification when deriving one class from another.

TABLE 2.1 VISIBILITY OF BASE CLASS MEMBERS IN DERIVED CLASS

Derivation Specification	Base Class Specification	Derived Class Access
private(default)	private	none
	protected	full access, private in derived class
	public	full access, public in derived class
public	private	none
	protected	full access, protected in derived class
	public	full access, public in derived class

REUSE OF CODE

C++ is also attractive for the extendibility of the programs written in it and for the reuse opportunity, thanks to the features in C++ such as inheritance and polymorphism mentioned earlier. A new programming project cannot only reuse classes that are created for some other program, if they are appropriate, but can extend another program with additional classes and functions as deemed necessary. You could inherit from an existing class hierarchy and only change functionality where you need to.

C++ COMPILERS

All of the programs in this book have been compiled and tested with Turbo C++, Borland C++, Microsoft C/C++, and Microsoft Visual C++. These are a few of the popular commercial C++ compilers available. You should be able to use most other commercial C++ compilers also. All of the programs should also port easily to other operating systems like Unix and the Mac, because they are all character-based programs.

WRITING C++ PROGRAMS

Before one starts writing a C++ program for a particular problem, one has to have a clear picture of the various parameters and variables that would be part of the problem definition and/or its solution. In addition, it should be clear as to what manipulations need to be performed during the solution process. Then one carefully determines what classes are needed and what relationships they have to each other in a hierarchy of classes. Think about is-a and has-a relationships to see where classes need to defined, and which classes could be derived from others. It would be far more clear to the programmer at this point in the program plan what the data and function access specifications should be and so on. The typical compilation error messages a programmer to C++ may encounter are stating that a particular function or data is not a member of a particular class, or that it is not accessible to a class, or that a constructor was not available for a particular class. When function arguments at declaration and at the place the function is called do not match, either for number or for types or both, the compiler thinks of them as two different functions. The compiler does not find the definition and/or declaration of one of the two and has reason to complain. This type of error in one line of code may cause the compiler to alert you that several other errors are also present, perhaps some in terms of improper punctuation. In that case, remedying the fundamental error that was pointed out would straighten many of the other argued matters.

The following list contains a few additional particulars you need to keep in mind when writing C++ programs.

- A member x of an object A is referred to with A.x just as done with structure elements in C.

- If you declare a class B, then the constructor function is also named B. B has no return type. If this constructor takes, say, one argument of type integer, you define the constructor using the syntax: **B::B(int){whatever the function does};**

- If you declare a member function C of class B, where return type of C is, say, float and C takes two arguments, one of type float, and the other int, then you define C with the syntax: **float B::C(float,int){whatever the function does};**

- If you declare a member function D of class B, where D does not return any value and takes no arguments, you define D using the syntax: **void B::D( ){whatever the function does};**

- If G is a class derived from, say, class B previously mentioned, you declare G using the syntax: **class G:B.** The constructor for G is defined using the syntax: **G::G(arguments of G):B(int){whatever the function does}.** If, on the other hand, G is derived from, say, class B as well as class T, then you declare G using the syntax: **class G:B,T.**

- If one class is declared as derived from more than one other class, that is, if there are more than one base class for it, the derivations specification can be different or the same. Thus the class may be derived from one class publicly and at the same time from another class privately.

- If you have declared a global variable *y* external to a class B, and if you also have a data member *y* in the class B, you can use the external *y* with the reference symbol ::. Thus ::*y* refers to the global variable, whereas y, within a member function of B, or B.y refers to the data member of B. This way polymorphic functions can also be distinguished from each other.

This is by no means a comprehensive list of features, but more a brief list of important constructs in C++. You will see examples of C++ usage in later chapters.

SUMMARY

A few highlights of the C++ language are presented.

- C++ is an object-oriented language with full compatibility with the C language.
- You create classes in C++ that encapsulate data and functions that operate on the data, and hiding data where a public interface is not needed.
- You can create hierarchies of classes with the facility of inheritance. Polymorphism is a feature that allows you to apply a function to a task according to the object the function is operating on.
- Another feature in C++ is overloading of operators, which allows you to create new functionality for existing operators in a different context.
- Overall, C++ is a powerful language fitting the object-oriented paradigm that enables software reuse and enhanced reliability.

CHAPTER 3

A Look at Fuzzy Logic

Crisp or Fuzzy Logic?

Logic deals with true and false. A *proposition* can be true on one occasion and false on another. "Apple is a red fruit" is such a proposition. If you are holding a Granny Smith apple that is green, the proposition that apple is a red fruit is false. On the other hand, if your apple is of a red delicious variety, it is a red fruit and the proposition in reference is true. If a proposition is true, it has a truth value of 1; if it is false, its truth value is 0. These are the only possible truth values. Propositions can be combined to generate other propositions, by means of logical operations.

When you say it will rain today or that you will have an outdoor picnic today, you are making statements with certainty. Of course your statements in this case can be either true or false. The truth values of your statements can be only 1, or 0. Your statements then can be said to be *crisp*.

On the other hand, there are statements you cannot make with such certainty. You may be saying that you think it will rain today. If pressed further, you may be able to say with a degree of certainty in your statement that it will rain today. Your level of certainty, however, is about 0.8, rather than 1. This type of situation is what *fuzzy logic* was developed to model. Fuzzy logic deals with propositions that can be true to a certain degree—somewhere from 0 to 1. Therefore, a proposition's truth value indicates the degree of certainty about which the proposition is true. The degree of certainity sounds like a probability (perhaps subjective probability), but it is not quite the same. Probabilities for mutually exclusive events cannot add up to more than 1, but their fuzzy values may. Suppose that the probability of a cup of coffee being hot is 0.8 and the probability of the cup of coffee being cold is 0.2. These probabilities must add up to 1.0. Fuzzy values do not need to add up to 1.0. The truth value of a proposition that a cup of coffee is hot is 0.8. The truth value of a proposition that the cup of coffee is cold can be 0.5. There is no restriction on what these truth values must add up to.

FUZZY SETS

Fuzzy logic is best understood in the context of *set membership*. Suppose you are assembling a set of rainy days. Would you put today in the set? When you deal only with crisp statements that are either true or false, your inclusion of today in the set of rainy days is based on certainty. When dealing with fuzzy logic, you would include today in the set of rainy days via an *ordered pair*, such as (today, 0.8). The first member in such an ordered pair is *a candidate for inclusion* in the set, and the second member is a value between 0 and 1, inclusive, called the *degree of membership* in the set. The inclusion of the degree of membership in the set makes it convenient for developers to come up with a set theory based on fuzzy logic, just as regular set theory is developed. Fuzzy sets are sets in which members are presented as ordered pairs that include information on degree of membership. A traditional set of, say, k elements, is a special case of a fuzzy set, where each of those k elements has 1 for the degree of membership, and every other element in the universal set has a degree of membership 0, for which reason you don't bother to list it.

FUZZY SET OPERATIONS

The usual operations you can perform on ordinary sets are *union*, in which you take all the elements that are in one set or the other; and *intersection*, in which you take the elements that are in both sets. In the case of fuzzy sets, taking a union is finding the degree of membership that an element should have in the new fuzzy set, which is the union of two fuzzy sets.

If a, b, c, and d are such that their degrees of membership in the fuzzy set A are 0.9, 0.4, 0.5, and 0, respectively, then the fuzzy set A is given by the *fit vector* (0.9, 0.4, 0.5, 0). The components of this fit vector are called *fit values* of a, b, c, and d.

Union of Fuzzy Sets

Consider a union of two traditional sets and an element that belongs to only one of those sets. Earlier you saw that if you treat these sets as fuzzy sets, this element has a degree of membership of 1 in one case and 0 in the other since it belongs to one set and not the other. Yet you are going to put this element in the union. The criterion you use in this action has to do with degrees of membership. You need to look at the two degrees of membership, namely, 0 and 1, and pick the higher value of the two, namely, 1. In other words, what you want for the degree of membership of an element when listed in the union of two fuzzy sets, is the maximum value of its degrees of membership within the two fuzzy sets forming a union.

If a, b, c, and d have the respective degrees of membership in fuzzy sets A, B as A = (0.9, 0.4, 0.5, 0) and B = (0.7, 0.6, 0.3, 0.8), then A ∪ B = (0.9, 0.6, 0.5, 0.8).

Intersection and Complement of Two Fuzzy Sets

Analogously, the degree of membership of an element in the intersection of two fuzzy sets is the *minimum*, or the smaller value of its degree of membership individually in the two sets forming the intersection. For example, if today has 0.8 for degree of membership in the set of rainy days and 0.5 for degree of membership in the set of days of work completion, then today belongs to the set of rainy days on which work is completed to a degree of 0.5, the smaller of 0.5 and 0.8.

Recall the fuzzy sets A and B in the previous example. A = (0.9, 0.4, 0.5, 0) and B = (0.7, 0.6, 0.3, 0.8). A∩B, which is the intersection of the fuzzy sets A and B, is obtained by taking, in each component, the smaller of the values found in that component in A and in B. Thus A∩B = (0.7, 0.4, 0.3, 0).

The idea of a universal set is implicit in dealing with traditional sets. For example, if you talk of the set of married persons, the universal set is the set of all persons. Every other set you consider in that context is a subset of the universal set. We bring up this matter of universal set because when you make the complement of a traditional set A, you need to put in every element in the universal set that is not in A. The complement of a fuzzy set, however, is obtained as follows. In the case of fuzzy sets, if the degree of membership is 0.8 for a member, then that member is not in that set to a degree of 1.0 − 0.8 = 0.2. So you can set the degree of membership in the complement fuzzy set to the complement with respect to 1. If we return to the scenario of having a degree of 0.8 in the set of rainy days, then today has to have 0.2 membership degree in the set of nonrainy or clear days.

Continuing with our example of fuzzy sets A and B, and denoting the complement of A by A', we have A' = (0.1, 0.6, 0.5, 1) and B' = (0.3, 0.4, 0.7, 0.2). Note that A' ∪ B' = (0.3, 0.6, 0.7, 1), which is also the complement of A ∩ B. You can similarly verify that the complement of A ∪ B is the same as A' ∩ B'. Furthermore, A ∪ A' = (0.9, 0.6, 0.5, 1) and A ∩ A' = (0.1, 0.4, 0.5, 0), which is not a vector of zeros only, as would be the case in conventional sets. In fact, A and A' will be equal in the sense that their fit vectors are the same, if each component in the fit vector is equal to 0.5.

Applications of Fuzzy Logic

Applications of fuzzy sets and fuzzy logic are found in many fields, including artificial intelligence, engineering, computer science, operations research, robotics, and pattern recognition. These fields are also ripe for applications for neural networks. So it seems natural that fuzziness should be introduced in neural networks themselves. Any area where humans need to indulge in making decisions, fuzzy sets can find a place, since information on which decisions are to be based may not always be complete and the reliability of the supposed values of the underlying parameters is not always certain.

Examples of Fuzzy Logic

Let us say five tasks have to be performed in a given period of time, and each task requires one person dedicated to it. Suppose there are six people capable of doing these tasks. As you have more than enough people, there is no problem in scheduling this work and getting it done. Of course who gets assigned to which task depends on some criterion, such as total time for completion, on which some optimization can be done. But suppose these six people are not necessarily available during the particular period of time in question. Suddenly, the equation is seen in less than crisp terms. The availability of the people is fuzzy-valued. Here is an example of an assignment problem where fuzzy sets can be used.

Commercial Applications

Many commercial uses of fuzzy logic exist today. A few examples are listed here:

- A subway in Sendai, Japan uses a fuzzy controller to control a subway car. This controller has outperformed human and conventional controllers in giving a smooth ride to passengers in all terrain and external conditions.
- Cameras and camcorders use fuzzy logic to adjust autofocus mechanisms and to cancel the jitter caused by a shaking hand.
- Some automobiles use fuzzy logic for different control applications. Nissan has patents on fuzzy logic braking systems, transmission controls, and fuel injectors. GM uses a fuzzy transmission system in its Saturn vehicles.

- FuziWare has developed and patented a fuzzy spreadsheet called *FuziCalc* that allows users to incorporate fuzziness in their data.

- Software applications to search and match images for certain pixel regions of interest have been developed. Avian Systems has a software package called *FullPixelSearch.*

- A stock market charting and research tool called *SuperCharts* from Omega Research, uses fuzzy logic in one of its modules to determine whether the market is bullish, bearish, or neutral.

FUZZINESS IN NEURAL NETWORKS

There are a number of ways fuzzy logic can be used with neural networks. Perhaps the simplest way is to use a *fuzzifier* function to preprocess or post-process data for a neural network. This is shown in Figure 3.1, where a neural network has a preprocessing fuzzifier that converts data into fuzzy data for application to a neural network.

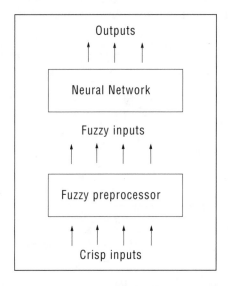

FIGURE 3.1 A NEURAL NETWORK WITH FUZZY PREPROCESSOR.

Let us build a simple fuzzifier based on an application to predict the direction of the stock market. Suppose that you wish to fuzzify one set of data used in the network, the Federal Reserve's fiscal policy, in one of four fuzzy categories: very accommodative, accommodative, tight, or very tight. Let us suppose that the raw data that we need to fuzzify is the discount rate and the interest rate that the Federal Reserve controls to set the fiscal policy. Now, a low discount rate usually indicates a loose fiscal policy, but this depends not only on the observer, but also on the political climate. There is a probability, for a given discount rate that you will find two people who offer different categories for the Fed fiscal policy. Hence, it is appropriate to fuzzify the data, so that the data we present to the neural network is like what an observer would see.

Figure 3.2 shows the fuzzy categories for different interest rates. Note that the category tight has the largest range. At any given interest rate level, you could have one possible category or several. If only one interest rate is present on the graph, this indicates that membership in that fuzzy set is 1.0. If you have three possible fuzzy sets, there is a requirement that membership add up to 1.0. For an interest rate of 8%, you have some chance of finding this in the tight category or the accommodative category. To find out the percentage probability from the graph, take the height of each curve at a given interest rate and normalize this to a one-unit length. At 8%, the tight category is about 0.8 unit in height, and accommodative is about 0.3 unit in height. The total is about 1.1 units, and the probability of the value being tight is then 0.8/1.1 = 0.73, while the probability of the value being accommodative is 0.27.

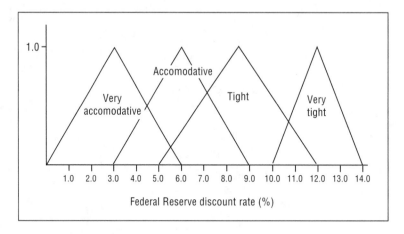

FIGURE 3.2 FUZZY CATEGORIES FOR FEDERAL RESERVE POLICY BASED ON THE FED DISCOUNT RATE.

CODE FOR THE FUZZIFIER

Let's develop C++ code to create a simple fuzzifier. A class called *category* is defined in Listing 3.1. This class encapsulates the data that we need to define, the categories in Figure 3.2. There are three private data members called *lowval, midval,* and *highval.* These represent the values on the graph that define the category triangle. In the tight category, the lowval is 5.0, the midval is 8.5, and the highval is 12.0. The category class allows you to instantiate a category object and assign parameters to it to define it. Also, there is a string called **name** that identifies the category, e.g. "tight." Various member functions are used to interface to the private data members. There is **setval(),** for example, which lets you set the value of the three parameters, while **gethighval()** returns the value of the parameter highval. The function **getshare()** returns the relative value of membership in a category given an input. In the example discussed earlier, with the number 8.0 as the Fed discount rate and the category tight defined according to the graph in Figure 3.2, **getshare()** would return 0.8. Note that this is not yet normalized. Following this example, the **getshare()** value from the accommodative category would also be used to determine the membership weights. These weights define a probability in a given category. A random number generator is used to define a value that is used to select a fuzzy category based on the probabilities defined.

LISTING 3.1 FUZZFIER.H

```
// fuzzfier.h V. Rao, H. Rao
// program to fuzzify data


class category
{
private:
        char name[30];
        float   lowval,highval,midval;

public:
        category(){};
        void setname(char *);
        char * getname();
        void setval(float&,float&,float&);
        float getlowval();
        float getmidval();
        float gethighval();

        float getshare(const float&);
```

```
        ~category(){};

};

int randnum(int);
```

Let's look at the implementation file in Listing 3.2.

Listing 3.2 fuzzfier.cpp

```
// fuzzfier.cpp        V. Rao, H. Rao
// program to fuzzify data

#include <iostream.h>
#include <stdlib.h>
#include <time.h>
#include <string.h>
#include <fuzzfier.h>

void category::setname(char *n)
{
strcpy(name,n);
}

char * category::getname()
{
return name;
}

void category::setval(float &h, float &m, float &l)
{
highval=h;
midval=m;
lowval=l;
}

float category::getlowval()
{
return lowval;
```

```
}

float category::getmidval()
{
return midval;
}

float category::gethighval()
{
return highval;
}

float category::getshare(const float & input)
{
// this member function returns the relative membership
// of an input in a category, with a maximum of 1.0

float output;
float midlow, highmid;

midlow=midval-lowval;
highmid=highval-midval;

// if outside the range, then output=0
if ((input <= lowval) || (input >= highval))
        output=0;

else

        {

        if (input > midval)

                output=(highval-input)/highmid;


        else

                if (input==midval)
```

```
                    output=1.0;

            else

                    output=(input-lowval)/midlow;

        }
return output;
}

int randomnum(int maxval)
{
// random number generator
// will return an integer up to maxval

srand ((unsigned)time(NULL));
return rand() % maxval;
}

void main()
{
// a fuzzifier program that takes category information:
// lowval, midval and highval and category name
// and fuzzifies an input based on
// the total number of categories and the membership
// in each category

int i=0,j=0,numcat=0,randnum;
float l,m,h, inval=1.0;

char input[30]="                    ";
category * ptr[10];
float relprob[10];
float total=0, runtotal=0;

//input the category information; terminate with 'done';

while (1)
```

```
        {

        cout << "\nPlease type in a category name, e.g. Cool\n";
        cout << "Enter one word without spaces\n";
        cout << "When you are done, type 'done' :\n\n";

        ptr[i]= new category;
        cin >> input;


        if ((input[0]=='d' && input[1]=='o' &&
                input[2]=='n' && input[3]=='e')) break;

        ptr[i]->setname(input);

        cout << "\nType in the lowval, midval and highval\n";
        cout << "for each category, separated by spaces\n";
        cout << " e.g. 1.0 3.0 5.0 :\n\n";

        cin >> l >> m >> h;
        ptr[i]->setval(h,m,l);

        i++;

        }

numcat=i; // number of categories

// Categories set up: Now input the data to fuzzify
cout <<"\n\n";
cout << "===================================\n";
cout << "==Fuzzifier is ready for data==\n";
cout << "===================================\n";

while (1)

        {
        cout << "\ninput a data value, type 0 to terminate\n";
```

```
cin >> inval;

if (inval == 0) break;

// calculate relative probabilities of
//      input being in each category

total=0;

for (j=0;j<numcat;j++)
        {
        relprob[j]=100*ptr[j]->getshare(inval);
        total+=relprob[j];
        }

if (total==0)
        {
        cout << "data out of range\n";
        exit(1);
        }

randnum=randomnum((int)total);

j=0;
runtotal=relprob[0];

while ((runtotal<randnum)&&(j<numcat))
        {
        j++;
        runtotal += relprob[j];
        }

cout << "\nOutput fuzzy category is ==> " <<
        ptr[j]->getname()<<"<== \n";

        cout <<"category\t"<<"membership\n";
        cout <<"————-\n";

for (j=0;j<numcat;j++)
```

```
        {
        cout << ptr[j]->getname()<<"\t\t"<<
                (relprob[j]/total) <<"\n";
        }


    }


cout << "\n\nAll done. Have a fuzzy day !\n";


    }
```

This program first sets up all the categories you define. These could be for the example we choose or any example you can think of. After the categories are defined, you can start entering data to be fuzzified. As you enter data you see the probability aspect come into play. If you enter the same value twice, you may end up with different categories! You will see sample output shortly, but first a technical note on how the weighted probabilities are set up. The best way to explain it is with an example. Suppose that you have defined three categories, A, B, and C. Suppose that category A has a relative membership of 0.8, category B of 0.4, and category C of 0.2. In the program, these numbers are first multiplied by 100, so you end up with A=80, B=40, and C=20. Now these are stored in a vector with an index j initialized to point to the first category. Let's say that these three numbers represent three adjacent number bins that are joined together. Now pick a random number to index into the bin that has its maximum value of (80+40+20). If the number is 100, then it is greater than 80 and less than (80+40), you end up in the second bin that represents B. Does this scheme give you weighted probabilities? Yes it does, since the size of the bin (given a uniform distribution of random indexes into it) determines the probability of falling into the bin. Therefore, the probability of falling into bin A is 80/(80+40+20).

Sample output from the program is shown below. Our input is in italic; computer output is not. The categories defined by the graph in Figure 3.2 are entered in this example. Once the categories are set up, the first data entry of 4.0 gets fuzzified to the accommodative category. Note that the memberships are also presented in each category. The same value is entered again, and this time it gets fuzzified to the very accommodative category. For the last data entry of 12.5, you see that only the very tight category holds membership for this value. In all cases you will note that the memberships add up to 1.0.

fuzzfier

Please type in a category name, e.g. Cool
Enter one word without spaces
When you are done, type 'done' :

v.accommodative

Type in the lowval, midval and highval
for each category, separated by spaces
 e.g. 1.0 3.0 5.0 :

0 3 6

Please type in a category name, e.g. Cool
Enter one word without spaces
When you are done, type 'done' :

accommodative

Type in the lowval, midval and highval
for each category, separated by spaces
 e.g. 1.0 3.0 5.0 :

3 6 9

Please type in a category name, e.g. Cool
Enter one word without spaces
When you are done, type 'done' :

tight

Type in the lowval, midval and highval
for each category, separated by spaces
 e.g. 1.0 3.0 5.0 :

5 8.5 12

Please type in a category name, e.g. Cool

Enter one word without spaces
When you are done, type 'done' :

v.tight

Type in the lowval, midval and highval
for each category, separated by spaces
 e.g. 1.0 3.0 5.0 :

10 12 14

Please type in a category name, e.g. Cool
Enter one word without spaces
When you are done, type 'done' :

done

```
====================================
==Fuzzifier is ready for data==
====================================
```

input a data value, type 0 to terminate
4.0

Output fuzzy category is ==> accommodative<==
category   membership
_____-

v.accommodative   0.666667
accommodative   0.333333
tight   0
v.tight   0

input a data value, type 0 to terminate
4.0

Output fuzzy category is ==> v.accommodative<==
category   membership
_____-

v.accommodative   0.666667

```
accommodative           0.333333
tight           0
v.tight         0
```

input a data value, type 0 to terminate
7.5

Output fuzzy category is ==> accommodative<==
```
category        membership
_____
v.accommodative         0
accommodative           0.411765
tight           0.588235
v.tight         0
```

input a data value, type 0 to terminate
11.0

        Output fuzzy category is ==> tight<==
```
category        membership
_____
v.accommodative         0
accommodative           0
tight           0.363636
v.tight         0.636364
```

input a data value, type 0 to terminate
12.5

Output fuzzy category is ==> v.tight<==
```
category        membership
_____
v.accommodative         0
accommodative           0
tight           0
v.tight         1
```

input a data value, type 0 to terminate

```
0
All done. Have a fuzzy day !
```

FUZZY CONTROL SYSTEMS

The most widespread use of fuzzy logic today is in *fuzzy control* applications. You can use fuzzy logic to make your air conditioner cool your room. Or you can design a subway system to use fuzzy logic to control the braking system for smooth and accurate stops. A control system is a closed-loop system that typically controls a machine to achieve a particular desired response, given a number of environmental inputs. A fuzzy control system is a closed-loop system that uses the process of fuzzification, as shown in the Federal Reserve policy program example, to generate fuzzy inputs to an *inference engine*, which is a knowledge base of actions to take. The inverse process, called *defuzzification*, is also used in a fuzzy control system to create crisp, real values to apply to the machine or process under control. In Japan, fuzzy controllers have been used to control many machines, including washing machines and camcorders.

Figure 3.3 shows a diagram of a fuzzy control system. The major parts of this closed-loop system are:

- **machine under control**—this is the machine or process that you are controlling, for example, a washing machine
- **outputs**—these are the measured response behaviors of your machine, for example, the temperature of the water
- **fuzzy outputs**—these are the same outputs passed through a fuzzifier, for example, hot or very cold
- **inference engine/fuzzy rule base**—an inference engine converts fuzzy outputs to actions to take by accessing fuzzy rules in a fuzzy rule base. An example of a fuzzy rule: IF the output is very cold, THEN increase the water temperature setting by a very large amount
- **fuzzy inputs**—these are the fuzzy actions to perform, such as increase the water temperature setting by a very large amount
- **inputs**—these are the (crisp) dials on the machine to control its behavior, for example, water temperature setting = 3.423, converted from fuzzy inputs with a defuzzifier

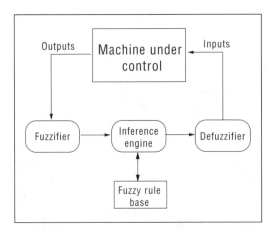

FIGURE 3.3 DIAGRAM OF A FUZZY CONTROL SYSTEM.

The key to development of a fuzzy control system is to iteratively construct a fuzzy rule base that yields the desired response from your machine. You construct these fuzzy rules from knowledge about the problem. In many cases this is very intuitive and gives you a robust control system in a very short amount of time.

FUZZINESS IN NEURAL NETWORKS

Fuzziness can enter neural networks to define the weights from fuzzy sets. A comparison between expert systems and fuzzy systems is important to understand in the context of neural networks. Expert systems are based on crisp rules. Such crisp rules may not always be available. Expert systems have to consider an exhaustive set of possibilities. Such sets may not be known beforehand. When crisp rules are not possible, and when it is not known if the possibilities are exhaustive, the expert systems approach is not a good one.

Some neural networks, through the features of training and learning, can function in the presence of unexpected situations. Therein neural networks have an advantage over expert systems, and they can manage with far less information than expert systems need.

One form of fuzziness in neural networks is called a *fuzzy cognitive map*. A fuzzy cognitive map is like a dynamic state machine with fuzzy states. A traditional *state machine* is a machine with defined states and outputs associated with each state. Transitions from state to state take place according to input events or *stimuli*. A fuzzy cognitive map looks like a state machine but has fuzzy states (not just 1 or 0). You have a set of weights along each transition path, and these weights can be learned from a set of training data.

Our treatment of fuzziness in neural networks is with the discussion of the fuzzy associative memory, abbreviated as *FAM*, which, like the fuzzy cognitive map, was developed by Bart Kosko. The FAM and the C++ implementation are discussed in Chapter 9.

NEURAL-TRAINED FUZZY SYSTEMS

So far we have considered how fuzzy logic plays a role in neural networks. The converse relationship, neural networks in fuzzy systems, is also an active area of research. In order to build a fuzzy system, you must have a set of membership rules for fuzzy categories. It is sometimes difficult to deduce these membership rules with a given set of complex data. Why not use a neural network to define the fuzzy rules for you? A neural network is good at discovering relationships and patterns in data and can be used to pre-process data in a fuzzy system. Further, a neural network that can learn new relationships with new input data can be used to refine fuzzy rules to create a fuzzy adaptive system. Neural trained fuzzy systems are being used in many commercial applications, especially in Japan:

- The Laboratory for International Fuzzy Engineering Research (LIFE) in Yokohama, Japan has a backpropagation neural network that derives fuzzy rules and membership functions. The LIFE system has been successfully applied to a foreign-exchange trade support system with approximately 5000 fuzzy rules.

- Ford Motor Company has developed trainable fuzzy systems for automobile idle-speed control.

- National Semiconductor Corporation has a software product called *NeuFuz* that supports the generation of fuzzy rules with a neural network for control applications.

- A number of Japanese consumer and industrial products use neural networks with fuzzy systems, including vacuum cleaners, rice cookers, washing machines, and photocopying machines.

- AEG Corporation of Germany uses a neural-network-trained fuzzy control system for its water- and energy-conserving washing machine. After the machine is loaded with laundry, it measures the water level with a pressure sensor and infers the amount of laundry in the machine by the speed and volume of water. A total of 157 rules were generated by a neural network that was trained on data correlating the amount of laundry with the measurement of water level on the sensor.

Summary

In this chapter, you read about fuzzy logic, fuzzy sets, and simple operations on fuzzy sets. Fuzzy logic, unlike Boolean logic, has more than two on or off categories to describe behavior of systems. You use membership values for data in fuzzy categories, which may overlap. In this chapter, you also developed a fuzzifier program in C++ that takes crisp values and converts them to fuzzy values, based on categories and memberships that you define. For use with neural networks, fuzzy logic can serve as a post-processing or pre-processing filter. Kosko developed neural networks that use fuzziness and called them *fuzzy associative memories*, which will be discussed in later chapters. You also read about how neural networks can be used in fuzzy systems to define membership functions and fuzzy rules.

CHAPTER 4

CONSTRUCTING A NEURAL NETWORK

FIRST EXAMPLE FOR C++ IMPLEMENTATION

The neural network we presented in Chapter 1 is an example of a Hopfield network with a single layer. Now we present a C++ implementation of this network. Suppose we place four neurons, all connected to one another on this layer, as shown in Figure 4.1. Some of these connections have a positive weight and the rest have a negative weight. You may recall from the earlier presentation of this example, that we used two input patterns to determine the weight matrix. The network recalls them when the inputs are presented to the network, one at a time. These inputs are binary and orthogonal so that their stable recall is assured. Each component of a binary input pattern is either a 0 or a 1. Two vectors are orthogonal when their *dot product*—the sum of the products of their corresponding components—is zero. An example of a binary input pattern is 1 0 1 0 0. An example of a pair of orthogonal vectors is (0, 1, 0, 0, 1) and (1, 0, 0, 1, 0). An example of a pair of vectors that are not orthogonal is (0, 1, 0, 0, 1) and (1, 1, 0, 1, 0). These last two vectors have a dot product of 1, different from 0.

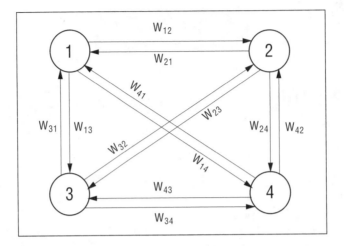

FIGURE 4.1 LAYOUT OF A HOPFIELD NETWORK

The two patterns we want the network to have stable recall for are $A = (1, 0, 1, 0)$ and $B = (0, 1, 0, 1)$. The weight matrix **W** is given as follows:

$$
W = \begin{array}{cccc}
0 & -3 & 3 & -3 \\
-3 & 0 & -3 & 3 \\
3 & -3 & 0 & -3 \\
-3 & 3 & -3 & 0
\end{array}
$$

The positive links (values with positive signs) tend to encourage agreement in a stable configuration, whereas negative links (values with negative signs) tend to discourage agreement in a stable configuration.

NOTE

We need a **threshold** function also, and we define it using a threshold value, θ, as follows:

$$
f(t) = \begin{cases} 1 & \text{if } t >= \theta \\ 0 & \text{if } t < \theta \end{cases}
$$

The threshold value θ is used as a cut-off value for the activation of a neuron to enable it to fire. The activation should equal or exceed the threshold value for the neuron to fire, meaning to have output 1. For our Hopfield network, θ is taken as 0. There are four neurons in the only layer in this network. The first node's output is the output of the **threshold** function. The argument for the **threshold** function is the activation of the node. And the activation of the node is the dot product of the input vector and the first column of the weight matrix. So if the input vector is A, the dot product becomes 3, and $f(3) = 1$. And the dot products of the second, third, and fourth nodes become -6, 3, and -6, respectively. The corresponding outputs therefore are 0, 1, and 0. This means that the output of the network is the vector $(1, 0, 1, 0)$, which is the same as the input pattern. Therefore, the network has recalled the pattern as presented. When B is presented, the dot product obtained at the first node is -6 and the output is 0. The activations of all the four nodes together with the **threshold** function give $(0, 1, 0, 1)$ as output from the network, which means that the network recalled B as well. The weight matrix worked well with both input patterns, and we do not need to modify it.

Classes in C++ Implementation

In our C++ implementation of this network, there are the following classes: a **network** class, and a **neuron** class. In our implementation, we create the network with four neurons, and these four neurons are all connected to one another. A neuron is not self-connected, though. That is, there is no edge in the directed graph representing the network, where the edge is from one node to itself. But for simplicity, we could pretend that such a connection exists carrying a weight of 0, so that the weight matrix has 0's in its principal diagonal.

The functions that determine the neuron activations and the network output are declared **public**. Therefore they are visible and accessible without restriction. The activations of the neurons are calculated with functions defined in the neuron class. When there are more than one layer in a neural network, the outputs of neurons in one layer become the inputs for neurons in the next layer. In order to facilitate passing the outputs from one layer as inputs to another layer, our C++ implementations compute the neuron outputs in the **network** class. For this reason the **threshold** function is made a member of the **network** class. We do this for the Hopfield network as well. To see if the network has achieved correct recall, you make comparisons between the presented pattern and the network output, component by component.

C++ PROGRAM FOR A HOPFIELD NETWORK

For convenience every C++ program has two components: One is the header file with all of the class declarations and lists of include library files; the other is the source file that includes the header file and the detailed descriptions of the member functions of the classes declared in the header file. You also put the function **main** in the source file. Most of the computations are done by class member functions, when class objects are created in the function **main**, and calls are made to the appropriate functions. The header file has an **.h** (or **.hpp**) extension, as you know, and the source file has a **.cpp** extension, to indicate that it is a C++ code file. It is possible to have the contents of the header file written at the beginning of the **.cpp** file and work with one file only, but separating the declarations and implementations into two files allows you to change the implementation of a class(**.cpp**) without changing the interface to the class (**.h**).

Header File for C++ Program for Hopfield Network

Listing 4.1 contains Hop.h, the header file for the C++ program for the Hopfield network. The include files listed in it are the stdio.h, iostream.h, and math.h. The iostream.h file contains the declarations and details of the C++ streams for input and output. A **network** class and a **neuron** class, are declared in Hop.h. The data members and member functions are declared within each class, and their accessibility is specified by the keywords *protected* or *public*.

LISTING 4.1 HEADER FILE FOR C++ PROGRAM FOR HOPFIELD NETWORK.

```
//Hop.h       V. Rao, H. Rao
//Single layer Hopfield Network with 4 neurons

#include <stdio.h>
#include <iostream.h>
#include <math.h>

class neuron
{
protected:
        int activation;
        friend class network;
public:
        int weightv[4];
        neuron() {};
        neuron(int *j) ;
```

```
        int act(int, int*);
};


class network
{
public:
        neuron   nrn[4];
        int output[4];
        int threshld(int) ;
        void activation(int j[4]);
        network(int*,int*,int*,int*);


};
```

Notes on the Header File Hop.h

Notice that the data item **activation** in the **neuron** class is declared as protected. In order to make the member **activation** of the neuron class accessible to the **network** class, the network is declared a friend class in the class **neuron.** Also, there are two constructors for the class **neuron.** One of them creates the object neuron without initializing any data members. The other creates the object neuron and initializes the connection weights.

Source Code for the Hopfield network

Listing 4.2 contains the source code for the C++ program for a Hopfield network in the file Hop.cpp. The member functions of the classes declared in Hop.h are implemented here. The function **main** contains the input patterns, values to initialize the weight matrix, and calls to the constructor of **network** class and other member functions of the **network** class.

Listing 4.2 SOURCE CODE FOR C++ PROGRAM FOR HOPFIELD NETWORK.

```
//Hop.cpp      V. Rao, H. Rao
//Single layer Hopfield Network with 4 neurons

#include "hop.h"

neuron::neuron(int *j)
```

```
{
int i;
for(i=0;i<4;i++)
        {
        weightv[i]= *(j+i);
        }


}

int neuron::act(int m, int *x)
{
int i;
int a=0;

for(i=0;i<m;i++)
        {
        a += x[i]*weightv[i];
        } .
return a;
}

int network::threshld(int k)
{
if(k>=0)
        return (1);
else
        return (0);
}

network::network(int a[4],int b[4],int c[4],int d[4])
{
nrn[0] = neuron(a) ;
nrn[1] = neuron(b) ;
nrn[2] = neuron(c) ;
nrn[3] = neuron(d) ;
}

void network::activation(int *patrn)
{
```

```
int i,j;

for(i=0;i<4;i++)
        {
        for(j=0;j<4;j++)
                {
                cout<<"\n nrn["<<i<<"].weightv["<<j<<"] is "
                        <<nrn[i].weightv[j];
                }
        nrn[i].activation = nrn[i].act(4,patrn);
        cout<<"\nactivation is "<<nrn[i].activation;
        output[i]=threshld(nrn[i].activation);
        cout<<"\noutput value is  "<<output[i]<<"\n";
        }
}

void main ()
{
int patrn1[]= {1,0,1,0},i;
int wt1[]= {0,-3,3,-3};
int wt2[]= {-3,0,-3,3};
int wt3[]= {3,-3,0,-3};
int wt4[]= {-3,3,-3,0};

cout<<"\nTHIS PROGRAM IS FOR A HOPFIELD NETWORK WITH A SINGLE LAYER OF";
cout<<"\n4 FULLY INTERCONNECTED NEURONS. THE NETWORK SHOULD RECALL THE";
cout<<"\nPATTERNS 1010 AND 0101 CORRECTLY.\n";

//create the network by calling its constructor.
// the constructor calls neuron constructor as many times as the number
of
// neurons in the network.
network h1(wt1,wt2,wt3,wt4);

//present a pattern to the network and get the activations of the neurons
h1.activation(patrn1);

//check if the pattern given is correctly recalled and give message
for(i=0;i<4;i++)
```

```
        {
        if (h1.output[i] == patrn1[i])
                cout<<"\n pattern= "<<patrn1[i]<<
                "   output = "<<h1.output[i]<<"  component matches";
        else
                cout<<"\n pattern= "<<patrn1[i]<<
                "   output = "<<h1.output[i]<<
                "   discrepancy occurred";
        }
cout<<"\n\n";
int patrn2[]= {0,1,0,1};
h1.activation(patrn2);
for(i=0;i<4;i++)
        {
        if (h1.output[i] == patrn2[i])
                cout<<"\n pattern= "<<patrn2[i]<<
                "   output = "<<h1.output[i]<<"  component matches";
        else
                cout<<"\n pattern= "<<patrn2[i]<<
                "   output = "<<h1.output[i]<<
                "   discrepancy occurred";

        }


}
```

Comments on the C++ Program for Hopfield Network

Note the use of the output stream operator **cout<<** to output text strings or numerical output. C++ has **istream** and **ostream** classes from which the **iostream** class is derived. The standard input and output streams are **cin** and **cout**, respectively, used, correspondingly, with the operators >> and <<. Use of **cout** for the output stream is much simpler than the use of the C function **printf**. As you can see, there is no formatting suggested for output. However, there is a provision that allows you to format the output, while using **cout**.

Also note the way comments are introduced in the program. The line with comments should start with a double slash //. Unlike C, the comment does not have to end with a double slash. Of course, if the comments extend to subsequent lines, each such line should have a double slash at the start. You can still use the pair, /* at the beginning with */ at the end of lines of comments, as you do in C. If the comment continues through many lines, the C facility will be handier to delimit the comments.

The neurons in the network are members of the network class and are identified by the abbreviation **nrn**. The two patterns, 1010 and 0101, are presented to the network one at a time in the program.

Output from the C++ Program for Hopfield Network

The output from this program is as follows and is self-explanatory. When you run this program, you're likely to see a lot of output whiz by, so in order to leisurely look at the output, use redirection. Type **Hop > filename**, and your output will be stored in a file, which you can edit with any text editor or list by using the **type filename | more** command.

```
THIS PROGRAM IS FOR A HOPFIELD NETWORK WITH A SINGLE LAYER OF 4 FULLY
INTERCONNECTED NEURONS. THE NETWORK SHOULD RECALL THE PATTERNS 1010 AND
0101 CORRECTLY.

    nrn[0].weightv[0] is  0
    nrn[0].weightv[1] is  -3
    nrn[0].weightv[2] is  3
    nrn[0].weightv[3] is  -3
activation is 3
output value is  1

    nrn[1].weightv[0] is  -3
    nrn[1].weightv[1] is  0
    nrn[1].weightv[2] is  -3
    nrn[1].weightv[3] is  3
activation is -6
output value is  0

    nrn[2].weightv[0] is  3
    nrn[2].weightv[1] is  -3
    nrn[2].weightv[2] is  0
    nrn[2].weightv[3] is  -3
activation is 3
output value is  1

    nrn[3].weightv[0] is  -3
```

```
     nrn[3].weightv[1] is   3
     nrn[3].weightv[2] is  -3
     nrn[3].weightv[3] is   0
activation is -6
output value is   0


     pattern= 1  output = 1  component matches
     pattern= 0  output = 0. component matches
     pattern= 1  output = 1  component matches
     pattern= 0  output = 0  component matches


     nrn[0].weightv[0] is   0
     nrn[0].weightv[1] is  -3
     nrn[0].weightv[2] is   3
     nrn[0].weightv[3] is  -3
activation is -6
output value is   0

     nrn[1].weightv[0] is  -3
     nrn[1].weightv[1] is   0
     nrn[1].weightv[2] is  -3
     nrn[1].weightv[3] is   3
activation is 3
output value is   1

     nrn[2].weightv[0] is   3
     nrn[2].weightv[1] is  -3
     nrn[2].weightv[2] is   0
     nrn[2].weightv[3] is  -3
activation is -6
output value is   0

     nrn[3].weightv[0] is  -3
     nrn[3].weightv[1] is   3
     nrn[3].weightv[2] is  -3
     nrn[3].weightv[3] is   0
activation is 3
output value is   1
```

```
pattern= 0   output = 0   component matches
pattern= 1   output = 1   component matches
pattern= 0   output = 0   component matches
pattern= 1   output = 1   component matches
```

Further Comments on the Program and Its Output

Let us recall our previous discussion of this example in Chapter 1. What does the network give as output if we present a pattern different from both A and B? If C = (0, 1, 0, 0) is the input pattern, the activation (dot products) would be −3, 0, −3, 3 making the outputs (next state) of the neurons 0,1,0,1, so that B would be recalled. This is quite interesting, because if we intended to input B, and we made a slight error and ended up presenting C instead, the network would recall B. You can run the program by changing the pattern to 0, 1, 0, 0 and compiling again, to see that the B pattern is recalled.

Another element about the example in Chapter 1 is that the weight matrix W is not the only weight matrix that would enable the network to recall the patterns A and B correctly. If we replace the 3 and −3 in the matrix with 2 and −2, respectively, the resulting matrix would facilitate the same performance from the network. One way for you to check this is to change the wt1, wt2, wt3, wt4 given in the program accordingly, and compile and run the program again. The reason why both of the weight matrices work is that they are closely related. In fact, one is a scalar (constant) multiple of the other, that is, if you multiply each element in the matrix by the same scalar, namely 2/3, you get the corresponding matrix in cases where 3 and −3 are replaced with 2 and −2, respectively.

A NEW WEIGHT MATRIX TO RECALL MORE PATTERNS

Let's continue to discuss this example. Suppose we are interested in having the patterns E = (1, 0, 0, 1) and F = (0, 1, 1, 0) also recalled correctly, in addition to the patterns A and B. In this case we would need to train the network and come up with a learning algorithm, which we will discuss in more detail later in the book. We come up with the matrix $W1$, which follows.

$$W_1 = \begin{matrix} 0 & -5 & 4 & 4 \\ -5 & 0 & 4 & 4 \\ 4 & 4 & 0 & -5 \\ 4 & 4 & -5 & 0 \end{matrix}$$

Try to use this modification of the weight matrix in the source program, and then compile and run the program to see that the network successfully recalls all four patterns A, B, E, and F.

The C++ implementation shown does not include the asynchronous update feature mentioned in Chapter 1, which is not necessary for the patterns presented. The coding of this feature is left as an exercise for the reader.

Weight Determination

You may be wondering about how these weight matrices were developed in the previous example, since so far we've only discussed how the network does its job, and how to implement the model. You have learned that the choice of weight matrix is not necessarily unique. But you want to be assured that there is some established way besides trial and error, in which to construct a weight matrix. You can go about this in the following way.

Binary to Bipolar Mapping

Let's look at the previous example. You have seen that by replacing each 0 in a binary string with a −1, you get the corresponding bipolar string. If you keep all 1's the same and replace each 0 with a −1, you will have a formula for the above option. You can apply the following function to each bit in the string:

$$f(x) = 2x - 1$$

When you give the binary bit x, you get the corresponding bipolar character $f(x)$

For inverse mapping, which turns a bipolar string into a binary string, you use the following function:

$$f(x) = (x + 1) / 2$$

NOTE

When you give the bipolar character x, you get the corresponding binary bit $f(x)$

Pattern's Contribution to Weight

Next, we work with the bipolar versions of the input patterns. You take each pattern to be recalled, one at a time, and determine its contribution to the weight matrix of the network. The contribution of each pattern is itself a matrix. The size of such a matrix is the same as the weight matrix of the network. Then add these contributions, in the way matrices are added, and you end up with the weight matrix for the network, which is also referred to as the *correlation matrix.* Let us find the contribution of the pattern $A = (1, 0, 1, 0)$:

First, we notice that the binary to bipolar mapping of $A = (1, 0, 1, 0)$ gives the vector $(1, -1, 1, -1)$.

Then we take the transpose, and multiply, the way matrices are multiplied, and we see the following:

$$
\begin{array}{l}
1 \\
1 \\
1 \\
1
\end{array}
\begin{bmatrix} 1 & -1 & 1 & -1 \end{bmatrix}
\quad = \quad
\begin{array}{rrrr}
1 & -1 & 1 & -1 \\
-1 & 1 & -1 & 1 \\
1 & -1 & 1 & -1 \\
-1 & 1 & -1 & 1
\end{array}
$$

Now subtract 1 from each element in the main diagonal (that runs from top left to bottom right). This operation gives the same result as subtracting the identity matrix from the given matrix, obtaining 0's in the main diagonal. The resulting matrix, which is given next, is the contribution of the pattern $(1, 0, 1, 0)$ to the weight matrix.

$$
\begin{array}{rrrr}
0 & -1 & 1 & -1 \\
-1 & 0 & -1 & 1 \\
1 & -1 & 0 & -1 \\
-1 & 1 & -1 & 0
\end{array}
$$

Similarly, we can calculate the contribution from the pattern $B = (0, 1, 0, 1)$ by verifying that pattern B's contribution is the same matrix as pattern A's contribution. Therefore, the matrix of weights for this exercise is the matrix W shown here.

$$W \quad = \quad \begin{matrix} 0 & -2 & 2 & -2 \\ -2 & 0 & -2 & 2 \\ 2 & -2 & 0 & -2 \\ -2 & 2 & -2 & 0 \end{matrix}$$

You can now optionally apply an arbitrary scalar multiplier to all the entries of the matrix if you wish. This is how we had previously obtained the +/- 3 values instead of +/- 2 values shown above.

AUTOASSOCIATIVE NETWORK

The Hopfield network just shown has the feature that the network associates an input pattern with itself in recall. This makes the network an *autoassociative* network. The patterns used for determining the proper weight matrix are also the ones that are autoassociatively recalled. These patterns are called the *exemplars*. A pattern other than an exemplar may or may not be recalled by the network. Of course, when you present the pattern 0 0 0 0, it is stable, even though it is not an exemplar pattern.

ORTHOGONAL BIT PATTERNS

You may be wondering how many patterns the network with four nodes is able to recall. Let us first consider how many different bit patterns are orthogonal to a given bit pattern. This question really refers to bit patterns in which at least one bit is equal to 1. A little reflection tells us that if two bit patterns are to be orthogonal, they cannot both have 1's in the same position, since the dot product would need to be 0. In other words, a bitwise logical **AND** operation of the two bit patterns has to result in a 0. This suggests the following. If a pattern P has k, less than 4, bit positions with 0 (and so 4-k bit positions with 1), and if pattern Q is to be orthogonal to P, then Q can have 0 or 1 in those k positions, but it must have only 0 in the rest 4-k positions. Since there are two choices for each of the k positions, there are 2^k possible patterns orthogonal to P. This number 2^k of patterns includes the pattern with all zeroes. So there really are 2^k-1 non-zero patterns orthogonal to P. Some of these 2^k-1 patterns are not orthogonal to each other. As an example, P can be the pattern 0 1 0 0, which has $k = 3$ positions with 0. There are $2^3-1=7$ nonzero patterns orthogonal to 0 1 0 0. Among these are patterns 1 0 1 0 and 1 0 0 1, which are not orthogonal to each other, since their dot product is 1 and not 0.

NETWORK NODES AND INPUT PATTERNS

Since our network has four neurons in it, it also has four nodes in the *directed graph* that represents the network. These are laterally connected because connections are established from node to node. They are lateral because the nodes are all in the same layer. We started with the patterns A = (1, 0, 1, 0) and B = (0, 1, 0, 1) as the exemplars. If we take any other nonzero pattern that is orthogonal to A, it will have a 1 in a position where B also has a 1. So the new pattern will not be orthogonal to B. Therefore, the orthogonal set of patterns that contains A and B can have only those two as its elements. If you remove B from the set, you can get (at most) two others to join A to form an orthogonal set. They are the patterns (0, 1, 0, 0) and (0, 0, 0, 1).

If you follow the procedure described earlier to get the correlation matrix, you will get the following weight matrix:

$$W = \begin{matrix} 0 & -1 & 3 & -1 \\ -1 & 0 & -1 & -1 \\ 3 & -1 & 0 & -1 \\ -1 & -1 & -1 & 0 \end{matrix}$$

With this matrix, pattern A is recalled, but the zero pattern (0, 0, 0, 0) is obtained for the two patterns (0, 1, 0, 0) and (0, 0, 0, 1). Once the zero pattern is obtained, its own recall will be stable.

SECOND EXAMPLE FOR C++ IMPLEMENTATION

Recall the cash register game from the show *The Price is Right*, used as one of the examples in Chapter 1. This example led to the description of the Perceptron neural network. We will now resume our discussion of the Perceptron model and follow up with its C++ implementation. Keep the cash register game example in mind as you read the following C++ implementation of the Perceptron model. Also note that the input signals in this example are not necessarily binary, but they may be real numbers. It is because the prices of the items the contestant has to choose are real numbers (dollars and cents). A Perceptron has one layer of input neurons and one layer of output neurons. Each input layer neuron is connected to each neuron in the output layer.

C++ Implementation of Perceptron Network

In our C++ implementation of this network, we have the following classes: we have separate classes for input neurons and output neurons. The **ineuron** class is for the input neurons. This class has weight and activation as data members. The **oneuron** class is similar and is for the output neuron. It is declared as a **friend** class in the **ineuron** class. The **output neuron** class has also a data member called **output**. There is a **network** class, which is a **friend** class in the **oneuron** class. An instance of the **network** class is created with four input neurons. These four neurons are all connected with one output neuron.

The member functions of the **ineuron** class are: (1) a default constructor, (2) a second constructor that takes a real number as an argument, and (3) a function that calculates the output of the input neuron. The constructor taking one argument uses that argument to set the value of the weight on the connection between the input neuron and the output neuron. The functions that determine the neuron activations and the network output are declared public. The activations of the neurons are calculated with functions defined in the neuron classes. A threshold value is used by a member function of the output neuron to determine if the neuron's activation is large enough for it to fire, giving an output of 1.

Header File

Listing 4.3 contains **percept.h**, the header file for the C++ program for the Perceptron network. **percept.h** contains the declarations for three classes, one for input neurons, one for output neurons, and one for network.

LISTING 4.3 THE PERCEPT.H HEADER FILE.

```
//percept.h        V. Rao, H. Rao
// Perceptron model

#include <stdio.h>
#include <iostream.h>
#include <math.h>
```

```
class ineuron
{
protected:
        float weight;
        float activation;
        friend class oneuron;
public:
        ineuron() {};
        ineuron(float j) ;
        float act(float x);
};


class oneuron
{
protected:
        int output;
        float activation;
        friend class network;
public:
        oneuron() { };
        void actvtion(float x[4], ineuron *nrn);
        int outvalue(float j) ;
};


class network
{
public:
        ineuron    nrn[4];
        oneuron    onrn;
        network(float,float,float,float);

};
```

Implementation of Functions

The network is designed to have four neurons in the input layer. Each of them is an object of class **ineuron**, and these are member classes in the class network. There is one explicitly defined output neuron of the class **oneuron**. The network constructor also invokes the neuron constructor for each input layer neuron in the network by providing it with the initial weight for its connection to the neuron in the output layer. The constructor for the output neuron is also invoked by the network constructor, at the same time initializing the output and activation data members of the output neuron each to zero. To make sure there is access to needed information and functions, the output neuron is declared a **friend** class in the class **ineuron**. The network is declared as a **friend** class in the class **oneuron**.

Source Code for Perceptron Network

Listing 4.4 contains the source code in **percept.cpp** for the C++ implementation of the Perceptron model previously discussed.

Listing 4.4 Source code for Perceptron model.

```
//percept.cpp  V. Rao, H. Rao
//Perceptron model

#include "percept.h"
#include "stdio.h"
#include "stdlib.h"

ineuron::ineuron(float j)
{
weight= j;
}

float ineuron::act(float x)
{
float a;

a = x*weight;

return a;
}

void oneuron::actvtion(float *inputv, ineuron *nrn)
{
int i;
```

```
activation = 0;

for(i=0;i<4;i++)
        {
        cout<<"\nweight for neuron "<<i+1<<" is  "<<nrn[i].weight;
        nrn[i].activation = nrn[i].act(inputv[i]);
        cout<<"            activation is "<<nrn[i].activation;
        activation += nrn[i].activation;
        }

cout<<"\n\nactivation is  "<<activation<<"\n";
}


int oneuron::outvalue(float j)
{
if(activation>=j)
        {
        cout<<"\nthe output neuron activation \
exceeds the threshold value of "<<j<<"\n";
        output = 1;
        }
else
        {
        cout<<"\nthe output neuron activation \
is smaller than the threshold value of "<<j<<"\n";
        output = 0;
        }

cout<<" output value is "<< output;
return (output);
}


network::network(float a,float b,float c,float d)
{
nrn[0] = ineuron(a) ;
nrn[1] = ineuron(b) ;
nrn[2] = ineuron(c) ;
nrn[3] = ineuron(d) ;
onrn = oneuron();
```

```
onrn.activation = 0;
onrn.output = 0;
}
void main (int argc, char * argv[])
{

float inputv1[]= {1.95,0.27,0.69,1.25};
float wtv1[]= {2,3,3,2}, wtv2[]= {3,0,6,2};
FILE * wfile, * infile;
int num=0, vecnum=0, i;
float threshold = 7.0;

if (argc < 2)
        {
        cerr << "Usage: percept Weightfile Inputfile";
        exit(1);
        }
// open  files

wfile= fopen(argv[1], "r");
infile= fopen(argv[2], "r");

if ((wfile == NULL) || (infile == NULL))
        {
        cout << " Can't open a file\n";
        exit(1);
        }

cout<<"\nTHIS PROGRAM IS FOR A PERCEPTRON NETWORK WITH AN INPUT LAYER
OF";
cout<<"\n4 NEURONS, EACH CONNECTED TO THE OUTPUT NEURON.\n";
cout<<"\nTHIS EXAMPLE TAKES REAL NUMBERS AS INPUT SIGNALS\n";

//create the network by calling its constructor.
//the constructor calls neuron constructor as many times as the number of
//neurons in input layer of the network.

cout<<"please enter the number of weights/vectors \n";
cin >> vecnum;
```

```
for (i=1;i<=vecnum;i++)
        {
        fscanf(wfile,"%f %f %f %f\n",
&wtv1[0],&wtv1[1],&wtv1[2],&wtv1[3]);
        network h1(wtv1[0],wtv1[1],wtv1[2],wtv1[3]);
        fscanf(infile,"%f %f %f %f \n",
                &inputv1[0],&inputv1[1],&inputv1[2],&inputv1[3]);
        cout<<"this is vector # " << i << "\n";
        cout << "please enter a threshold value, eg 7.0\n";
        cin >> threshold;

        h1.onrn.actvtion(inputv1, h1.nrn);
        h1.onrn.outvalue(threshold);
        cout<<"\n\n";
        }

fclose(wfile);
fclose(infile);
}
```

Comments on Your C++ Program

Notice the use of input stream operator **cin>>** in the C++ program, instead of the C function **scanf** in several places. The **iostream** class in C++ was discussed earlier in this chapter. The program works like this:

First, the network input neurons are given their connection weights, and then an input vector is presented to the input layer. A threshold value is specified, and the output neuron does the weighted sum of its inputs, which are the outputs of the input layer neurons. This weighted sum is the activation of the output neuron, and it is compared with the threshold value, and the output neuron fires (output is 1) if the threshold value is not greater than its activation. It does not fire (output is 0) if its activation is smaller than the threshold value. In this implementation, neither supervised nor unsupervised training is incorporated.

Input/Output for percept.cpp

There are two data files used in this program. One is for setting up the weights, and the other for setting up the input vectors. On the command line, you enter the program name followed by the weight file name and the input file name. For this discussion (also on the accompanying disk for this book) create a file called weight.dat, which contains the following data:

2.0 3.0 3.0 2.0
3.0 0.0 6.0 2.0

These are two weight vectors. Create also an input file called input.dat with the two data vectors below:

1.95 0.27 0.69 1.25
0.30 1.05 0.75 0.19

During the execution of the program, you are first prompted for the number of vectors that are used (in this case, 2), then for a threshold value for the input/weight vectors (use 7.0 in both cases). You will then see the following output. Note that the user input is in italic.

percept weight.dat input.dat

THIS PROGRAM IS FOR A PERCEPTRON NETWORK WITH AN INPUT LAYER OF 4 NEU-
RONS, EACH CONNECTED TO THE OUTPUT
NEURON.

THIS EXAMPLE TAKES REAL NUMBERS AS INPUT SIGNALS
please enter the number of weights/vectors
2
this is vector # 1
please enter a threshold value, eg 7.0
7.0

weight for neuron 1 is 2	activation is 3.9
weight for neuron 2 is 3	activation is 0.81
weight for neuron 3 is 3	activation is 2.07
weight for neuron 4 is 2	activation is 2.5

activation is 9.28

the output neuron activation exceeds the threshold value of 7
 output value is 1

```
this is vector # 2
please enter a threshold value, eg 7.0
7.0


weight for neuron 1 is   3          activation is 0.9
weight for neuron 2 is   0          activation is 0
weight for neuron 3 is   6          activation is 4.5
weight for neuron 4 is   2          activation is 0.38


activation is   5.78


the output neuron activation is smaller than the threshold value of 7
output value is 0
```

Finally, try adding a data vector of (1.4, 0.6, 0.35, 0.99) to the data file. Add a weight vector of (2, 6, 8, 3) to the weight file and use a threshold value of 8.25 to see the result. You can use other values to experiment also.

NETWORK MODELING

So far, we have considered the construction of two networks, the Hopfield memory and the Perceptron. What are other considerations (which will be discussed in more depth in the chapters to follow) that you should keep in mind ?

Some of the considerations that go into the modeling of a neural network for an application are:

nature of inputs

      fuzzy

             binary

             analog

     crisp

             binary

             analog

number of inputs

nature of outputs

      fuzzy

            binary

            analog

      crisp

            binary

            analog

number of outputs

nature of the application

         to complete patterns (recognize corrupted patterns)

         to classify patterns

         to do an optimization

         to do approximation

         to perform data clustering

         to compute functions

dynamics

      adaptive

          learning

             training

                with exemplars

                without exemplars

          self-organizing

      nonadaptive

          learning

training
    with exemplars
    without exemplars
self-organizing

hidden layers

    number
        fixed
        variable
    sizes
        fixed
        variable

processing

    additive
    multiplicative
    hybrid
        additive and multiplicative
        combining other approaches
            expert systems
            genetic algorithms

Hybrid models, as indicated above, could be of the variety of combining neural network approach with expert system methods or of combining additive and multiplicative processing paradigms.

Decision support systems are amenable to approaches that combine neural networks with expert systems. An example of a hybrid model that combines different modes of processing by neurons is the *Sigma Pi* neural network, wherein one layer of neurons uses summation in aggregation and the next layer of neurons uses multiplicative processing.

A hidden layer, if only one, in a neural network is a layer of neurons that operates in between the input layer and the output layer of the network.

Neurons in this layer receive inputs from those in the input layer and supply their outputs as the inputs to the neurons in the output layer. When a hidden layer comes in between other hidden layers, it receives input and supplies input to the respective hidden layers.

In modeling a network, it is often not easy to determine how many, if any, hidden layers, and of what sizes, are needed in the model. Some approaches, like genetic algorithms—which are paradigms competing with neural network approaches in many situations but nevertheless can be cooperative, as here—are at times used to make a determination on the needed or optimum, as the case may be, numbers of hidden layers and/or the neurons in those hidden layers. In what follows, we outline one such application.

TIC-TAC-TOE ANYONE?

David Fogel describes evolutionary general problem solving and uses the familiar game of Tic-Tac-Toe as an example. The idea is to come up with optimal strategies in playing this game. The first player's marker is an X, and the second player's marker is an O. Whoever gets three of his or her markers in a row or a column or a diagonal before the other player does, wins. Shrewd players manage a draw position, if their equally shrewd opponent thwarts their attempts to win. A draw position is one where neither player has three of his or her markers in a row, or a column, or a diagonal.

The board can be described by a vector of nine components, each of which is a three-valued number. Imagine the squares of the board for the game as taken in sequence row by row from top to bottom. Allow a 1 to show the presence of an X in that square, a 0 to indicate a blank there, and a -1 to correspond to an O. This is an example of a coding for the status of the board. For example, (-1, 0, 1, 0, -1, 0, 1, 1, -1) is a winning position for the second player, because it corresponds to the board looking as below.

```
O           X

        O

X       X       O
```

A neural network for this problem will have an input layer with nine neurons, as each input pattern has nine components. There would be some hidden layers. But the example is with one hidden layer. The output layer also contains nine neurons, so that one cycle of operation of the network shows what the best configuration of the board is to be, given a particular input. Of course, during this cycle of operation, all that needs to be determined is which blank space, indicated by a 0 in the input, should be changed to 1, if strategy is being worked for player 1. None of the 1's and -1's is to be changed.

In this particular example, the neural network architecture itself is dynamic. The network expands or contracts according to some rules, which are described next.

Fogel describes the network as an evolving network in the sense that the number of neurons in the hidden layer changed with a probability of 0.5. A node was equally likely to be added or deleted. Since the number of unmarked squares dwindles after each play, this kind of approach with varying numbers of neurons in the network seems to be reasonable, and interesting.

The initial set of weights is random values between -0.5 and 0.5, inclusive, according to a uniform distribution. Bias and threshold values also come from this distribution. The sigmoid function:

$$1/(1 + e^x)$$

is used also, to determine the outputs.

Weights and biases were changed during the network operation training cycles. Thus, the network had a learning phase. (You will read more on learning in Chapter 6.) This network is adaptive, since it changes its architecture. Other forms of adaptation in neural networks are in changing parameter values for a fixed architecture. (See Chapter 6.) The results of the experiment Fogel describes show that you need nine neurons in the hidden layer also, for your network to be the best for this problem. They also purged any strategy that was likely to lose.

Fogel's emphasis is on the evolutionary aspect of an adaptive process or experiment. Our interest in this example is primarily due to the fact that an adaptive neural network is used.

The choice of Tic-Tac-Toe, while being a simple and all too familiar game, is in the genre of much more complicated games. These games ask a player to place a marker in some position in a given array, and as players take turns doing so, some criterion determines if it is a draw, or who won. Unlike in Tic-Tac-Toe, the criterion by which one wins may not be known to the players.

STABILITY AND PLASTICITY

We discuss now a few other considerations in neural network modeling by introducing short-term memory and long-term memory concepts. Neural network training is usually done in an *iterative* way, meaning that the procedure is repeated a certain number of times. These iterations are referred to as *cycles.* After each cycle, the input used may remain the same or change, or the weights may remain the same or change. Such change is based on the output of a completed cycle. If the number of cycles is not preset, and the network is allowed to go through cycles until some other criterion is met, the question of whether or not the termination of the iterative process occurs eventually, arises naturally.

Stability for a Neural Network

Stability refers to such convergence that facilitates an end to the iterative process. For example, if any two consecutive cycles result in the same output for the network, then there may be no need to do more iterations. In this case, convergence has occurred, and the network has stabilized in its operation. If weights are being modified after each cycle, then convergence of weights would constitute stability for the network.

In some situations, it takes many more iterations than you desire, to have output in two consecutive cycles to be the same. Then a tolerance level on the convergence criterion can be used. With a tolerance level, you accomplish early but satisfactory termination of the operation of the network.

Plasticity for a Neural Network

Suppose a network is trained to learn some patterns, and in this process the weights are adjusted according to an algorithm. After learning these patterns and encountering a new pattern, the network may modify the weights in order to learn the new pattern. But what if the new weight structure is not responsive to the new pattern? Then the network does not possess *plasticity*—the ability to deal satisfactorily with new short-term memory (STM) while retaining long-term memory (LTM). Attempts to endow a network with plasticity may have some adverse effects on the stability of your network.

SHORT-TERM MEMORY AND LONG-TERM MEMORY

We alluded to short-term memory (STM) and long-term memory (LTM) in the previous paragraph. STM is basically the information that is currently and perhaps temporarily being processed. It is manifested in the patterns that the network encounters. LTM, on the other hand, is information that is already stored and is not being currently processed. In a neural network, STM is usually characterized by patterns and LTM is characterized by the connections' weights. The weights determine how an input is processed in the network to yield output. During the cycles of operation of a network, the weights may change. After convergence, they represent LTM, as the weight levels achieved are stable.

SUMMARY

You saw in this chapter, the C++ implementations of a simple Hopfield network and of a simple Perceptron network. What have not been included in them is an automatic iteration and a learning algorithm. They were not necessary for the examples that were used in this chapter to show C++ implementation, the emphasis was on the method of implementation. In a later chapter, you will read about the learning algorithms and examples of how to implement some of them.

Considerations in modeling a neural network are presented in this chapter along with an outline of how Tic-Tac-Toe is used as an example of an adaptive neural network model.

You also were introduced to the following concepts: stability, plasticity, short-term memory, and long-term memory (discussed further in later chapters). Much more can be said about them, in terms of the so-called noise-saturation dilemma, or stability–plasticity dilemma and what research has developed to address them (for further reading, see References).

CHAPTER 5

A Survey of Neural Network Models

Neural Network Models

You were introduced in the preceding pages to the Perceptron model, the Feedforward network, and the Hopfield network. You learned that the differences between the models lie in their architecture, encoding, and recall. We aim now to give you a comprehensive picture of these and other neural network models. We will show details and implementations of some networks in later chapters.

The models we briefly review in this chapter are the Perceptron, Hopfield, Adaline, Feed-Forward Backpropagation, Bidirectional Associative Memory, Brain-State-in-a-Box, Neocognitron, Fuzzy Associative Memory, ART1, and ART2. C++ implementations of some of these and the role of fuzzy logic in some will be treated in the subsequent chapters. For now, our discussion will be about the distinguishing characteristics of a neural network. We will follow it with the description of some of the models.

Layers in a Neural Network

A neural network has its neurons divided into subgroups, or fields, and elements in each subgroup are placed in a row, or a column, in the diagram depicting the network. Each subgroup is then referred to as a layer of neurons in the network. A great many models of neural networks have two layers, quite a few have one layer, and some have three or more layers. A number of additional, so-called hidden layers are possible in some networks, such as the Feed-forward backpropagation network. When the network has a single layer, the input signals are received at that layer, processing is done by its neurons, and output is generated at that layer. When more than one layer is present, the first field is for the neurons that supply the input signals for the neurons in the next layer.

Every network has a layer of input neurons, but in most of the networks, the sole purpose of these neurons is to feed the input to the next layer of neurons. However, there are feedback connections, or recurrent connections in some networks, so that the neurons in the input layer may also do some processing. In the Hopfield network you saw earlier, the input and output layers are the same. If any layer is present between the input and output layers, it may be referred to as a hidden layer in general, or as a layer with a special name taken after the researcher who proposed its inclusion to achieve certain performance from the network. Examples are the Grossberg and the Kohonen layers. The number of hidden layers is not limited except by the scope of the problem being addressed by the neural network.

A layer is also referred to as a *field*. Then the different layers can be designated as field A, field B, and so on, or shortly, F_A, F_B.

Single-Layer Network

A neural network with a single layer is also capable of processing for some important applications, such as integrated circuit implementations or assembly line control. The most common capability of the different models of neural networks is pattern recognition. But one network, called the *Brain-State-in-a-Box*, which is a single-layer neural network, can do pattern completion. *Adaline* is a network with A and B fields of neurons, but *aggregation* or processing of input signals is done only by the field B neurons.

The Hopfield network is a single-layer neural network. The Hopfield network makes an association between different patterns (heteroassociation) or associates a pattern with itself (autoassociation). You may characterize this as being able to recognize a given pattern. The idea of viewing it as a case of pattern recognition becomes more relevant if a pattern is presented with some noise, meaning that there is some slight deformation in the pattern, and if the network is able to relate it to the correct pattern.

The Perceptron technically has two layers, but has only one group of weights. We therefore still refer to it as a single-layer network. The second layer consists solely of the output neuron, and the first layer consists of the neurons that receive input(s). Also, the neurons in the same layer, the input layer in this case, are not interconnected, that is, no connections are made between two neurons in that same layer. On the other hand, in the Hopfield network, there is no separate output layer, and hence, it is strictly a single-layer network. In addition, the neurons are all fully connected with one another.

Let us spend more time on the single-layer Perceptron model and discuss its limitations, and thereby motivate the study of multilayer networks.

XOR FUNCTION AND THE PERCEPTRON

The ability of a Perceptron in evaluating functions was brought into question when Minsky and Papert proved that a simple function like XOR (the logical function exclusive or) could not be correctly evaluated by a Perceptron. The XOR logical function, f(A,B), is as follows:

A	B	f(A,B)= XOR(A,B)
0	0	0
0	1	1
1	0	1
1	1	0

To summarize the behavior of the XOR, if both inputs are the same value, the output is 0, otherwise the output is 1.

Minsky and Papert showed that it is impossible to come up with the proper set of weights for the neurons in the single layer of a simple Perceptron to evaluate the XOR function. The reason for this is that such a Perceptron, one with a single layer of neurons, requires the function to be evaluated, to be linearly separable by means of the function values. The concept of *linear separability* is explained next. But let us show you first why the simple perceptron fails to compute this function.

Since there are two arguments for the XOR function, there would be two neurons in the input layer, and since the function's value is one number, there would be one output neuron. Therefore, you need two weights w_1 and w_2 ,and a threshold value θ. Let us now look at the conditions to be satisfied by the w's and the θ so that the outputs corresponding to given inputs would be as for the XOR function.

First the output should be 0 if inputs are 0 and 0. The activation works out as 0. To get an output of 0, you need $0 < \theta$. This is your first condition. Table 5.1 shows this and two other conditions you need, and why.

Table 5.1 Conditions on Weights

Input	Activation	Output	Needed Condition
0, 0	0	0	$0 < \theta$
1, 0	w_1	1	$w_1 > \theta$
0, 1	w_2	1	$w_2 > \theta$
1, 1	$w_1 + w_2$	0	$w_1 + w_2 < \theta$

From the first three conditions, you can deduce that the sum of the two weights has to be greater than θ, which has to be positive itself. Line 4 is inconsistent with lines 1, 2, and 3, since line 4 requires the sum of the two weights to be less than θ. This affirms the contention that it is not possible to compute the XOR function with a simple perceptron.

Geometrically, the reason for this failure is that the inputs (0, 1) and (1, 0) with which you want output 1, are situated diagonally opposite each other, when plotted as points in the plane, as shown below in a diagram of the output (1=T, 0=F):

F   T

T   F

You can't separate the T's and the F's with a straight line. This means that you cannot draw a line in the plane in such a way that neither (1, 1) ->F nor (0, 0)->F is on the same side of the line as (0, 1) ->T and (1, 0)-> T.

Linear Separability

What linearly separable means is, that a type of a linear barrier or a separator—a line in the plane, or a plane in the three-dimensional space, or a hyperplane in higher dimensions—should exist, so that the set of inputs that give rise to one value for the function all lie on one side of this barrier, while on the other side lie the inputs that do not yield that value for the function. A *hyperplane* is a surface in a higher dimension, but with a linear equation defining it much the same way a line in the plane and a plane in the three-dimensional space are defined.

To make the concept a little bit clearer, consider a problem that is similar but, let us emphasize, not the same as the XOR problem.

Imagine a cube of 1-unit length for each of its edges and lying in the positive octant in a xyz-rectangular coordinate system with one corner at the origin. The other corners or vertices are at points with coordinates (0, 0, 1), (0, 1, 0), (0, 1, 1), (1, 0, 0), (1, 0, 1), (1, 1, 0), and (1, 1, 1). Call the origin O, and the seven points listed as A, B, C, D, E, F, and G, respectively. Then any two faces opposite to each other are linearly separable because you can define the separating plane as the plane halfway between these two faces and also parallel to these two faces.

For example, consider the faces defined by the set of points O, A, B, and C and by the set of points D, E, F, and G. They are parallel and 1 unit apart, as you can see in Figure 5.1. The separating plane for these two faces can be seen to be one of many possible planes—any plane in between them and parallel to them. One example, for simplicity, is the plane that passes through the points (1/2, 0, 0), (1/2, 0, 1), (1/2, 1, 0), and (1/2, 1, 1). Of course, you need only specify three of those four points because a plane is uniquely determined by three points that are not all on the same line. So if the first set of points corresponds to a value of say, +1 for the function, and the second set to a value of −1, then a single-layer Perceptron can determine, through some training algorithm, the correct weights for the connections, even if you start with the weights being initially all 0.

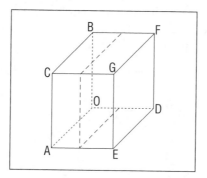

FIGURE 5.1 SEPARATING PLANE.

Consider the set of points O, A, F, and G. This set of points cannot be linearly separated from the other vertices of the cube. In this case, it would be impossible for the single-layer Perceptron to determine the proper weights for the neurons in evaluating the type of function we have been discussing.

A Second Look at the XOR Function: Multilayer Perceptron

By introducing a set of cascaded Perceptrons, you have a Perceptron network, with an input layer, middle or hidden layer, and an output layer. You will see that the multilayer Perceptron can evaluate the XOR function as well as other logic functions (AND, OR, MAJORITY, etc.). The absence of the separability that we talked about earlier is overcome by having a second stage, so to speak, of connection weights.

You need two neurons in the input layer and one in the output layer. Let us put a hidden layer with two neurons. Let w_{11}, w_{12}, w_{21}, and w_{22}, be the weights on connections from the input neurons to the hidden layer neurons. Let v_1, v_2, be the weights on the connections from the hidden layer neurons to the outout neuron.

We will select the w's (weights) and the threshold values θ_1, and θ_2 at the hidden layer neurons, so that the input (0, 0) generates the output vector (0, 0), and the input vector (1, 1) generates (1, 1), while the inputs (1, 0) and (0, 1) generate (0, 1) as the hidden layer output. The inputs to the output layer neurons would be from the set {(0, 0), (1, 1), (0, 1)}. These three vectors are separable, with (0, 0), and (1, 1) on one side of the separating line, while (0, 1) is on the other side.

We will select the v's (weights) and τ, the threshold value at the output neuron, so as to make the inputs (0, 0) and (1, 1) cause an output of 0 for the network, and an output of 1 is caused by the input (0, 1). The network layout within the labels of weights and threshold values inside the nodes representing hidden layer and output neurons is shown in Figure 5.1a. Table 5.2 gives the results of operation of this network.

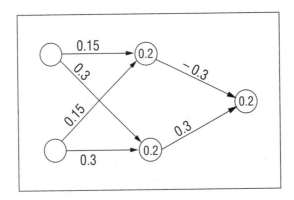

FIGURE 5.1A EXAMPLE NETWORK.

TABLE 5.2 RESULTS FOR THE PERCEPTRON WITH ONE HIDDEN LAYER.

Input	Hidden Layer Activations	Hidden Layer Outputs	Output Neuron activation	Output of network
(0, 0)	(0, 0)	(0, 0)	0	0
(1, 1)	(0.3, 0.6)	(1, 1)	0	0
(0, 1)	(0.15, 0.3)	(0, 1)	0.3	1
(1, 0)	(0.15, 0.3)	(0, 1)	0.3	1

It is clear from Table 5.2, that the above perceptron with a hidden layer does compute the XOR function successfully.

The activation should exceed the threshold value for a neuron to fire. Where the output of a neuron is shown to be 0, it is because the internal activation of that neuron fell short of its threshold value.

N O T E

Example of the Cube Revisited

Let us return to the example of the cube with vertices at the origin O, and the points labeled A, B, C, D, E, F, and G. Suppose the set of vertices O, A, F, and G give a value of 1 for the function to be evaluated, and the other vertices give a –1. The two sets are not linearly separable as mentioned before. A simple Perceptron cannot evaluate this function.

Can the addition of another layer of neurons help? The answer is yes. What would be the role of this additional layer? The answer is that it will do the final processing for the problem after the previous layer has done some preprocessing. This can do two separations in the sense that the set of eight vertices can be separated—or partitioned—into three separable subsets. If this partitioning can also help collect within each subset, like vertices, meaning those that map onto the same value for the function, the network will succeed in its task of evaluating the function when the aggregation and thresholding is done at the output neuron.

Strategy

So the strategy is first to consider the set of vertices that give a value of +1 for the function and determine the minimum number of subsets that can be identified to be each separable from the rest of the vertices. It is evident that since the vertices O and A lie on one edge of the cube, they can form one subset that is separable. The other two vertices, viz., F and one for G, which correspond to the value +1 for the function, can form a second subset that is separable, too. We need not bother with the last four vertices from the point of view of further partitioning that subset. It is clear that one new layer of three neurons, one of which fires for the inputs corresponding to the vertices O and A, one for F, and G, and the third for the rest, will then facilitate the correct evaluation of the function at the output neuron.

Details

Table 5.3 lists the vertices and their coordinates, together with a flag that indicates to which subset in the partitioning the vertex belongs. Note that you can think of the action of the Multilayer Perceptron as that of evaluating the intersection and union of linearly separable subsets.

TABLE 5.3 PARTITIONING OF VERTICES OF A CUBE

Vertex	Coordinates	Subset
O	(0,0,0)	1
A	(0,0,1)	1
B	(0,1,0)	2
C	(0,1,1)	2
D	(1,0,0)	2
E	(1,0,1)	2
F	(1,1,0)	3
G	(1,1,1)	3

The network, which is a two-layer Perceptron, meaning two layers of weights, has three neurons in the first layer and one output neuron in the second layer. Remember that we are counting those layers in which the neurons do the aggregation of the signals coming into them using the connection weights. The first layer with the three neurons is what is generally described as the hidden layer, since the second layer is not hidden and is at the extreme right in the layout of the neural network. Table 5.4 gives an example of the weights you can use for the connections between the input neurons and the hidden layer neurons. There are three input neurons, one for each coordinate of the vertex of the cube.

TABLE 5.4 WEIGHTS FOR CONNECTIONS BETWEEN INPUT NEURONS AND HIDDEN LAYER NEURONS.

Input Neuron #	Hidden Layer Neuron #	Connection Weight
1	1	1
1	2	0.1
1	3	-1
2	1	1
2	2	-1
2	3	-1
3	1	0.2
3	2	0.3
3	3	0.6

Now we give, in Table 5.5, the weights for the connections between the three hidden-layer neurons and the output neuron.

TABLE 5.5 WEIGHTS FOR CONNECTION BETWEEN THE HIDDEN-LAYER NEURONS AND THE OUTPUT NEURON

Hidden Layer Neuron #	Connection Weight
1	0.6
3	0.3
3	0.6

It is not apparent whether or not these weights will do the job. To determine the activations of the hidden-layer neurons, you need these weights, and you also need the threshold value at each neuron that does processing. A hidden-layer neuron will fire, that is, will output a 1, if the weighted sum of the signals it receives is greater than the threshold value. If the output neuron fires, the function value is taken as +1, and if it does not fire, the function value is −1. Table 5.6 gives the threshold values. Figure 5.1b shows the neural network with connection weights and threshold values.

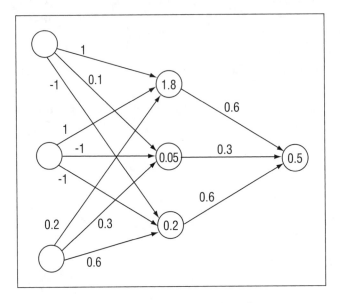

FIGURE 5.1B NEURAL NETWORK FOR CUBE EXAMPLE

TABLE 5.6 THRESHOLD VALUES

Layer	Neuron	Threshold Value
hidden	1	1.8
hidden	2	0.05
hidden	3	-0.2
output	1	0.5

Performance of the Perceptron

When you input the coordinates of the vertex G, which has 1 for each coordinate, the first hidden-layer neuron aggregates these inputs and gets a value of 2.2. Since 2.2 is more than the threshold value of the first neuron in the hidden layer, that neuron fires, and its output of 1 becomes an input to the output neuron on the connection with weight 0.6. But you need the activations of the other hidden-layer neurons as well. Let us describe the performance with coordinates of G as the inputs to the network. Table 5.7 describes this.

TABLE 5.7 RESULTS WITH COORDINATES OF VERTEX G AS INPUT

Vertex/ Coordinates	Hidden Layer	Weighted Sum	Comment	Activation	Contribution to Output	Sum
G:1,1,1	1	2.2	>1.8	1	0.6	
	2	-0.8	<0.05	0	0	
	3	-1.4	<-0.2	0	0	0.6

The weighted sum at the output neuron is 0.6, and it is greater than the threshold value 0.5. Therefore, the output neuron fires, and at the vertex G, the function is evaluated to have a value of +1.

Table 5.8 shows the performance of the network with the rest of the vertices of the cube. You will notice that the network computes a value of +1 at the vertices, O, A, F, and G, and a −1 at the rest.

TABLE 5.8 RESULTS WITH OTHER INPUTS

	Hidden Layer Neuron#	Weighted Sum	Comment	Activation	Contribution to output	Sum
O :0, 0, 0	1	0	<1.8	0	0	
	2	0	<0.05	0	0	
	3	0	>-0.2	1	0.6	0.6 *
A :0, 0, 1	1	0.2	<1.8	0	0	
	2	0.3	>0.05	1	0.3	
	3	0.6	>-0.2	1	0.6	0.9*
B :0, 1, 0	1	1	<1.8	0	0	
	2	-1	<0.05	0	0	
	3	-1	<-0.2	0	0	0
C :0, 1, 1	1	1.2	<1.8	0	0	
	2	0.2	>0.05	1	0.3	
	3	-0.4	<-0.2	0	0	0.3

D :1, 0, 0	1	1	<1.8	0	0	
	2	.1	>0.05	1	0.3	
	3	-1	<-0.2	0	0	0.3
E :1, 0, 1	1	1.2	<1.8	0	0	
	2	0.4	>0.05	1	0.3	
	3	-0.4	<-0.2	0	0	0.3
F :1, 1, 0	1	2	>1.8	1	0.6	
	2	-0.9	<0.05	0	0	
	3	-2	<-0.2	0	0	0.6*

*The output neuron fires, as this value is greater than 0.5 (the threshold value); the function value is +1.

Other Two-Layer Networks

Many important neural network models have two layers. The Feedforward backpropagation network, in its simplest form, is one example. Grossberg and Carpenter's ART1 paradigm uses a two-layer network. The Counterpropagation network has a Kohonen layer followed by a Grossberg layer. Bidirectional Associative Memory, (BAM), Boltzman Machine, Fuzzy Associative Memory, and Temporal Associative Memory are other two-layer networks. For autoassociation, a single-layer network could do the job, but for heteroassociation or other such mappings, you need at least a two-layer network. We will give more details on these models shortly.

Many Layer Networks

Kunihiko Fukushima's Neocognitron, noted for identifying handwritten characters, is an example of a network with several layers. Some previously mentioned networks can also be multilayered from the addition of more hidden layers. It is also possible to combine two or more neural networks into one network by creating appropriate connections between layers of one subnetwork to those of the others. This would certainly create a multilayer network.

CONNECTIONS BETWEEN LAYERS

You have already seen some difference in the way connections are made between neurons in a neural network. In the Hopfield network, every neuron was connected to every other in the one layer that was present in the network. In the Perceptron, neurons within the same layer were not connected with one another, but the connections were between the neurons in one layer and those in the next layer. In the former case, the connections are described as being lateral. In the latter case, the connections are forward and the signals are fed forward within the network.

Two other possibilities also exist. All the neurons in any layer may have extra connections, with each neuron connected to itself. The second possibility is that there are connections from the neurons in one layer to the neurons in a previous layer, in which case there is both forward and backward signal feeding. This occurs, if feedback is a feature for the network model. The type of layout for the network neurons and the type of connections between the neurons constitute the architecture of the particular model of the neural network.

INSTAR AND OUTSTAR

Outstar and *instar* are terms defined by Stephen Grossberg for ways of looking at neurons in a network. A neuron in a web of other neurons receives a large number of inputs from outside the neuron's boundaries. This is like an inwardly radiating star, hence, the term instar. Also, a neuron may be sending its output to many other destinations in the network. In this way it is acting as an outstar. Every neuron is thus simultaneously both an instar and an outstar. As an instar it receives stimuli from other parts of the network or from outside the network. Note that the neurons in the input layer of a network primarily have connections away from them to the neurons in the next layer, and thus behave mostly as outstars. Neurons in the output layer have many connections coming to it and thus behave mostly as instars. A neural network performs its work through the constant interaction of instars and outstars.

A layer of instars can constitute a competitive layer in a network. An outstar can also be described as a source node with some associated sink nodes that the source feeds to. Grossberg identifies the source input with a conditioned stimulus and the sink inputs with unconditioned stimuli. Robert Hecht-Nielsen's Counterpropagation network is a model built with instars and outstars.

WEIGHTS ON CONNECTIONS

Weight assignments on connections between neurons not only indicate the strength of the signal that is being fed for aggregation but also the type of interaction between the two neurons. The type of interaction is one of cooperation or of competition. The cooperative type is suggested by a positive weight, and the competition by a negative weight, on the connection. The positive weight connection is meant for what is called *excitation,* while the negative weight connection is termed an *inhibition.*

Initialization of Weights

Initializing the network weight structure is part of what is called the *encoding phase* of a network operation. The encoding algorithms are several, differing by model and by application. You may have gotten the impression that the weight matrices used in the examples discussed in detail thus far have been arbitrarily determined; or if there is a method of setting them up, you are not told what it is.

It is possible to start with randomly chosen values for the weights and to let the weights be adjusted appropriately as the network is run through successive iterations. This would make it easier also. For example, under supervised training, if the error between the desired and computed output is used as a criterion in adjusting weights, then one may as well set the initial weights to zero and let the training process take care of the rest. The small example that follows illustrates this point.

A Small Example

Suppose you have a network with two input neurons and one output neuron, with forward connections between the input neurons and the output neuron, as shown in Figure 5.2. The network is required to output a 1 for the input patterns (1, 0) and (1, 1), and the value 0 for (0, 1) and (0, 0). There are only two connection weights w_1 and w_2.

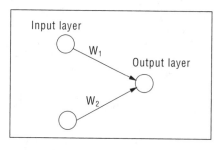

FIGURE 5.2 NEURAL NETWORK WITH FORWARD CONNECTIONS.

Let us set initially both weights to 0, but you need a threshold function also. Let us use the following threshold function, which is slightly different from the one used in a previous example:

$$f(x) = \begin{cases} 1 \text{ if } x > 0 \\ \\ 0 \text{ if } x \leq 0 \end{cases}$$

The reason for modifying this function is that if $f(x)$ has value 1 when $x = 0$, then no matter what the weights are, the output will work out to 1 with input $(0, 0)$. This makes it impossible to get a correct computation of any function that takes the value 0 for the arguments $(0, 0)$.

Now we need to know by what procedure we adjust the weights. The procedure we would apply for this example is as follows.

- If the output with input pattern (a, b) is as desired, then do not adjust the weights.
- If the output with input pattern (a, b) is smaller than what it should be, then increment each of w_1 and w_2 by 1.
- If the output with input pattern (a, b) is greater than what it should be, then subtract 1 from w_1 if the product aw_1 is smaller than 1, and adjust w_2 similarly.

Table 5.9 shows what takes place when we follow these procedures, and at what values the weights settle.

TABLE 5.9 ADJUSTMENT OF WEIGHTS

step	w_1	w_2	a	b	activation	output	comment
1	0	0	1	1	0	0	desired output is 1; increment both w's
2	1	1	1	1	2	1	output is what it should be
3	1	1	1	0	1	1	output is what it should be
4	1	1	0	1	1	1	output is 1; it should be 0.
5							subtract 1 from w_2
6	1	0	0	1	0	0	output is what it should be
7	1	0	0	0	0	0	output is what it should be
8	1	0	1	1	1	1	output is what it should be
9	1	0	1	0	1	1	output is what it should be

Table 5.9 shows that the network weight vector changed from an initial vector (0, 0) to the final weight vector (1, 0) in eight iterations. This example is not of a network for pattern matching. If you think about it, you will realize that the network is designed to fire if the first digit in the pattern is a 1, and not otherwise. An analogy for this kind of a problem is determining if a given image contains a specific object in a specific part of the image, such as a dot should occur in the letter i.

If the initial weights are chosen somewhat prudently and to make some particular relevance, then the speed of operation can be increased in the sense of convergence being achieved with fewer iterations than otherwise. Thus, encoding algorithms are important. We now present some of the encoding algorithms.

Initializing Weights for Autoassociative Networks

Consider a network that is to associate each input pattern with itself and which gets binary patterns as inputs. Make a bipolar mapping on the input pattern. That is, replace each 0 by −1. Call the mapped pattern the vector **x**, when written as a column vector. The transpose, the same vector written as a row vector, is **x**^{T}. You will get a matrix of order the size of **x** when you form the product **xx**^{T}. Obtain similar matrices for the other patterns you want the network to store. Add these matrices to give you the matrix of weights to be used initially, as we did in Chapter 4. This process can be described with the following equation:

$$W = \Sigma_i\, x_i x_i{}^T$$

Weight Initialization for Heteroassociative Networks

Consider a network that is to associate one input pattern with another pattern and that gets binary patterns as inputs. Make a bipolar mapping on the input pattern. That is, replace each 0 by −1. Call the mapped pattern the vector **x** when written as a column vector. Get a similar bipolar mapping for the corresponding associated pattern. Call it **y**. You will get a matrix of size **x** by size **y** when you form the product **xy**^{T}. Obtain similar matrices for the other patterns you want the network to store. Add these matrices to give you the matrix of weights to be used initially. The following equation restates this process:

$$W = \Sigma_i\, x_i y_i{}^T$$

ON CENTER, OFF SURROUND

In one of the many interesting paradigms you encounter in neural network models and theory, is the strategy *winner takes all*. Well, if there should be one winner emerging from a crowd of neurons in a particular layer, there needs to be competition. Since everybody is for himself in such a competition, in this case every neuron for itself, it would be necessary to have lateral connections that indicate this circumstance. The lateral connections from any neuron to the others should have a negative weight. Or, the neuron with the highest activation is considered the winner and only its weights are modified in the training process, leaving the weights of others the same. Winner takes all means that only one neuron in that layer fires and the others do not. This can happen in a hidden layer or in the output layer.

In another situation, when a particular category of input is to be identified from among several groups of inputs, there has to be a subset of the neurons that are dedicated to seeing it happen. In this case, inhibition increases for distant neurons, whereas excitation increases for the neighboring ones, as far as such a subset of neurons is concerned. The phrase *on center, off surround* describes this phenomenon of distant inhibition and near excitation.

Weights also are the prime components in a neural network, as they reflect on the one hand the memory stored by the network, and on the other hand the basis for learning and training.

INPUTS

You have seen that mutually orthogonal or almost orthogonal patterns are required as stable stored patterns for the Hopfield network, which we discussed before for pattern matching. Similar restrictions are found also with other neural networks. Sometimes it is not a restriction, but the purpose of the model makes natural a certain type of input. Certainly, in the context of pattern classification, binary input patterns make problem setup simpler. Binary, bipolar, and analog signals are the varieties of inputs. Networks that accept analog signals as inputs are for continuous models, and those that require binary or bipolar inputs are for discrete models. Binary inputs can be fed to networks for continuous models, but analog signals cannot be input to networks for discrete models (unless they are fuzzified). With input possibilities being discrete or analog, and the model possibilities being discrete or continuous, there are potentially four situations, but one of them where analog inputs are considered for a discrete model is untenable.

An example of a continuous model is where a network is to adjust the angle by which the steering wheel of a truck is to be turned to back up the truck into a parking space. If a network is supposed to recognize characters of the alphabet, a means of discretization of a character allows the use of a discrete model.

What are the types of inputs for problems like image processing or handwriting analysis? Remembering that artificial neurons, as processing elements, do aggregation of their inputs by using connection weights, and that the output neuron uses a **threshold** function, you know that the inputs have to be numerical. A handwritten character can be superimposed on a grid, and the input can consist of the cells in each row of the grid where a part of the character is present. In other words, the input corresponding to one character will be a set of binary or gray-scale sequences containing one sequence for each row of the grid. A 1 in a particular position in the sequence for a row shows that the corresponding pixel is present(black) in that part of the grid, while 0 shows it is not. The size of the grid has to be big enough to accommodate the largest character under study, as well as the most complex features.

OUTPUTS

The output from some neural networks is a spatial pattern that can include a bit pattern, in some a binary function value, and in some others an analog signal. The type of mapping intended for the inputs determines the type of outputs, naturally. The output could be one of classifying the input data, or finding associations between patterns of the same dimension as the input.

The **threshold** functions do the final mapping of the activations of the output neurons into the network outputs. But the outputs from a single cycle of operation of a neural network may not be the final outputs, since you would iterate the network into further cycles of operation until you see convergence. If convergence seems possible, but is taking an awful lot of time and effort, that is, if it is too slow to learn, you may assign a tolerance level and settle for the network to achieve near convergence.

THE THRESHOLD FUNCTION

The output of any neuron is the result of thresholding, if any, of its internal activation, which, in turn, is the weighted sum of the neuron's inputs. Thresholding sometimes is done for the sake of scaling down the activation and mapping it into a meaningful output for the problem, and sometimes for adding a bias. Thresholding (scaling) is important for multilayer networks to preserve a meaningful range across each layer's operations. The most often used **threshold** function is the **sigmoid** function. A **step** function or a **ramp** function or just a **linear** function can be used, as when you simply add the bias to the activation. The **sigmoid** function accomplishes mapping the activation into the interval [0, 1]. The equations are given as follows for the different threshold functions just mentioned.

The Sigmoid Function

More than one function goes by the name *sigmoid* function. They differ in their formulas and in their ranges. They all have a graph similar to a stretched letter **s**. We give below two such functions. The first is the hyperbolic tangent function with values in (−1, 1). The second is the logistic function and has values between 0 and 1. You therefore choose the one that fits the range you want. The graph of the **sigmoid logistic** function is given in Fig. 5.3.

1. $f(x) = \tanh(x) = (e^x - e^{-x}) / (e^x + e^{-x})$
2. $f(x) = 1 / (1 + e^{-x})$

Note that the first function here, the hyperbolic tangent function, can also be written, as $1 - 2e^{-x} / (e^x + e^{-x})$ after adding and also subtracting e^{-x} to the numerator, and then simplifying. If now you multiply in the second term both numerator and denominator by e^x, you get $1 - 2/ (e^{2x} + 1)$. As x approaches -∞, this function goes to -1, and as x approaches +∞, it goes to +1. On the other hand, the second function here, the **sigmoid logistic** function, goes to 0 as x approaches -∞, and to +1 as x approaches +∞. You can see this if you rewrite $1 / (1 + e^{-x})$ as $1 - 1 / (1 + e^x)$, after manipulations similar to those above.

You can think of equation 1 as the bipolar equivalent of binary equation 2. Both functions have the same shape.

Figure 5.3 is the graph of the **sigmoid logistic** function (number 2 of the preceding list).

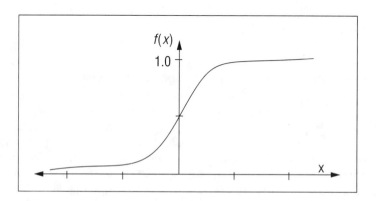

FIGURE 5.3 THE SIGMOID FUNCTION.

The Step Function

The **step** function is also frequently used as a **threshold** function. The function is 0 to start with and remains so to the left of some **threshold** value θ. A jump to 1 occurs for the value of the function to the right of θ, and the function then remains at the level 1. In general, a **step** function can have a finite number of points at which jumps of equal or unequal size occur. When the jumps are equal and at many points, the graph will resemble a staircase. We are interested in a **step** function that goes from 0 to 1 in one step, as soon as the argument exceeds the **threshold** value θ. You could also have two values other than 0 and 1 in defining the range of values of such a **step** function. A graph of the **step** function follows in Figure 5.4.

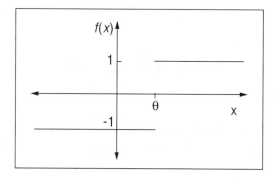

FIGURE 5.4 THE STEP FUNCTION.

You can think of a **sigmoid** function as a **fuzzy step** function.

NOTE

The Ramp Function

To describe the **ramp** function simply, first consider a step function that makes a jump from 0 to 1 at some point. Instead of letting it take a sudden jump like that at one point, let it gradually gain in value, along a straight line (looks like a ramp), over a finite interval reaching from an initial 0 to a final 1. Thus, you get a **ramp** function. You can think of a **ramp** function as a piecewise linear approximation of a sigmoid. The graph of a **ramp** function is illustrated in Figure 5.5.

102

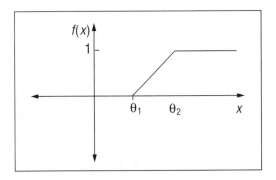

FIGURE 5.5 GRAPH OF A RAMP FUNCTION.

Linear Function

A **linear** function is a simple one given by an equation of the form:

$$f(x) = \alpha x + \beta$$

When $\alpha = 1$, the application of this **threshold** function amounts to simply adding a bias equal to β to the sum of the inputs.

APPLICATIONS

As briefly indicated before, the areas of application generally include auto- and heteroassociation, pattern recognition, data compression, data completion, signal filtering, image processing, forecasting, handwriting recognition, and optimization. The type of connections in the network, and the type of learning algorithm used must be chosen appropriate to the application. For example, a network with lateral connections can do autoassociation, while a feed-forward type can do forecasting.

SOME NEURAL NETWORK MODELS

Adaline and Madaline

Adaline is the acronym for *adaptive linear element*, due to Bernard Widrow and Marcian Hoff. It is similar to a Perceptron. Inputs are real numbers in the interval [−1,+1], and learning is based on the criterion of minimizing the average squared error. Adaline has a high capacity to store patterns. *Madaline* stands for many Adalines and is a neural network that is widely used. It is composed of field A and field B neurons, and there is one connection from each field A neuron to each field B neuron. Figure 5.6 shows a diagram of the Madaline.

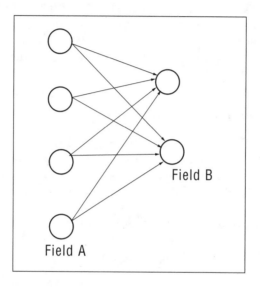

FIGURE 5.6 THE MADALINE MODEL.

Backpropagation

The *Backpropagation* training algorithm for training feed-forward networks was developed by Paul Werbos, and later by Parker, and Rummelhart and McClelland. This type of network configuration is the most common in use, due to its ease of training. It is estimated that over 80% of all neural network projects in development use backpropagation. In backpropagation, there are two phases in its learning cycle, one to propagate the input pattern through the network and the other to adapt the output, by changing the weights in the network. It is the error signals that are backpropagated in the network operation to the hidden layer(s). The portion of the error signal that a hidden-layer neuron receives in this process is an estimate of the contribution of a particular neuron to the output error. Adjusting on this basis the weights of the connections, the squared error, or some other metric, is reduced in each cycle and finally minimized, if possible.

Figure for Backpropagation Network

You will find in Figure 7.1 in Chapter 7, the layout of the nodes that represent the neurons in a feedforward Backpropagation network and the connections between them. For now, you try your hand at drawing this layout based on the following description, and compare your drawing with Figure 7.1. There are three fields of neurons. The connections are forward and are from each neuron in a layer to every neuron in the next layer. There are no lateral or recurrent connections. Labels on connections indicate weights. Keep in mind that the number of neurons is not necessarily the same in different layers, and this fact should be evident in the notation for the weights.

Bidirectional Associative Memory

Bidirectional Associative Memory, (BAM), and other models described in this section were developed by Bart Kosko. BAM is a network with feedback connections from the output layer to the input layer. It associates a member of the set of input patterns with a member of the set of output patterns that is the closest, and thus it does heteroassociation. The patterns can be with binary or bipolar values. If all possible input patterns are known, the matrix of connection weights can be determined as the sum of matrices obtained by taking the matrix product of an input vector (as a column vector) with its transpose (written as a row vector).

The pattern obtained from the output layer in one cycle of operation is fed back at the input layer at the start of the next cycle. The process continues until the network stabilizes on all the input patterns. The stable state so achieved is described as *resonance*, a concept used in the Adaptive Resonance Theory.

You will find in Figure 8.1 in Chapter 8, the layout of the nodes that represent the neurons in a BAM network and the connections between them. There are two fields of neurons. The network is fully connected with feedback connections and forward connections. There are no lateral or recurrent connections.

Fuzzy Associative memories are similar to Bidirectional Associative memories, except that association is established between fuzzy patterns. Chapter 9 deals with Fuzzy Associative memories.

Temporal Associative Memory

Another type of associative memory is *temporal associative memory*. Amari, a pioneer in the field of neural networks, constructed a Temporal

Associative Memory model that has feedback connections between the input and output layers. The forte of this model is that it can store and retrieve spatiotemporal patterns. An example of a spatiotemporal pattern is a waveform of a speech segment.

Brain-State-in-a-Box

Introduced by James Anderson and others, this network differs from the single-layer fully connected Hopfield network in that *Brain-State-in-a-Box* uses what we call recurrent connections as well. Each neuron has a connection to itself. With target patterns available, a modified *Hebbian learning* rule is used. The adjustment to a connection weight is proportional to the product of the desired output and the error in the computed output. You will see more on Hebbian learning in Chapter 6. This network is adept at noise tolerance, and it can accomplish pattern completion. Figure 5.7 shows a Brain-State-in-a-Box network.

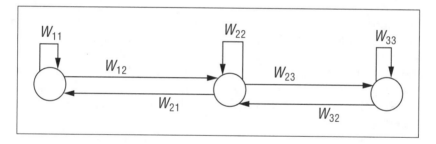

FIGURE 5.7 A BRAIN-STATE-IN-A-BOX, NETWORK.

What's in a Name?

More like what's in the box? Suppose you find the following: there is a square box and its corners are the locations for an entity to be. The entity is not at one of the corners, but is at some point inside the box. The next position for the entity is determined by working out the change in each coordinate of the position, according to a weight matrix, and a squashing function. This process is repeated until the entity settles down at some position. The choice of the weight matrix is such that when the entity reaches a corner of the square box, its position is stable and no more movement takes place. You would perhaps guess that the entity finally settles at the corner nearest to the initial position of it within the box. It is said that this kind of an example is the reason for the name Brain-State-in-a-Box for the model. Its forte is that it represents linear transformations. Some type of association of patterns can be achieved with this model. If an incomplete pattern is associated with a completed pattern, it would be an example of autoassociation.

Counterpropagation

This is a neural network model developed by Robert Hecht-Nielsen, that has one or two additional layers between the input and output layers. If it is one, the middle layer is a Grossberg layer with a bunch of outstars. In the other case, a Kohonen layer, or a self-organizing layer, follows the input layer, and in turn is followed by a Grossberg layer of outstars. The model has the distinction of considerably reducing training time. With this model, you gain a tool that works like a look-up table.

Neocognitron

Compared to all other neural network models, Fukushima's Neocognitron is more complex and ambitious. It demonstrates the advantages of a multi-layered network. The Neocognitron is one of the best models for recognizing handwritten symbols. Many pairs of layers called the S layer, for *simple layer*, and C layer, for *complex layer*, are used. Within each S layer are several planes containing simple cells. Similarly, there are within each C layer, an equal number of planes containing complex cells. The input layer does not have this arrangement and is like an input layer in any other neural network.

The number of planes of simple cells and of complex cells within a pair of S and C layers being the same, these planes are paired, and the complex plane cells process the outputs of the simple plane cells. The simple cells are trained so that the response of a simple cell corresponds to a specific portion of the input image. If the same part of the image occurs with some distortion, in terms of scaling or rotation, a different set of simple cells responds to it. The complex cells output to indicate that some simple cell they correspond to did fire. While simple cells respond to what is in a contiguous region in the image, complex cells respond on the basis of a larger region. As the process continues to the output layer, the C-layer component of the output layer responds, corresponding to the entire image presented in the beginning at the input layer.

Adaptive Resonance Theory

ART1 is the first model for adaptive resonance theory for neural networks developed by Gail Carpenter and Stephen Grossberg. This theory was developed to address the stability–plasticity dilemma. The network is supposed to be plastic enough to learn an important pattern. But at the same time it should remain stable when, in short-term memory, it encounters some distorted versions of the same pattern.

ART1 model has A and B field neurons, a gain, and a reset as shown in Figure 5.8. There are top-down and bottom-up connections between neurons of fields A and B. The neurons in field B have lateral connections as well as recurrent connections. That is, every neuron in this field is connected to every other neuron in this field, including itself, in addition to the connections to the neurons in field A. The external input (or bottom-up signal), the top-down signal, and the gain constitute three elements of a set, of which at least two should be a +1 for the neuron in the A field to fire. This is what is termed the *two-thirds rule*. Initially, therefore, the gain would be set to +1. The idea of a single winner is also employed in the B field. The gain would not contribute in the top-down phase; actually, it will inhibit. The two-thirds rule helps move toward stability once *resonance,* or equilibrium, is obtained. A *vigilance parameter* ρ is used to determine the parameter reset. Vigilance parameter corresponds to what degree the resonating category can be predicted. The part of the system that contains gain is called the *attentional subsystem*, whereas the rest, the part that contains reset, is termed the *orienting subsystem*. The top-down activity corresponds to the orienting subsystem, and the bottom-up activity relates to the attentional subsystem.

In ART1, classification of an input pattern in relation to stored patterns is attempted, and if unsuccessful, a new stored classification is generated. Training is unsupervised. There are two versions of training: slow and fast. They differ in the extent to which the weights are given the time to reach their eventual values. Slow training is governed by differential equations, and fast training by algebraic equations.

ART2 is the analog counterpart of ART1, which is for discrete cases. These are self-organizing neural networks, as you can surmise from the fact that training is present but unsupervised. The ART3 model is for recognizing a coded pattern through a parallel search, and is developed by Carpenter and

Grossberg. It tries to emulate the activities of chemical transmitters in the brain during what can be construed as a parallel search for pattern recognition.

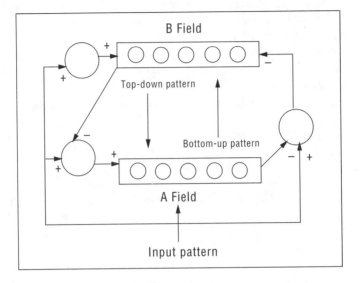

FIGURE 5.8 THE ART1 NETWORK.

SUMMARY

The basic concepts of neural network layers, connections, weights, inputs, and outputs have been discussed. An example of how adding another layer of neurons in a network can solve a problem that could not be solved without it is given in detail. A number of neural network models are introduced briefly. Learning and training, which form the basis of neural network behavior has not been included here, but will be discussed in the following chapter.

CHAPTER 6

LEARNING AND TRAINING

In the last chapter, we presented an overview of different neural network models. In this chapter, we continue the broad discussion of neural networks with two important topics: Learning and Training. Here are key questions that we would like to answer:

- How do neural networks learn?
- What does it mean for a network to learn ?
- What differences are there between supervised and unsupervised learning ?
- What training regimens are in common use for neural networks?

OBJECTIVE OF LEARNING

There are many varieties of neural networks. In the final analysis, as we have discussed briefly in Chapter 4 on network modeling, all neural networks do one or more of the following :

- Pattern classification
- Pattern completion
- Optimization
- Data clustering
- Approximation
- Function evaluation

A neural network, in any of the previous tasks, maps a set of inputs to a set of outputs. This nonlinear mapping can be thought of as a multidimensional *mapping surface*. The objective of learning *is to mold the mapping surface according to a desired response*, either with or without an explicit training process.

LEARNING AND TRAINING

A network can learn when training is used, or the network can learn also in the absence of training. The difference between *supervised* and *unsupervised training* is that, in the former case, external prototypes are used as target outputs for specific inputs, and the network is given a learning algorithm to follow and calculate new connection weights that bring the output closer to the target output. Unsupervised learning is the sort of learning that takes place without a teacher. For example, when you are finding your way out of a labyrinth, no teacher is present. You learn from the responses or events that develop as you try to feel your way through the maze. For neural networks, in the unsupervised case, a learning algorithm may be given but target outputs are not given. In such a case, data input to the network gets clustered together; similar input stimuli cause similar responses.

When a neural network model is developed and an appropriate learning algorithm is proposed, it would be based on the theory supporting the model. Since the dynamics of the operation of the neural network is under study, the learning equations are initially formulated in terms of differential equations. After solving the differential equations, and using any initial conditions that are available, the algorithm could be simplified to consist of an algebraic equation for the changes in the weights. These simple forms of learning equations are available for your neural networks.

At this point of our discussion you need to know what learning algorithms are available, and what they look like. We will now discuss two main rules for learning—*Hebbian learning,* used with unsupervised learning and the *delta rule,* used with supervised learning. Adaptations of these by simple modifications to suit a particular context generate many other learning rules in use today. Following the discussion of these two rules, we present variations for each of the two classes of learning: supervised learning and unsupervised learning.

Hebb's Rule

Learning algorithms are usually referred to as *learning rules.* The foremost such rule is due to Donald Hebb. Hebb's rule is a statement about how the firing of one neuron, which has a role in the determination of the activation of another neuron, affects the first neuron's influence on the activation of the second neuron, especially if it is done in a repetitive manner. As a learning rule, Hebb's observation translates into a formula for the difference in a connection weight between two neurons from one iteration to the next, as a constant μ times the product of activations of the two neurons. How a connection weight is to be modified is what the learning rule suggests. In the case of Hebb's rule, it is adding the quantity $\mu a_i a_j$, where a_i is the activation of the ith neuron, and a_j is the activation of the jth neuron to the connection weight between the ith and jth neurons. The constant μ itself is referred to as the *learning rate*. The following equation using the notation just described, states it succinctly:

$$\Delta w_{ij} = \mu a_i a_j$$

As you can see, the learning rule derived from Hebb's rule is quite simple and is used in both simple and more involved networks. Some modify this rule by replacing the quantity a_i with its deviation from the average of all as and, similarly, replacing a_j by a corresponding quantity. Such rule variations can yield rules better suited to different situations.

For example, the output of a neural network being the activations of its output layer neurons, the Hebbian learning rule in the case of a perceptron takes the form of adjusting the weights by adding μ times the difference between the output and the target. Sometimes a situation arises where some unlearning is required for some neurons. In this case a reverse Hebbian rule is used in which the quantity $\mu a_i a_j$ is subtracted from the connection weight under question, which in effect is employing a negative learning rate.

In the Hopfield network of Chapter 1, there is a single layer with all neurons fully interconnected. Suppose each neuron's output is either $a + 1$ or $a - 1$. If we take $\mu = 1$ in the Hebbian rule, the resulting modification of the connection weights can be described as follows: add 1 to the weight, if both neuron outputs match, that is, both are +1 or −1. And if they do not match (meaning one of them has output +1 and the other has −1), then subtract 1 from the weight.

Delta Rule

The delta rule is also known as the *least mean squared* error rule (LMS). You first calculate the square of the errors between the target or desired values and computed values, and then take the average to get the mean squared error. This quantity is to be minimized. For this, realize that it is a function of the weights themselves, since the computation of output uses them. The set of values of weights that minimizes the mean squared error is what is needed for the next cycle of operation of the neural network. Having worked this out mathematically, and having compared the weights thus found with the weights actually used, one determines their difference and gives it in the delta rule, each time weights are to be updated. So the delta rule, which is also the rule used first by Widrow and Hoff, in the context of learning in neural networks, is stated as an equation defining the change in the weights to be affected.

Suppose you fix your attention to the weight on the connection between the ith neuron in one layer and the jth neuron in the next layer. At time t, this weight is $w_{ij}(t)$. After one cycle of operation, this weight becomes $w_{ij}(t + 1)$. The difference between the two is $w_{ij}(t + 1) - w_{ij}(t)$, and is denoted by Δw_{ij}. The delta rule then gives Δw_{ij} as :

$$\Delta w_{ij} = 2\mu x_i(\text{desired output value} - \text{computed output value})_j$$

Here, μ is the learning rate, which is positive and much smaller than 1, and x_i is the ith component of the input vector.

Supervised Learning

Supervised neural network paradigms to be discussed include :

- Perceptron
- Adaline
- Feedforward Backpropagation network
- Statistical trained networks (Boltzmann/Cauchy machines)
- Radial basis function networks

The Perceptron and the Adaline use the delta rule; the only difference is that the Perceptron has binary output, while the Adaline has continuous valued output. The Feedforward Backpropagation network uses the *generalized delta rule*, which is described next.

Generalized Delta Rule

While the delta rule uses local information on error, the generalized delta rule uses error information that is not local. It is designed to minimize the total of the squared errors of the output neurons. In trying to achieve this minimum, the steepest descent method, which uses the gradient of the weight surface, is used. (This is also used in the delta rule.) For the next error calculation, the algorithm looks at the gradient of the error surface, which gives the direction of the largest slope on the error surface. This is used to determine the direction to go to try to minimize the error. The algorithm chooses the negative of this gradient, which is the direction of steepest descent. Imagine a very hilly error surface, with peaks and valleys that have a wide range of magnitude. Imagine starting your search for minimum error at an arbitrary point. By choosing the negative gradient on all iterations, you eventually end up at a valley. You cannot know, however, if this valley is the global minimum or a local minimum. Getting stuck in a local minimum is one well-known potential problem of the steepest descent method. You will see more on the generalized delta rule in the chapter on backpropagation (Chapter 7).

Statistical Training and Simulated Annealing

The Boltzmann machine (and Cauchy machine) uses probabilities and statistical theory, along with an energy function representing temperature. The learning is probabilistic and is called *simulated annealing*. At different temperature levels, a different number of iterations in processing are used, and this constitutes an annealing schedule. Use of probability distributions is for the goal of reaching a state of global minimum of energy. Boltzmann distribution and Cauchy distribution are probability distributions used in this process. It is obviously desirable to reach a global minimum, rather than settling down at a local minimum.

Figure 6.1 clarifies the distinction between a local minimum and a global minimum. In this figure you find the graph of an energy function and points A and B. These points show that the energy levels there are smaller than the energy levels at any point in their vicinity, so you can say they represent points of minimum energy. The overall or global minimum, as you can see, is at point B, where the energy level is smaller than that even at point A, so A corresponds only to a local minimum. It is desirable to get to B and not get stopped at A itself, in the pursuit of a minimum for the energy function. If point C is reached, one would like the further movement to be toward B and not A. Similarly, if a point near A is reached, the subsequent movement should avoid reaching or settling at A but carry on to B. Perturbation techniques are useful for these considerations.

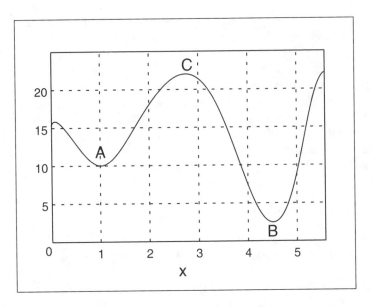

FIGURE 6.1 LOCAL AND GLOBAL MINIMA.

Clamping Probabilities

Sometimes in simulated annealing, first a subset of the neurons in the network are associated with some inputs, and another subset of neurons are associated with some outputs, and these are clamped with probabilities, which are not changed in the learning process. Then the rest of the network is subjected to adjustments. Updating is not done for the clamped units in the network. This training procedure of Geoffrey Hinton and Terrence Sejnowski provides an extension of the Boltzmann technique to more general networks.

Radial Basis-Function Networks

Although details of radial basis functions are beyond the scope of this book, it is worthwhile to contrast the learning characteristics for this type of neural network model. Radial basis-function networks in topology look similar to feedforward networks. Each neuron has an output to input characteristic that resembles a radial function (for two inputs, and thus two dimensions). Specifically, the output $h(x)$ is as follows:

$$h(\mathbf{x}) = \exp\left[\,(\mathbf{x} - \mathbf{u})^2 / 2\sigma^2\,\right]$$

Here, **x** is the input vector, **u** is the mean, and σ is the standard deviation of the output response curve of the neuron. Radial basis function (RBF) networks have rapid training time (orders of magnitude faster than backpropagation) and do not have problems with local minima as backpropagation does. RBF networks are used with supervised training, and typically only the output layer is trained. Once training is completed, a RBF network may be slower to use than a feedforward Backpropagation network, since more computations are required to arrive at an output.

UNSUPERVISED NETWORKS

Unsupervised neural network paradigms to be discussed include:

- Hopfield Memory
- Bidirectional associative memory
- Fuzzy associative memory
- Learning vector quantizer
- Kohonen self-organizing map
- ART1

Self-Organization

Unsupervised learning and self-organization are closely related. Unsupervised learning was mentioned in Chapter 1, along with supervised learning. Training in supervised learning takes the form of external exemplars being provided. The network has to compute the correct weights for the connections for neurons in some layer or the other. Self-organization implies unsupervised learning. It was described as a characteristic of a neural network model, ART1, based on adaptive resonance theory (to be covered in Chapter 10). With the winner-take-all criterion, each neuron of field B learns a distinct classification. The winning neuron in a layer, in this case the field B, is the one with the largest activation, and it is the only neuron in that layer that is allowed to fire. Hence, the name winner take all.

Self-organization means self-adaptation of a neural network. Without target outputs, the closest possible response to a given input signal is to be generated. Like inputs will cluster together. The connection weights are modified through different iterations of network operation, and the network capable of self-organizing creates on its own the closest possible set of outputs for the given inputs. This happens in the model in Kohonen's self-organizing map.

Kohonen's *Linear Vector Quantizer* (LVQ) described briefly below is later extended as a self-organizing feature map. Self-organization is also learning, but without supervision; it is a case of self-training. Kohonen's topology preserving maps illustrate self-organization by a neural network. In these cases, certain subsets of output neurons respond to certain subareas of the inputs, so that the firing within one subset of neurons indicates the presence of the corresponding subarea of the input. This is a useful paradigm in applications such as speech recognition. The winner-take-all strategy used in ART1 also facilitates self-organization.

LEARNING VECTOR QUANTIZER

Suppose the goal is the classification of input vectors. Kohonen's Vector Quantization is a method in which you first gather a finite number of vectors of the dimension of your input vector. Kohonen calls these *codebook vectors*. You then assign groups of these codebook vectors to the different classes under the classification you want to achieve. In other words, you make a correspondence between the codebook vectors and classes, or, partition the set of codebook vectors by classes in your classification.

Now examine each input vector for its distance from each codebook vector, and find the nearest or closest codebook vector to it. You identify the input vector with the class to which the codebook vector belongs.

Codebook vectors are updated during training, according to some algorithm. Such an algorithm strives to achieve two things: (1), a codebook vector closest to the input vector is brought even closer to it; and (two), a codebook vector indicating a different class is made more distant from the input vector.

For example, suppose (2, 6) is an input vector, and (3, 10) and (4, 9) are a pair of codebook vectors assigned to different classes. You identify (2, 6) with the class to which (4, 9) belongs, since (4, 9) with a distance of $\sqrt{13}$ is closer to it than (3, 10) whose distance from (2, 6) is $\sqrt{17}$. If you add 1 to each component of (3, 10) and subtract 1 from each component of (4, 9), the new distances of these from (2, 6) are $\sqrt{29}$ and $\sqrt{5}$, respectively. This shows that (3, 10) when changed to (4, 11) becomes more distant from your input vector than before the change, and (4, 9) is changed to (3, 8), which is a bit closer to (2, 6) than (4, 9) is.

Training continues until all input vectors are classified. You obtain a stage where the classification for each input vector remains the same as in the previous cycle of training. This is a process of self-organization.

The Learning Vector Quantizer (LVQ) of Kohonen is a self-organizing net-work. It classifies input vectors on the basis of a set of stored or reference vectors. The B field neurons are also called *grandmother cells*, each of which represents a specific class in the reference vector set. Either supervised or unsupervised learning can be used with this network. (See Figure 6.2.)

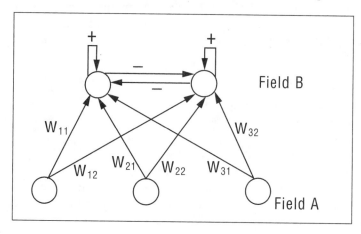

FIGURE 6.2 LAYOUT FOR LEARNING VECTOR QUANTIZER.

ASSOCIATIVE MEMORY MODELS AND ONE-SHOT LEARNING

The Hopfield memory, Bidirectional Associative memory and Fuzzy Associative memory are all unsupervised networks that perform pattern completion, or pattern association. That is, with corrupted or missing infor-mation, these memories are able to recall or complete an expected output. Gallant calls the training method used in these networks as *one-shot* learn-ing, since you determine the weight matrix as a function of the completed patterns you wish to recall just once. An example of this was shown in Chapter 4 with determination of weights for the Hopfield memory.

LEARNING AND RESONANCE

ART1 is the first neural network model based on adaptive resonance theory of Carpenter and Grossberg. When you have a pair of patterns such that when one of them is input to a neural network the output turns out to be the other pattern in the pair, and if this happens consistently in both directions, then you may describe it as resonance. We discuss in Chapter 8 bidirectional associative memories and resonance. By the time training is completed, and learning is through, many other pattern pairs would have been presented to the network as well. If changes in the short-term memory do not disturb or affect the long-term memory, the network shows adaptive resonance. The ART1 model is designed to maintain it. Note that this discussion relates largely to stability.

LEARNING AND STABILITY

Learning, convergence, and stability are matters of much interest. As *learning* is taking place, you want to know if the process is going to halt at some appropriate point, which is a question of *convergence*. Is what is learned *stable*, or will the network have to learn all over again, as each new event occurs? These questions have their answers within a mathematical model with differential equations developed to describe a learning algorithm. Proofs showing stability are part of the model inventor's task. One particular tool that aids in the process of showing convergence is the idea of state energy, or cost, to describe whether the direction the process is taking can lead to convergence.

The *Lyapunov* function, discussed later in this chapter, is found to provide the right energy function, which can be minimized during the operation of the neural network. This function has the property that the value gets smaller with every change in the state of the system, thus assuring that a minimum will be reached eventually. The Lyapunov function is discussed further because of its significant utility for neural network models, but briefly because of the high level of mathematics involved. Fortunately, simple forms are derived and put into learning algorithms for neural networks. The high-level mathematics is used in making the proofs to show the viability of the models.

Alternatively, temperature relationships can be used, as in the case of the Boltzmann machine, or any other well-suited cost function such as a function of distances used in the formulation of the *Traveling Salesman Problem*, in which the total distance for the tour of the traveling salesman is to be minimized, can be employed. The Traveling Salesman Problem is important and well-known. A set of cities is to be visited by the salesman, each only once, and the aim is to devise a tour that minimizes the total distance traveled. The search continues for an efficient algorithm for this problem. Some algorithms solve the problem in a large number but not all of the situations. A neural network formulation can also work for the Traveling Salesman Problem. You will see more about this in Chapter 15.

TRAINING AND CONVERGENCE

Suppose you have a criterion such as energy to be minimized or cost to be decreased, and you know the optimum level for this criterion. If the network achieves the optimum value in a finite number of steps, then you have convergence for the operation of the network. Or, if you are making pairwise associations of patterns, there is the prospect of convergence if after each cycle of the network operation, the number of errors is decreasing.

It is also possible that convergence is slow, so much so that it may seem to take forever to achieve the convergence state. In that case, you should specify a tolerance value and require that the criterion be achieved within that tolerance, avoiding a lot of computing time. You may also introduce a *momentum* parameter to further change the weight and thereby speed up the convergence. One technique used is to add a portion of the previous change in weight.

Instead of converging, the operation may result in oscillations. The weight structure may keep changing back and forth; learning will never cease. Learning algorithms need to be analyzed in terms of convergence as being an essential algorithm property.

LYAPUNOV FUNCTION

Neural networks are dynamic systems in the learning and training phase of their operation, and convergence is an essential feature, so it was necessary for the researchers developing the models and their learning algorithms to find a provable criterion for convergence in a dynamic system. The Lyapunov function, mentioned previously, turned out to be the most convenient and appropriate function. It is also referred to as the energy function. The function decreases as the system states change. Such a function needs to be found and watched as the network operation continues from cycle to cycle. Usually it involves a quadratic form. The least mean squared error is an example of such a function. Lyapunov function usage assures a system stability that cannot occur without convergence. It is convenient to have one value, that of the Lyapunov function specifying the system behavior. For example, in the Hopfield network, the energy function is a constant times the sum of products of outputs of different neurons and the connection weight between them. Since pairs of neuron outputs are multiplied in each term, the entire expression is a quadratic form.

Other Training Issues

Besides the applications for which a neural network is intended, and depending on these applications, you need to know certain aspects of the model. The length of encoding time and the length of learning time are among the important considerations. These times could be long but should not be prohibitive. It is important to understand how the network behaves with new inputs; some networks may need to be trained all over again, but some tolerance for distortion in input patterns is desirable, where relevant. Restrictions on the format of inputs should be known.

An advantage of neural networks is that they can deal with nonlinear functions better than traditional algorithms can. The ability to store a number of patterns, or needing more and more neurons in the output field with an increasing number of input patterns are the kind of aspects addressing the capabilities of a network and also its limitations.

Adaptation

Sometimes neural networks are used as *adaptive filters*, the motivation for such an architecture being selectivity. You want the neural network to classify each input pattern into its appropriate category. Adaptive models involve changing of connection weights during all their operations, while nonadaptive ones do not alter the weights after the phase of learning with exemplars. The Hopfield network is often used in modeling a neural network for optimization problems, and the Backpropagation model is a popular choice in most other applications. Neural network models are distinguishable sometimes by their architecture, sometimes by their adaptive methods, and sometimes both. Methods for adaptation, where adaptation is incorporated, assume great significance in the description and utility of a neural network model.

For adaptation, you can modify parameters in an architecture during training, such as the learning rate in the backpropagation training method for example. A more radical approach is to modify the architecture itself during training. New neural network paradigms change the number or layers and the number of neurons in a layer during training. These node adding or pruning algorithms are termed *constructive algorithms*. (See Gallant for more details.)

Generalization Ability

The analogy for a neural network presented at the beginning of the chapter was that of a multidimensional mapping surface that maps inputs to outputs. For each unseen input with respect to a training set, the *generalization* ability of a network determines how well the mapping surface renders the new input in the output space. A stock market forecaster must generalize well, otherwise you lose money in unseen market conditions. The opposite of generalization is *memorization*. A pattern recognition system for images of handwriting, should be able to generalize a letter A that is handwritten in several different ways by different people. If the system memorizes, then you will not recognize the letter A in all cases, but instead will categorize each letter A variation separately. The trick to achieve generalization is in network architecture, design, and training methodology. You do not want to overtrain your neural network on expected outcomes, but rather should accept a slightly worse than minimum error on your training set data. You will learn more about generalization in Chapter 14.

SUMMARY

Learning and training are important issues in applying neural networks. Two broad categories of network learning are supervised and unsupervised learning. Supervised learning provides example outputs to compare to while unsupervised learning does not. During supervised training, external prototypes are used as target outputs and the network is given a learning algorithm to follow and calculate new connection weights that bring the output closer to the target output. You can refer to networks using unsupervised learning as self-organizing networks, since no external information or guidance is used in learning. Several neural network paradigms were presented in this chapter along with their learning and training characteristics.

CHAPTER 7

BACKPROPAGATION

FEEDFORWARD BACKPROPAGATION NETWORK

The feedforward backpropagation network is a very popular model in neural networks. It does not have feedback connections, but errors are back-propagated during training. Least mean squared error is used. Many applications can be formulated for using a feedforward backpropagation network, and the methodology has been a model for most multilayer neural networks. Errors in the output determine measures of hidden layer output errors, which are used as a basis for adjustment of connection weights between the input and hidden layers. Adjusting the two sets of weights between the pairs of layers and recalculating the outputs is an iterative process that is carried on until the errors fall below a tolerance level. Learning rate parameters scale the adjustments to weights. A momentum parameter can also be used in scaling the adjustments from a previous iteration and adding to the adjustments in the current iteration.

Mapping

The feedforward backpropagation network maps the input vectors to output vectors. Pairs of input and output vectors are chosen to train the network first. Once training is completed, the weights are set and the network can be used to find outputs for new inputs. The dimension of the input vector determines the number of neurons in the input layer, and the number of neurons in the output layer is determined by the dimension of the outputs. If there are k neurons in the input layer and m neurons in the output layer, then this network can make a mapping from k-dimensional space to an m-dimensional space. Of course, what that mapping is depends on what pair of patterns or vectors are used as exemplars to train the network, which determine the network weights. Once trained, the network gives you the image of a new input vector under this mapping. Knowing what mapping you want the feedforward backpropagation network to be trained for implies the dimensions of the input space and the output space, so that you can determine the numbers of neurons to have in the input and output layers.

Layout

The architecture of a feedforward backpropagation network is shown in Figure 7.1. While there can be many hidden layers, we will illustrate this network with only one hidden layer. Also, the number of neurons in the input layer and that in the output layer are determined by the dimensions of the input and output patterns, respectively. It is not easy to determine how many neurons are needed for the hidden layer. In order to avoid cluttering the figure, we will show the layout in Figure 7.1 with five input neurons, three neurons in the hidden layer, and four output neurons, with a few representative connections.

The network has three fields of neurons: one for input neurons, one for hidden processing elements, and one for the output neurons. As already stated, connections are for feed forward activity. There are connections from every neuron in field A to every one in field B, and, in turn, from every neuron in field B to every neuron in field C. Thus, there are two sets of weights, those figuring in the activations of hidden layer neurons, and those that help determine the output neuron activations. In training, all of these weights are adjusted by considering what can be called a *cost function* in terms of the error in the computed output pattern and the desired output pattern.

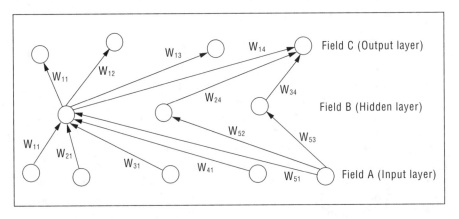

FIGURE 7.1 LAYOUT OF A FEEDFORWARD BACKPROPAGATION NETWORK.

Training

The feedforward backpropagation network undergoes supervised training, with a finite number of pattern pairs consisting of an input pattern and a desired or target output pattern. An input pattern is presented at the input layer. The neurons here pass the pattern activations to the next layer neurons, which are in a hidden layer. The outputs of the hidden layer neurons are obtained by using perhaps a bias, and also a threshold function with the activations determined by the weights and the inputs. These hidden layer outputs become inputs to the output neurons, which process the inputs using an optional **bias** and a **threshold** function. The final output of the network is determined by the activations from the output layer.

The computed pattern and the input pattern are compared, a function of this error for each component of the pattern is determined, and adjustment to weights of connections between the hidden layer and the output layer is computed. A similar computation, still based on the error in the output, is made for the connection weights between the input and hidden layers. The procedure is repeated with each pattern pair assigned for training the network. Each pass through all the training patterns is called a *cycle* or an *epoch*. The process is then repeated as many cycles as needed until the error is within a prescribed tolerance.

There can be more than one learning rate parameter used in training in a feedforward backpropagation network. You can use one with each set of weights between consecutive layers.

N O T E

ILLUSTRATION: ADJUSTMENT OF WEIGHTS OF CONNECTIONS FROM A NEURON IN THE HIDDEN LAYER

We will be as specific as is needed to make the computations clear. First recall that the activation of a neuron in a layer other than the input layer is the sum of products of its inputs and the weights corresponding to the connections that bring in those inputs. Let us discuss the jth neuron in the hidden layer. Let us be specific and say $j = 2$. Suppose that the input pattern is (1.1, 2.4, 3.2, 5.1, 3.9) and the target output pattern is (0.52, 0.25, 0.75, 0.97). Let the weights be given for the second hidden layer neuron by the vector (−0.33, 0.07, −0.45, 0.13, 0.37). The activation will be the quantity:

(-0.33 * 1.1) + (0.07 * 2.4) + (-0.45 * 3.2) + (0.13 * 5.1) + (0.37 * 3.9) = 0.471

Now add to this an optional bias of, say, 0.679, to give 1.15. If we use the **sigmoid** function given by:

1 / (1+ exp(-x)),

with $x = 1.15$, we get the output of this hidden layer neuron as 0.7595.

We are taking values to a few decimal places only for illustration, unlike the precision that can be obtained on a computer.

N O T E

We need the computed output pattern also. Let us say it turns out to be actual =(0.61, 0.41, 0.57, 0.53), while the desired pattern is desired =(0.52, 0.25, 0.75, 0.97). Obviously, there is a discrepancy between what is desired and what is computed. The component-wise differences are given in the vector, desired - actual = (-0.09, -0.16, 0.18, 0.44). We use these to form another vector where each component is a product of the error component, corresponding computed pattern component, and the complement of the latter with respect to 1. For example, for the first component, error is –0.09, computed pattern component is 0.61, and its complement is 0.39. Multiplying these together (0.61*0.39*-0.09), we get -0.02. Calculating the other components similarly, we get the vector (–0.02, –0.04, 0.04, 0.11). The desired–actual vector, which is the error vector multiplied by the actual output vector, gives you a value of error reflected back at the output of the hidden layer. This is scaled by a value of (1-output vector), which is the first derivative of the output activation function for numerical stability). You will see the formulas for this process later in this chapter.

The backpropagation of errors needs to be carried further. We need now the weights on the connections between the second neuron in the hidden layer that we are concentrating on, and the different output neurons. Let us say these weights are given by the vector (0.85, 0.62, –0.10, 0.21). The error of the second neuron in the hidden layer is now calculated as below, using its output.

error = 0.7595 * (1 - 0.7595) * ((0.85 * -0.02) + (0.62 * -0.04) + (-0.10 * 0.04) + (0.21 * 0.11)) = -0.0041.

Again, here we multiply the error (e.g., -0.02) from the output of the current layer, by the output value (0.7595) and the value (1-0.7595). We use the weights on the connections between neurons to work backwards through the network.

Next, we need the *learning rate parameter* for this layer; let us set it as 0.2. We multiply this by the output of the second neuron in the hidden layer, to get 0.1519. Each of the components of the vector (−0.02, −0.04, 0.04, 0.11) is multiplied now by 0.1519, which our latest computation gave. The result is a vector that gives the adjustments to the weights on the connections that go from the second neuron in the hidden layer to the output neurons. These values are given in the vector (−0.003, −0.006, 0.006, 0.017). After these adjustments are added, the weights to be used in the next cycle on the connections between the second neuron in the hidden layer and the output neurons become those in the vector (0.847, 0.614, −0.094, 0.227).

ILLUSTRATION: ADJUSTMENT OF WEIGHTS OF CONNECTIONS FROM A NEURON IN THE INPUT LAYER

Let us look at how adjustments are calculated for the weights on connections going from the ith neuron in the input layer to neurons in the hidden layer. Let us take specifically $i = 3$, for illustration.

Much of the information we need is already obtained in the previous discussion for the second hidden layer neuron. We have the errors in the computed output as the vector (−0.09, −0.16, 0.18, 0.44), and we obtained the error for the second neuron in the hidden layer as −0.0041, which was not used above. Just as the error in the output is propagated back to assign errors for the neurons in the hidden layer, those errors can be propagated to the input layer neurons.

To determine the adjustments for the weights on connections between the input and hidden layers, we need the errors determined for the outputs of hidden layer neurons, a learning rate parameter, and the activations of the input neurons, which are just the input values for the input layer. Let us take the learning rate parameter to be 0.15. Then the weight adjustments for the connections from the third input neuron to the hidden layer neurons are obtained by multiplying the particular hidden layer neuron's output error by the learning rate parameter and by the input component from the input neuron. The adjustment for the weight on the connection from the third input neuron to the second hidden layer neuron is 0.15 * 3.2 * −0.0041, which works out to −0.002.

127

If the weight on this connection is, say, −0.45, then adding the adjustment of -0.002, we get the modified weight of −0.452, to be used in the next iteration of the network operation. Similar calculations are made to modify all other weights as well.

ADJUSTMENTS TO THRESHOLD VALUES OR BIASES

The bias or the threshold value we added to the activation, before applying the **threshold** function to get the output of a neuron, will also be adjusted based on the error being propagated back. The needed values for this are in the previous discussion.

The adjustment for the threshold value of a neuron in the output layer is obtained by multiplying the calculated error (not just the difference) in the output at the output neuron and the learning rate parameter used in the adjustment calculation for weights at this layer. In our previous example, we have the learning rate parameter as 0.2, and the error vector as (−0.02, −0.04, 0.04, 0.11), so the adjustments to the threshold values of the four output neurons are given by the vector (−0.004, −0.008, 0.008, 0.022). These adjustments are added to the current levels of threshold values at the output neurons.

The adjustment to the threshold value of a neuron in the hidden layer is obtained similarly by multiplying the learning rate with the computed error in the output of the hidden layer neuron. Therefore, for the second neuron in the hidden layer, the adjustment to its threshold value is calculated as 0.15 * −0.0041, which is −0.0006. Add this to the current threshold value of 0.679 to get 0.6784, which is to be used for this neuron in the next training pattern for the neural network.

ANOTHER EXAMPLE OF BACKPROPAGATION CALCULATIONS

You have seen, in the preceding sections, the details of calculations for one particular neuron in the hidden layer in a feedforward backpropagation network with five input neurons and four neurons in the output layer, and two neurons in the hidden layer.

You are going to see all the calculations in the C++ implementation later in this chapter. Right now, though, we present another example and give the complete picture of the calculations done in one completed iteration or cycle of backpropagation.

Consider a feedforward backpropagation network with three input neurons, two neurons in the hidden layer, and three output neurons. The weights on connections from the input neurons to the neurons in the hidden layer are given in Matrix M-1, and those from the neurons in the hidden layer to output neurons are given in Matrix M-2.

We calculate the output of each neuron in the hidden and output layers as follows. We add a bias or threshold value to the activation of a neuron (call this result x) and use the **sigmoid** function below to get the output.

$$f(x) = 1/ (1 + e^{-x})$$

Learning parameters used are 0.2 for the connections between the hidden layer neurons and output neurons and 0.15 for the connections between the input neurons and the neurons in the hidden layer. These values as you recall are the same as in the previous illustration, to make it easy for you to follow the calculations by comparing them with similar calculations in the preceding sections.

The input pattern is (0.52, 0.75, 0.97), and the desired output pattern is (0.24, 0.17, 0.65). The initial weight matrices are as follows:

M-1 Matrix of weights from input layer to hidden layer

```
   0.6     - 0.4
   0.2       0.8
 - 0.5       0.3
```

M-2 Matrix of weights from hidden layer to output layer

```
 -0.90      0.43    0.25
  0.11    - 0.67 - 0.75
```

The threshold values (or bias) for neurons in the hidden layer are 0.2 and 0.3, while those for the output neurons are 0.15, 0.25, and 0.05, respectively.

Table 7.1 presents all the results of calculations done in the first iteration. You will see modified or new weight matrices and threshold values. You will use these and the original input vector and the desired output vector to carry out the next iteration.

The top row in the table gives headings for the columns. They are, Item, I-1, I-2, I-3 (I-k being for input layer neuron k); H-1, H-2 (for hidden layer neurons); and O-1, O-2, O-3 (for output layer neurons).

In the first column of the table, M-1 and M-2 refer to weight matrices as above. Where an entry is appended with -H, like in Output -H, the information refers to the hidden layer. Similarly, -O refers to the output layer, as in Activation + threshold -O.

Table 7.1 Backpropagation Calculations

Item	I-1	I-2	I-3	H-1	H-2	O-1	O-2	O-3
Input	0.52	0.75	0.97					
Desired Output						0.24	0.17	0.65
M-1 Row 1				0.6	- 0.4			
M-1 Row 2				0.2	0.8			
M-1 Row 3				- 0.5	0.3			
M-2 Row 1						- 0.90	0.43	0.25
M-2 Row 2						0.11	- 0.67	- 0.75
Threshold				0.2	0.3	0.15	0.25	0.05
Activation - H				- 0.023	0.683			
Activation + Threshold -H				0.177	0.983			
Output -H				0.544	0.728			
Complement				0.456	0.272			
Activation -O						- 0.410	- 0.254	- 0.410
Activation + Threshold -O						- 0.260	- 0.004	- 0.360
Output -O						0.435	0.499	0.411
Complement						0.565	0.501	0.589
Diff. from Target						- 0.195	- 0.329	0.239
Computed Error -O						- 0.048	- 0.082	0.058
Computed Error -H				0.0056	0.0012			
Adjustment to Threshold				0.0008	0.0002	- 0.0096	- 0.0164	0.0116

Adjustment to M-2 Column 1	- 0.0005	- 0.0070			
Adjustment to M-2 Column 2	0.0007	0.0008			
Adjustment to M-2 Column 3	0.0008	0.0011			
New Matrix M-2 Row 1			- 0.91	0.412	0.262
New Matrix M-2 Row 2			0.096	- 0.694	- 0.734
New Threshold Values -O			0.1404	0.2336	0.0616
Adjustment to M-1 Row 1	0.0004	- 0.0001			
Adjustment to M-1 Row 2	0.0006	0.0001			
Adjustment to M-1 Row 3	0.0008	0.0002			
New Matrix M-1 Row 1	0.6004	- 0.4			
New Matrix M-1 Row 2	0.2006	0.8001			
New Matrix M-1 Row 3	- 0.4992	0.3002			
New Threshold Values -H	0.2008	0.3002			

The next iteration uses the following information from the previous iteration, which you can identify from Table 7.1. The input pattern is (0.52, 0.75, 0.97), and the desired output pattern is (0.24, 0.17, 0.65). The current weight matrices are as follows:

M-1 Matrix of weights from input layer to hidden layer:

```
    0.6004      - 0.4
    0.2006        0.8001
  - 0.4992        0.3002
```

M-2 Matrix of weights from hidden layer to output layer:

```
-0.910        0.412        0.262
 0.096       -0.694       -0.734
```

The threshold values (or bias) for neurons in the hidden layer are 0.2008 and 0.3002, while those for the output neurons are 0.1404, 0.2336, and 0.0616, respectively.

You can keep the learning parameters as 0.15 for connections between input and hidden layer neurons, and 0.2 for connections between the hidden layer neurons and output neurons, or you can slightly modify them. Whether or not to change these two parameters is a decision that can be made perhaps at a later iteration, having obtained a sense of how the process is converging.

If you are satisfied with the rate at which the computed output pattern is getting close to the target output pattern, you would not change these learning rates. If you feel the convergence is much slower than you would like, then the learning rate parameters can be adjusted slightly upwards. It is a subjective decision both in terms of when (if at all) and to what new levels these parameters need to be revised.

NOTATION AND EQUATIONS

You have just seen an example of the process of training in the feedforward backpropagation network, described in relation to one hidden layer neuron and one input neuron. There were a few vectors that were shown and used, but perhaps not made easily identifiable. We therefore introduce some notation and describe the equations that were implicitly used in the example.

Notation

Let us talk about two matrices whose elements are the weights on connections. One matrix refers to the interface between the input and hidden layers, and the second refers to that between the hidden layer and the output layer. Since connections exist from each neuron in one layer to every neuron in the next layer, there is a vector of weights on the connections going out from any one neuron. Putting this vector into a row of the matrix, we get as many rows as there are neurons from which connections are established.

Let M_1 and M_2 be these matrices of weights. Then what does $M_1[i][j]$ represent? It is the weight on the connection from the ith input neuron to the jth neuron in the hidden layer. Similarly, $M_2[i][j]$ denotes the weight on the connection from the ith neuron in the hidden layer and the jth output neuron.

Next, we will use x, y, z for the outputs of neurons in the input layer, hidden layer, and output layer, respectively, with a subscript attached to denote which neuron in a given layer we are referring to. Let P denote the desired output pattern, with p_i as the components. Let m be the number of input neurons, so that according to our notation, $(x1, x2, ..., xm)$ will denote the input pattern. If P has, say, r components, the output layer needs r neurons. Let the number of hidden layer neurons be n. Let β_h be the learning rate parameter for the hidden layer, and β_o, that for the output layer. Let θ with the appropriate subscript represent the threshold value or bias for a hidden layer neuron, and τ with an appropriate subscript refer to the threshold value of an output neuron.

Let the errors in output at the output layer be denoted by ejs and those at the hidden layer by t_i's. If we use a Δ prefix of any parameter, then we are looking at the change in or adjustment to that parameter. Also, the thresholding function we would use is the **sigmoid** function, $f(x) = 1 / (1 + \exp(-x))$.

Equations

Output of jth hidden layer neuron:

$$y_j = f((\Sigma_i x_i M_1[i][j]) + \theta j) \tag{7.1}$$

Output of jth output layer neuron:

$$z_j = f((\Sigma_i y_i M_2[i][j]) + \tau_j) \tag{7.2}$$

Ith component of vector of output differences:

$$\text{desired value - computed value} = P_i - z_i$$

Ith component of output error at the output layer:

$$e_i = (P_i - z_i) \tag{7.3}$$

Ith component of output error at the hidden layer:

$$t_i = y_i (1 - y_i) (\Sigma j M_2[i][j] e_j) \tag{7.4}$$

Adjustment for weight between ith neuron in hidden layer and jth output neuron:

$$\Delta M_2[\,i\,][\,j\,] = \beta_o\, y_i e_j \qquad\qquad (7.5)$$

Adjustment for weight between ith input neuron and jth neuron in hidden layer:

$$M_1[\,i\,][\,j\,] = \beta_h x_i t_j \qquad\qquad (7.6)$$

Adjustment to the threshold value or bias for the jth output neuron:

$$\Delta \tau_j = \beta_o\, e_j$$

Adjustment to the threshold value or bias for the jth hidden layer neuron:

$$\Delta \theta_j = \beta_h\, e_j$$

For use of momentum parameter α (more on this parameter in Chapter 13), instead of equations 7.5 and 7.6, use:

$$\Delta M_2[\,i\,][\,j\,]\,(\,t\,) = \beta_o\, y_i e_j + \alpha \Delta M_2[\,i\,][\,j\,]\,(\,t - 1\,) \qquad (7.7)$$

and

$$\Delta M_1[\,i\,][\,j\,]\,(\,t\,) = \beta_h\, x_i t_j + \alpha \Delta M_1[\,i\,][\,j\,]\,(t - 1) \qquad (7.8)$$

C++ IMPLEMENTATION OF A BACKPROPAGATION SIMULATOR

The backpropagation simulator of this chapter has the following design objectives:

1. Allow the user to specify the number and size of all layers.
2. Allow the use of one or more hidden layers.
3. Be able to save and restore the state of the network.
4. Run from an arbitrarily large training data set or test data set.
5. Query the user for key network and simulation parameters.

6. Display key information at the end of the simulation.

7. Demonstrate the use of some C++ features.

A Brief Tour of How to Use the Simulator

In order to understand the C++ code, let us have an overview of the functioning of the program.

There are two modes of operation in the simulator. The user is queried first for which mode of operation is desired. The modes are **Training** mode and **Nontraining** mode (**Test** mode).

Training Mode

Here, the user provides a training file in the current directory called training.dat. This file contains exemplar pairs, or patterns. Each pattern has a set of inputs followed by a set of outputs. Each value is separated by one or more spaces. As a convention, you can use a few extra spaces to separate the inputs from the outputs. Here is an example of a training.dat file that contains two patterns:

```
0.4 0.5 0.89      -0.4 -0.8
0.23 0.8 -0.3     0.6 0.34
```

In this example, the first pattern has inputs 0.4, 0.5, and 0.89, with an expected output of −0.4 and −0.8. The second pattern has inputs of 0.23, 0.8, and −0.3 and outputs of 0.6 and 0.34. Since there are three inputs and two outputs, the input layer size for the network must be three neurons and the output layer size must be two neurons. Another file that is used in training is the weights file. Once the simulator reaches the error tolerance that was specified by the user, or the maximum number of iterations, the simulator saves the state of the network, by saving all of its weights in a file called **weights.dat.** This file can then be used subsequently in another run of the simulator in Nontraining mode. To provide some idea of how the network has done, information about the total and average error is presented at the end of the simulation. In addition, the output generated by the network for the last pattern vector is provided in an output file called **output.dat.**

Nontraining Mode (Test Mode)

In this mode, the user provides test data to the simulator in a file called **test.dat.** This file contains only input patterns. When this file is applied to an already trained network, an output.dat file is generated, which contains the outputs from the network for all of the input patterns. The network goes through one cycle of operation in this mode, covering all the patterns in the test data file. To start up the network, the weights file, weights.dat is read to initialize the state of the network. The user must provide the same network size parameters used to train the network.

Operation

The first thing to do with your simulator is to train a network with an architecture you choose. You can select the number of layers and the number of hidden layers for your network. Keep in mind that the input and output layer sizes are dictated by the input patterns you are presenting to the network and the outputs you seek from the network. Once you decide on an architecture, perhaps a simple three-layer network with one hidden layer, you prepare training data for it and save the data in the training.dat file. After this you are ready to train. You provide the simulator with the following information:

- The mode (select 1 for training)
- The values for the error tolerance and the learning rate parameter, lambda or beta
- The maximum number of cycles, or passes through the training data you'd like to try
- The number of layers (between three and five, three implies one hidden layer, while five implies three hidden layers)
- The size for each layer, from the input to the output

The simulator then begins training and reports the current cycle number and the average error for each cycle. *You should watch the error to see that it is on the whole decreasing with time.* If it is not, you should restart the simulation, because this will start with a brand new set of random weights and give you another, possibly better, solution. Note that there will be legitimate periods where the error *may* increase for some time. Once the simulation is done you will see information about the number of cycles and patterns used, and the total and average error that resulted. The weights are saved in the weights.dat file. You can rename this file to use this particular state of the network later. You can infer the size and number of layers from the information in this file, as will be shown in the next section for the **weights.dat** file format. You can have a peek at the **output.dat** file to see the kind of training result you have achieved. To get a full-blown accounting of each pattern and the match to that pattern, copy the training file to the test file and delete the output information from it. You can then run **Test** mode to get a full list of all the input stimuli and responses in the output.dat file.

Summary of Files Used in the Backpropagation Simulator

Here is a list of the files for your reference, as well as what they are used for.

- **weights.dat** You can look at this file to see the weights for the network. It shows the layer number followed by the weights that feed into the layer. The first layer, or input layer, layer zero, does not have any weights associated with it. An example of the weights.dat file is shown as follows for a network with three layers of sizes 3, 5, and 2. Note that the row width for layer n matches the column length for layer $n + 1$:

  1 -0.199660 -0.859660 -0.339660 -0.25966 0.520340

  1 0.292860 -0.487140 0.212860 -0.967140 -0.427140

  1 0.542106 -0.177894 0.322106 -0.977894 0.562106

  2 -0.175350 -0.835350

  2 -0.330167 -0.250167

  2 0.503317 0.283317

  2 -0.477158 0.222842

  2 -0.928322 -0.388322

  In this weights file the row width for layer 1 is 5, corresponding to the output of that (middle) layer. The input for the layer is the column length, which is 3, just as specified. For layer 2, the output size is the row width, which is 2, and the input size is the column length, 5, which is the same as the output for the middle layer. You can read the weights file to find out how things look.

- **training.dat** This file contains the input patterns for training. You can have as large a file as you'd like without degrading the performance of the simulator. The simulator caches data in memory for processing. This is to improve the speed of the simulation since disk accesses are expensive in time. A data buffer, which has a maximum size specified in a **#define** statement in the program, is filled with data from the training.dat file whenever data is needed. The format for the training.dat file has been shown in the **Training** mode section.

- **test.dat** The test.dat file is just like the training.dat file but without expected outputs. You use this file with a trained neural network in **Test** mode to see what responses you get for untrained data.

- **output.dat** The output.dat file contains the results of the simulation. In **Test** mode, the input and output vectors are shown for all pattern vectors. In the **Simulator** mode, the expected output is also shown, but only the last vector in the training set is presented, since the training set is usually quite large.

  Shown here is an example of an output file in **Training** mode:

  for input vector:
  0.400000 -0.400000
  output vector is:
  0.880095
  expected output vector is:
  0.900000

C++ Classes and Class Hierarchy

So far, you have learned how we address most of the objectives outlined for this program. The only objective left involves the demonstration of some C++ features. In this program we use a class hierarchy with the **inheritance** feature. Also, we use **polymorphism** with dynamic binding and function overloading with static binding.

First let us look at the class hierarchy used for this program (see Figure 7.2). An *abstract* class is a class that is never meant to be instantiated as an object, but serves as a base class from which others can inherit functionality and interface definitions. The layer class is such a class. You will see shortly that one of its functions is set = zero, which indicates that this class is an *abstract base class*. From the layer class are two branches. One is the **input_layer** class, and the other is the **output_layer** class. The middle layer class is very much like the output layer in function and so inherits from the **output_layer** class.

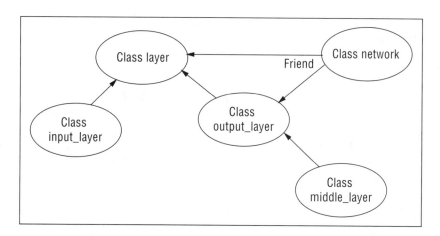

FIGURE 7.2 CLASS HIERARCHY USED IN THE BACKPROPAGATION SIMULATOR.

Function overloading can be seen in the definition of the **calc_error()** function. It is used in the **input_layer** with no parameters, while it is used in the **output_layer** (which the **input_layer** inherits from) with one parameter. Using the same function name is not a problem, and this is referred to as *overloading*. Besides function overloading, you may also have operator overloading, which is using an operator that performs some familiar function like + for addition, for another function, say, vector addition.

When you have overloading with the same parameters and the keyword *virtual*, then you have the potential for *dynamic binding*, which means that you determine which overloaded function to execute at run time and not at compile time. Compile time binding is referred to as *static binding*. If you put a bunch of C++ objects in an array of pointers to the **base** class, and then go through a loop that indexes each pointer and executes an overloaded virtual function that pointer is pointing to, then you will be using dynamic binding. This is exactly the case in the function **calc_out()**, which is declared with the virtual keyword in the **layer** base class. Each descendant of layer can provide a version of **calc_out()**, which differs in functionality from the base class, and the correct function will be selected at run time based on the object's identity. In this case **calc_out()**, which is a function to calculate the outputs for each layer, is different for the input layer than for the other two types of layers.

Let's look at some details in the header file in Listing 7.1:

Listing 7.1 Header file for the backpropagation simulator

```
// layer.h          V.Rao, H. Rao
// header file for the layer class hierarchy and
// the network class

#define MAX_LAYERS    5
#define MAX_VECTORS   100


class network;

class layer
{

protected:

        int num_inputs;
        int num_outputs;
        float *outputs;// pointer to array of outputs
        float *inputs; // pointer to array of inputs, which
                       // are outputs of some other layer

        friend network;

public:

        virtual void calc_out()=0;
};

class input_layer: public layer
{

private:
```

```
public:

        input_layer(int, int);
        ~input_layer();
        virtual void calc_out();

};

class middle_layer;

class output_layer:    public layer
{
protected:

        float * weights;
        float * output_errors; // array of errors at output
        float * back_errors; // array of errors back-propagated
        float * expected_values;      // to inputs
    friend network;

public:

        output_layer(int, int);
        ~output_layer();
        virtual void calc_out();
        void calc_error(float &);
        void randomize_weights();
        void update_weights(const float);
        void list_weights();
        void write_weights(int, FILE *);
        void read_weights(int, FILE *);
        void list_errors();
        void list_outputs();
};

class middle_layer:    public output_layer
{

private:
```

```
public:
    middle_layer(int, int);
    ~middle_layer();
        void calc_error();
};


class network

{

private:

    layer *layer_ptr[MAX_LAYERS];
    int number_of_layers;
    int layer_size[MAX_LAYERS];
    float *buffer;
    fpos_t position;
    unsigned training;

public:
    network();
    ~network();
        void set_training(const unsigned &);
        unsigned get_training_value();
        void get_layer_info();
        void set_up_network();
        void randomize_weights();
        void update_weights(const float);
        void write_weights(FILE *);
        void read_weights(FILE *);
        void list_weights();
        void write_outputs(FILE *);
        void list_outputs();
        void list_errors();
        void forward_prop();
```

```
void backward_prop(float &);
int fill_IObuffer(FILE *);
void set_up_pattern(int);

};
```

143

Details of the Backpropagation Header File

At the top of the file, there are two **#define** statements, which are used to set the maximum number of layers that can be used, currently five, and the maximum number of training or test vectors that can be read into an I/O buffer. This is currently 100. You can increase the size of the buffer for better speed at the cost of increased memory usage.

The following are definitions in the **layer** base class. Note that the number of inputs and outputs are *protected* data members, which means that they can be accessed freely by descendants of the class.

```
int num_inputs;
int num_outputs;
float *outputs;// pointer to array of outputs
float *inputs; // pointer to array of inputs, which
                      // are outputs of some other layer
friend network;
```

There are also two pointers to arrays of floats in this class. They are the pointers to the outputs in a given layer and the inputs to a given layer. To get a better idea of what a layer encompasses, Figure 7.3 shows you a small feedforward backpropagation network, with a dotted line that shows you the three layers for that network. A layer contains neurons and weights. The layer is responsible for calculating its output (**calc_out()**), stored in the **float * outputs** array, and errors (**calc_error()**) for each of its respective neurons. The errors are stored in another array called **float * output_errors** defined in the **output** class. Note that the **input** class does not have any weights associated with it and therefore is a special case. It does not need to provide any data members or function members related to errors or backpropagation. The only purpose of the input layer is to store data to be forward propagated to the next layer.

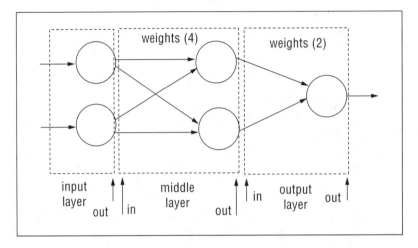

FIGURE 7.3 ORGANIZATION OF LAYERS FOR BACKPROPAGATION PROGRAM.

With the output layer, there are a few more arrays present. First, for storing backpropagated errors, there is an array called **float * back_errors**. There is a weights array called **float * weights**, and finally, for storing the expected values that initiate the error calculation process, there is an array called **float * expected_values**. Note that the middle layer needs almost all of these arrays and inherits them by being a derived class of the **output_layer** class.

There is one other class besides the **layer** class and its descendants defined in this header file, and that is the **network** class, which is used to set up communication channels between layers and to feed and remove data from the network. The **network** class performs the interconnection of layers by setting the pointer of an input array of a given layer to the output array of a previous layer.

This is a fairly extensible scheme that can be used to create variations on the feedforward backpropagation network with feedback connections, for instance.

N O T E

Another connection that the **network** class is responsible for is setting the pointer of an **output_error** array to the **back_error** array of the next layer (remember, errors flow in reverse, and the **back_error** array is the output error of the layer reflected at its inputs).

The **network** class stores an array of pointers to layers and an array of layer sizes for all the layers defined. These layer objects and arrays are dynamically allocated on the heap with the **New** and **Delete** functions in

C++. There is some minimal error checking for file I/O and memory allocation, which can be enhanced, if desired.

As you can see, the feedforward backpropagation network can quickly
become a memory and CPU hog, with large networks and large training sets.
The size and topology of the network, or *architecture*, will largely dictate
both these characteristics.

Details of the Backpropagation Implementation File

The implementation of the classes and methods is the next topic. Let's look

at the layer.cpp file in Listing 7.2.

LISTING 7.2 LAYER.CPP IMPLEMENTATION FILE FOR THE BACKPROPAGATION SIMULATOR

```
// layer.cpp           V.Rao, H.Rao
// compile for floating point hardware if available
#include <stdio.h>
#include <iostream.h>
#include <stdlib.h>
#include <math.h>
#include <time.h>
#include "layer.h"



inline float squash(float input)
// squashing function
// use sigmoid - can customize to something
// else if desired; can add a bias term too
//
{
if (input < -50)
        return 0.0;
else    if (input > 50)
                return 1.0;
        else return (float)(1/(1+exp(-(double)input)));

}


inline float randomweight(unsigned init)
```

```
{
int num;
// random number generator
// will return a floating point
// value between -1 and 1

if (init==1)    // seed the generator
        srand ((unsigned)time(NULL));

num=rand() % 100;

return 2*(float(num/100.00))-1;
}

// the next function is needed for Turbo C++
// and Borland C++ to link in the appropriate
// functions for fscanf floating point formats:
static void force_fpf()
{
        float x, *y;
        y=&x;
        x=*y;
}

// ————————————————-
//                              input layer
//—————————————
input_layer::input_layer(int i, int o)
{

num_inputs=i;
num_outputs=o;

outputs = new float[num_outputs];
if (outputs==0)
        {
        cout << "not enough memory\n";
        cout << "choose a smaller architecture\n";
        exit(1);
```

```
        }
}

input_layer::~input_layer()
{
delete [num_outputs] outputs;
}


void input_layer::calc_out()
{
//nothing to do, yet
}

// ——————————————-
//                                   output layer
//———————————————

output_layer::output_layer(int i, int o)
{

num_inputs=i;
num_outputs=o;
weights = new float[num_inputs*num_outputs];
output_errors = new float[num_outputs];
back_errors = new float[num_inputs];
outputs = new float[num_outputs];
expected_values = new float[num_outputs];
if ((weights==0)||(output_errors==0)||(back_errors==0)
        ||(outputs==0)||(expected_values==0))
        {
        cout << "not enough memory\n";
        cout << "choose a smaller architecture\n";
        exit(1);
        }

}

output_layer::~output_layer()
```

```
{
// some compilers may require the array
// size in the delete statement; those
// conforming to Ansi C++ will not
delete [num_outputs*num_inputs] weights;
delete [num_outputs] output_errors;
delete [num_inputs] back_errors;
delete [num_outputs] outputs;


}


void output_layer::calc_out()
{

int i,j,k;
float accumulator=0.0;


for (j=0; j<num_outputs; j++)
        {

        for (i=0; i<num_inputs; i++)

                {
                k=i*num_outputs;
                if (weights[k+j]*weights[k+j] > 1000000.0)
                        {
                        cout << "weights are blowing up\n";
                        cout << "try a smaller learning constant\n";
                        cout << "e.g. beta=0.02    aborting...\n";
                        exit(1);
                        }
                outputs[j]=weights[k+j]*(*(inputs+i));
                accumulator+=outputs[j];
                }
        // use the sigmoid squash function
        outputs[j]=squash(accumulator);
        accumulator=0;
```

```
        }

}


void output_layer::calc_error(float & error)
{
int i, j, k;
float accumulator=0;
float total_error=0;

for (j=0; j<num_outputs; j++)
    {
        output_errors[j] = expected_values[j]-outputs[j];
        total_error+=output_errors[j];
        }

error=total_error;

for (i=0; i<num_inputs; i++)
        {
        k=i*num_outputs;
        for (j=0; j<num_outputs; j++)
                {
                back_errors[i]=
                        weights[k+j]*output_errors[j];
                accumulator+=back_errors[i];
                }
        back_errors[i]=accumulator;
        accumulator=0;
        // now multiply by derivative of
        // sigmoid squashing function, which is
        // just the input*(1-input)
        back_errors[i]*=(*(inputs+i))*(1-(*(inputs+i)));
        }

 }
```

```
void output_layer::randomize_weights()
{
int i, j, k;
const unsigned first_time=1;

const unsigned not_first_time=0;
float discard;

discard=randomweight(first_time);

for (i=0; i< num_inputs; i++)
    {
    k=i*num_outputs;
    for (j=0; j< num_outputs; j++)
            weights[k+j]=randomweight(not_first_time);
    }
}

void output_layer::update_weights(const float beta)
{
int i, j, k;

// learning law: weight_change =
//              beta*output_error*input

for (i=0; i< num_inputs; i++)
    {
    k=i*num_outputs;
    for (j=0; j< num_outputs; j++)
            weights[k+j] +=
                    beta*output_errors[i]*(*(inputs+i));
    }

}

void output_layer::list_weights()
{
int i, j, k;
```

```
for (i=0; i< num_inputs; i++)
        {
        k=i*num_outputs;
        for (j=0; j< num_outputs; j++)
                cout << "weight["<<i<<","<<
                        j<<"] is: "<<weights[k+j];
        }


}


void output_layer::list_errors()
{
int i, j;

for (i=0; i< num_inputs; i++)
        cout << "backerror["<<i<<
                "] is : "<<back_errors[i]<<"\n";

for (j=0; j< num_outputs; j++)
        cout << "outputerrors["<<j<<
                        "] is: "<<output_errors[j]<<"\n";

}



void output_layer::write_weights(int layer_no,
                FILE * weights_file_ptr)
{
int i, j, k;

// assume file is already open and ready for
// writing

// prepend the layer_no to all lines of data
// format:
//              layer_no        weight[0,0] weight[0,1] ...
//              layer_no        weight[1,0] weight[1,1] ...
//              ...
```

```
for (i=0; i< num_inputs; i++)
        {
        fprintf(weights_file_ptr,"%i ",layer_no);
        k=i*num_outputs;
    for (j=0; j< num_outputs; j++)
        {
        fprintf(weights_file_ptr,"%f",
                    weights[k+j]);
        }
    fprintf(weights_file_ptr,"\n");
        }


}

void output_layer::read_weights(int layer_no,
            FILE * weights_file_ptr)
{
int i, j, k;

// assume file is already open and ready for
// reading

// look for the prepended layer_no
// format:
//          layer_no        weight[0,0] weight[0,1] ...
//          layer_no        weight[1,0] weight[1,1] ...
//          ...
while (1)

        {

        fscanf(weights_file_ptr,"%i",&j);
        if ((j==layer_no)|| (feof(weights_file_ptr)))
                break;
        else
                {
                while (fgetc(weights_file_ptr) != '\n')
                        {;}// get rest of line
                }
```

```
        }

if (!(feof(weights_file_ptr)))
        {
        // continue getting first line
        i=0;
        for (j=0; j< num_outputs; j++)
                    {

                        fscanf(weights_file_ptr,"%f",
                                    &weights[j]); // i*num_outputs = 0
                    }
        fscanf(weights_file_ptr,"\n");

        // now get the other lines
        for (i=1; i< num_inputs; i++)
                {
                fscanf(weights_file_ptr,"%i",&layer_no);
                k=i*num_outputs;
        for (j=0; j< num_outputs; j++)
                {
                fscanf(weights_file_ptr,"%f",
                        &weights[k+j]);
                }

                }
        fscanf(weights_file_ptr,"\n");
        }


else cout << "end of file reached\n";

}
void output_layer::list_outputs()
{
int j;

for (j=0; j< num_outputs; j++)
        {
```

```
            cout << "outputs["<<j
                <<"] is: "<<outputs[j]<<"\n";
            }


}


// ————————————————————-
//                              middle layer
//————————————————————

middle_layer::middle_layer(int i, int o):
        output_layer(i,o)
{

}

middle_layer::~middle_layer()
{
delete [num_outputs*num_inputs] weights;
delete [num_outputs] output_errors;
delete [num_inputs] back_errors;
delete [num_outputs] outputs;
}

void middle_layer::calc_error()
{
int i, j, k;
float accumulator=0;

for (i=0; i<num_inputs; i++)
        {
        k=i*num_outputs;
        for (j=0; j<num_outputs; j++)
                {
                back_errors[i]=
                        weights[k+j]*(*(output_errors+j));
                accumulator+=back_errors[i];
                }
        back_errors[i]=accumulator;
```

```
            accumulator=0;
            // now multiply by derivative of
            // sigmoid squashing function, which is
            // just the input*(1-input)
            back_errors[i]*=(*(inputs+i))*(1-(*(inputs+i)));
            }

    }


network::network()
{
position=0L;
}


network::~network()
{
int i,j,k;
i=layer_ptr[0]->num_outputs;// inputs
j=layer_ptr[number_of_layers-1]->num_outputs; //outputs
k=MAX_VECTORS;



delete [(i+j)*k]buffer;
}


void network::set_training(const unsigned & value)
{
training=value;
}


unsigned network::get_training_value()
{
return training;
}


void network::get_layer_info()
{
int i;
```

```
//————————————————
//
//        Get layer sizes for the network
//
// ————————————————-


cout << " Please enter in the number of layers for your network.\n";
cout << " You can have a minimum of 3 to a maximum of 5. \n";
cout << " 3 implies 1 hidden layer; 5 implies 3 hidden layers : \n\n";


cin >> number_of_layers;


cout << " Enter in the layer sizes separated by spaces.\n";
cout << " For a network with 3 neurons in the input layer,\n";
cout << " 2 neurons in a hidden layer, and 4 neurons in the\n";
cout << " output layer, you would enter: 3 2 4 .\n";
cout << " You can have up to 3 hidden layers,for five maximum entries
:\n\n";


for (i=0; i<number_of_layers; i++)
        {
        cin >> layer_size[i];
        }


// ————————————————
// size of layers:
//              input_layer           layer_size[0]
//              output_layer          layer_size[number_of_layers-1]
//              middle_layers         layer_size[1]
//                                    optional: layer_size[number_of_lay-
ers-3]
//                                    optional: layer_size[number_of_lay-
ers-2]
//————————————————-


}


void network::set_up_network()
{
int i,j,k;
```

```
//————————————————-
// Construct the layers
//
//————————————————-

layer_ptr[0] = new input_layer(0,layer_size[0]);

for (i=0;i<(number_of_layers-1);i++)
        {
        layer_ptr[i+1] =
        new middle_layer(layer_size[i],layer_size[i+1]);
        }

layer_ptr[number_of_layers-1] = new
output_layer(layer_size[number_of_layers-2],layer_size[number_of_layers-
1]);

for (i=0;i<(number_of_layers-1);i++)
        {
        if (layer_ptr[i] == 0)
                {
                cout << "insufficient memory\n";
                cout << "use a smaller architecture\n";
                exit(1);
                }
        }


//————————————————-
// Connect the layers
//
//————————————————-
// set inputs to previous layer outputs for all layers,
//      except the input layer

for (i=1; i< number_of_layers; i++)
        layer_ptr[i]->inputs = layer_ptr[i-1]->outputs;

// for back_propagation, set output_errors to next layer
//              back_errors for all layers except the output
```

```
//              layer and input layer


for (i=1; i< number_of_layers -1; i++)
        ((output_layer *)layer_ptr[i])->output_errors =
                ((output_layer *)layer_ptr[i+1])->back_errors;

// define the IObuffer that caches data from
// the datafile
i=layer_ptr[0]->num_outputs;// inputs
j=layer_ptr[number_of_layers-1]->num_outputs; //outputs
k=MAX_VECTORS;

buffer=new
        float[(i+j)*k];
if (buffer==0)
        cout << "insufficient memory for buffer\n";
}

void network::randomize_weights()
{
int i;

for (i=1; i<number_of_layers; i++)
        ((output_layer *)layer_ptr[i])
                ->randomize_weights();
}


void network::update_weights(const float beta)
{
int i;

for (i=1; i<number_of_layers; i++)
        ((output_layer *)layer_ptr[i])
                ->update_weights(beta);
}


void network::write_weights(FILE * weights_file_ptr)
```

```
{
int i;

for (i=1; i<number_of_layers; i++)
        ((output_layer *)layer_ptr[i])
                ->write_weights(i,weights_file_ptr);
}

void network::read_weights(FILE * weights_file_ptr)
{
int i;

for (i=1; i<number_of_layers; i++)
        ((output_layer *)layer_ptr[i])
                ->read_weights(i,weights_file_ptr);
}

void network::list_weights()
{
int i;

for (i=1; i<number_of_layers; i++)
        {
        cout << "layer number : " <<i<< "\n";
        ((output_layer *)layer_ptr[i])
                ->list_weights();
        }
}

void network::list_outputs()
{
int i;

for (i=1; i<number_of_layers; i++)
        {
        cout << "layer number : " <<i<< "\n";
        ((output_layer *)layer_ptr[i])
                ->list_outputs();
        }
```

```
}

void network::write_outputs(FILE *outfile)
{
int i, ins, outs;
ins=layer_ptr[0]->num_outputs;
outs=layer_ptr[number_of_layers-1]->num_outputs;
float temp;

fprintf(outfile,"for input vector:\n");

for (i=0; i<ins; i++)
        {
        temp=layer_ptr[0]->outputs[i];
        fprintf(outfile,"%f  ",temp);
        }

fprintf(outfile,"\noutput vector is:\n");

for (i=0; i<outs; i++)
        {
        temp=layer_ptr[number_of_layers-1]->
        outputs[i];
        fprintf(outfile,"%f  ",temp);

        }

if (training==1)
{
fprintf(outfile,"\nexpected output vector is:\n");

for (i=0; i<outs; i++)
        {
        temp=((output_layer *)(layer_ptr[number_of_layers-1]))->
        expected_values[i];
        fprintf(outfile,"%f  ",temp);

        }
}
```

```
fprintf(outfile,"\n————————\n");

}

void network::list_errors()
{
int i;

for (i=1; i<number_of_layers; i++)
        {
        cout << "layer number : " <<i<< "\n";
        ((output_layer *)layer_ptr[i])
                ->list_errors();
        }
}

int network::fill_IObuffer(FILE * inputfile)
{
// this routine fills memory with
// an array of input, output vectors
// up to a maximum capacity of
// MAX_INPUT_VECTORS_IN_ARRAY
// the return value is the number of read
// vectors


int i, k, count, veclength;

int ins, outs;

ins=layer_ptr[0]->num_outputs;

outs=layer_ptr[number_of_layers-1]->num_outputs;

if (training==1)
        veclength=ins+outs;
else
        veclength=ins;
```

```
count=0;
while  ((count<MAX_VECTORS)&&
              (!feof(inputfile)))
        {
        k=count*(veclength);
        for (i=0; i<veclength; i++)
                {
                fscanf(inputfile,"%f",&buffer[k+i]);
                }
        fscanf(inputfile,"\n");
        count++;
        }

if (!(ferror(inputfile)))
        return count;
else return -1; // error condition


}


void network::set_up_pattern(int buffer_index)
{
// read one vector into the network
int i, k;
int ins, outs;

ins=layer_ptr[0]->num_outputs;
outs=layer_ptr[number_of_layers-1]->num_outputs;
if (training==1)
        k=buffer_index*(ins+outs);
else
        k=buffer_index*ins;

for (i=0; i<ins; i++)
        layer_ptr[0]->outputs[i]=buffer[k+i];

if (training==1)
{
        for (i=0; i<outs; i++)
```

```
                ((output_layer *)layer_ptr[number_of_layers-1])->
                        expected_values[i]=buffer[k+i+ins];
}


}
```

```
void network::forward_prop()
{
int i;
for (i=0; i<number_of_layers; i++)
        {
        layer_ptr[i]->calc_out(); //polymorphic
                            // function
        }
}


void network::backward_prop(float & toterror)
{
int i;

// error for the output layer
((output_layer*)layer_ptr[number_of_layers-1])->
                        calc_error(toterror);

// error for the middle layer(s)
for (i=number_of_layers-2; i>0; i-)
        {
        ((middle_layer*)layer_ptr[i])->
                        calc_error();

        }


}
```

A Look at the Functions in the layer.cpp File

The following is a listing of the functions in the layer.cpp file along with a brief statement of each one's purpose.

- **void set_training(const unsigned &)** Sets the value of the private data member, training; use 1 for training mode, and 0 for test mode.

- **unsigned get_training_value()** Gets the value of the training constant that gives the mode in use.

- **void get_layer_info()** Gets information about the number of layers and layer sizes from the user.

- **void set_up_network()** This routine sets up the connections between layers by assigning pointers appropriately.

- **void randomize_weights()** At the beginning of the training process, this routine is used to randomize all of the weights in the network.

- **void update_weights(const float)** As part of training, weights are updated according to the learning law used in backpropagation.

- **void write_weights(FILE *)** This routine is used to write weights to a file.

- **void read_weights(FILE *)** This routine is used to read weights into the network from a file.

- **void list_weights()** This routine can be used to list weights while a simulation is in progress.

- **void write_outputs(FILE *)** This routine writes the outputs of the network to a file.

- **void list_outputs()** This routine can be used to list the outputs of the network while a simulation is in progress.

- **void list_errors()** Lists errors for all layers while a simulation is in progress.

- **void forward_prop()** Performs the forward propagation.

- **void backward_prop(float &)** Performs the backward error propagation.

- **int fill_IObuffer(FILE *)** This routine fills the internal IO buffer with data from the training or test data sets.

- **void set_up_pattern(int)** This routine is used to set up one pattern from the IO buffer for training.

- **inline float squash(float input)** This function performs the sigmoid function.

- **inline float randomweight (unsigned unit)** This routine returns a random weight between −1 and 1; use 1 to initialize the generator, and 0 for all subsequent calls.

Note that the functions **squash(float)** and **randomweight(unsigned)** are declared inline. This means that the function's source code is inserted wherever it appears. This increases code size, but also increases speed because a function call, which is expensive, is avoided.

N O T E

The final file to look at is the backprop.cpp file presented in Listing 7.3 .

LISTING 7.3 THE BACKPROP.CPP FILE FOR THE BACKPROPAGATION SIMULATOR

```
// backprop.cpp        V. Rao, H. Rao
#include "layer.cpp"


#define TRAINING_FILE  "training.dat"
#define WEIGHTS_FILE "weights.dat"
#define OUTPUT_FILE    "output.dat"
#define TEST_FILE      "test.dat"


void main()
{

float error_tolerance=0.1;
float total_error=0.0;
float avg_error_per_cycle=0.0;
float error_last_cycle=0.0;
float avgerr_per_pattern=0.0; // for the latest cycle
float error_last_pattern=0.0;
float learning_parameter=0.02;
unsigned temp, startup;
long int vectors_in_buffer;
long int max_cycles;
long int patterns_per_cycle=0;

long int total_cycles, total_patterns;
int i;

// create a network object
network backp;
```

```
FILE * training_file_ptr, * weights_file_ptr, * output_file_ptr;
FILE * test_file_ptr, * data_file_ptr;

// open output file for writing
if ((output_file_ptr=fopen(OUTPUT_FILE,"w"))==NULL)
                {
                cout << "problem opening output file\n";
                exit(1);
                }

// enter the training mode : 1=training on     0=training off
cout << "——————————————————————\n";
cout << " C++ Neural Networks and Fuzzy Logic \n";
cout << "        Backpropagation simulator \n";
cout << "              version 1 \n";
cout << "——————————————————————\n";
cout << "Please enter 1 for TRAINING on, or 0 for off: \n\n";
cout << "Use training to change weights according to your\n";
cout << "expected outputs. Your training.dat file should contain\n";
cout << "a set of inputs and expected outputs. The number of\n";
cout << "inputs determines the size of the first (input) layer\n";
cout << "while the number of outputs determines the size of the\n";
        cout << "last (output) layer :\n\n";

cin >> temp;
backp.set_training(temp);

if (backp.get_training_value() == 1)
        {
        cout << "—> Training mode is *ON*. weights will be saved\n";
        cout << "in the file weights.dat at the end of the\n";
        cout << "current set of input (training) data\n";
        }
else
        {
        cout << "—> Training mode is *OFF*. weights will be loaded\n";
        cout << "from the file weights.dat and the current\n";
        cout << "(test) data set will be used. For the test\n";
        cout << "data set, the test.dat file should contain\n";
```

```
            cout << "only inputs, and no expected outputs.\n";
            }

if (backp.get_training_value()==1)
        {
        // ————————————-
        //      Read in values for the error_tolerance,
        //      and the learning_parameter
        // ————————————-
        cout << " Please enter in the error_tolerance\n";
        cout << " -- between 0.001 to 100.0, try 0.1 to start \n";
        cout << "\n";
        cout << "and the learning_parameter, beta\n";
        cout << " -- between 0.01 to 1.0, try 0.5 to start - \n\n";
        cout << " separate entries by a space\n";
        cout << " example: 0.1 0.5 sets defaults mentioned :\n\n";

        cin >> error_tolerance >> learning_parameter;
        //————————————-
        // open training file for reading
        //————————————-
        if ((training_file_ptr=fopen(TRAINING_FILE,"r"))==NULL)
                {
                cout << "problem opening training file\n";
                exit(1);
                }
        data_file_ptr=training_file_ptr; // training on

        // Read in the maximum number of cycles
        // each pass through the input data file is a cycle
        cout << "Please enter the maximum cycles for the simula-\
        tion\n";
        cout << "A cycle is one pass through the data set.\n";
        cout << "Try a value of 10 to start with\n";

        cin >> max_cycles;

        }
else
```

```
        {
        if ((test_file_ptr=fopen(TEST_FILE,"r"))==NULL)
                {
                cout << "problem opening test file\n";
                exit(1);
                }

        data_file_ptr=test_file_ptr; // training off
        }




//
// training: continue looping until the total error is less than
//              the tolerance specified, or the maximum number of
//              cycles is exceeded; use both the forward signal propaga-
tion
//              and the backward error propagation phases. If the error
//              tolerance criteria is satisfied, save the weights in a
file.
// no training: just proceed through the input data set once in the
//              forward signal propagation phase only. Read the starting
//              weights from a file.
// in both cases report the outputs on the screen



// initialize counters
total_cycles=0; // a cycle is once through all the input data
total_patterns=0; // a pattern is one entry in the input data



// get layer information
backp.get_layer_info();

// set up the network connections
backp.set_up_network();

// initialize the weights
```

```
if (backp.get_training_value()==1)
        {
        // randomize weights for all layers; there is no
        // weight matrix associated with the input layer
        // weight file will be written after processing
        // so open for writing
        if ((weights_file_ptr=fopen(WEIGHTS_FILE,"w"))
                    ==NULL)
                {
                cout << "problem opening weights file\n";
                exit(1);
                }
        backp.randomize_weights();
        }
else
        {
        // read in the weight matrix defined by a
        // prior run of the backpropagation simulator
        // with training on
        if ((weights_file_ptr=fopen(WEIGHTS_FILE,"r"))
                    ==NULL)
                {
                cout << "problem opening weights file\n";
                exit(1);
                }
        backp.read_weights(weights_file_ptr);
        }



// main loop
// if training is on, keep going through the input data
//              until the error is acceptable or the maximum number of
//      cycles
//              is exceeded.
// if training is off, go through the input data once. report // outputs
// with inputs to file output.dat

startup=1;
```

```
vectors_in_buffer = MAX_VECTORS; // startup condition
total_error = 0;

while (                     ((backp.get_training_value()==1)
                    && (avgerr_per_pattern
                                    > error_tolerance)
                    && (total_cycles < max_cycles)
                    && (vectors_in_buffer !=0))
                    || ((backp.get_training_value()==0)
                    && (total_cycles < 1))
                    || ((backp.get_training_value()==1)
                    && (startup==1))
                    )
    {
    startup=0;
    error_last_cycle=0; // reset for each cycle
    patterns_per_cycle=0;
    // process all the vectors in the datafile
    // going through one buffer at a time
    // pattern by pattern



    while ((vectors_in_buffer==MAX_VECTORS))
            {

            vectors_in_buffer=
                    backp.fill_IObuffer(data_file_ptr); // fill buffer
                    if (vectors_in_buffer < 0)
                            {
                            cout << "error in reading in vectors, aborting\n";
                            cout << "check that there are no extra
linefeeds\n";
                            cout << "in your data file, and that the
number\n";
                            cout << "of layers and size of layers match
the\n";
                            cout << "the parameters provided.\n";
                            exit(1);
                            }
```

```
                  // process vectors
                  for (i=0; i<vectors_in_buffer; i++)
                        {
                        // get next pattern
                        backp.set_up_pattern(i);

                        total_patterns++;
                        patterns_per_cycle++;
                        // forward propagate

                        backp.forward_prop();

                        if (backp.get_training_value()==0)
                              backp.write_outputs(output_file_ptr);

                        // back_propagate, if appropriate
                        if (backp.get_training_value()==1)
                              {

                              backp.backward_prop(error_last_pattern);
                              error_last_cycle += error_last_pattern
 z                            *error_last_pattern;
                              backp.update_weights(learning_parameter);
                              // backp.list_weights();
                              // can
                              // see change in weights by
                              // using list_weights before and
                              // after back_propagation
                              }

                        }

            error_last_pattern = 0;
            }

avgerr_per_pattern=((float)sqrt((double)error_last_cycle
/patterns_per_cycle));
total_error += error_last_cycle;
total_cycles++;
```

```
// most character displays are 26 lines
// user will see a corner display of the cycle count
// as it changes

cout << "\n\n\n\n\n\n\n\n\n\n\n\n\n\n\n\n\n\n\n\n\n\n\n\n\n";
cout << total_cycles << "\t" << avgerr_per_pattern << "\n";

fseek(data_file_ptr, OL, SEEK_SET); // reset the file pointer
                        // to the beginning of
                        // the file
vectors_in_buffer = MAX_VECTORS; // reset


} // end main loop

cout << "\n\n\n\n\n\n\n\n\n\n\n";
cout << "————————————————————\n";
cout << "     done:   results in file output.dat\n";
cout << "             training: last vector only\n";
cout << "             not training: full cycle\n\n";
if (backp.get_training_value()==1)
        {
        backp.write_weights(weights_file_ptr);
        backp.write_outputs(output_file_ptr);
        avg_error_per_cycle = (float)sqrt((double)total_error/
total_cycles);
     error_last_cycle = (float)sqrt((double)error_last_cycle);



cout << "             weights saved in file weights.dat\n";
cout << "\n";
cout << "—>average error per cycle = " <<
avg_error_per_cycle << " <–-\n";
cout << "–>error last cycle= " << error_last_cycle << " <–\n";
cout << "->error last cycle per pattern= " << avgerr_per_pattern << " <--
\n";


        }
```

```
cout << "————>total cycles = " << total_cycles <<
" <--\n";
cout << "————>total patterns = " << total_patterns << " <--\n";
cout << "————————————————————\n";
// close all files
fclose(data_file_ptr);
fclose(weights_file_ptr);
fclose(output_file_ptr);



}
```

The backprop.cpp file implements the simulator controls. First, data is accepted from the user for network parameters. Assuming **Training** mode is used, the training file is opened and data is read from the file to fill the IO buffer. Then the main loop is executed where the network processes pattern by pattern to complete a cycle, which is one pass through the entire training data set. (The IO buffer is refilled as required during this process.) After executing one cycle, the file pointer is reset to the beginning of the file and another cycle begins. The simulator continues with cycles until one of the two fundamental criteria is met:

1. The maximum cycle count specified by the user is exceeded.
2. The average error per pattern for the latest cycle is less than the error tolerance specified by the user.

When either of these occurs, the simulator stops and reports out the error achieved, and saves weights in the weights.dat file and one output vector in the output.dat file.

In **Test** mode, exactly one cycle is processed by the network and outputs are written to the output.dat file. At the beginning of the simulation in **Test** mode, the network is set up with weights from the weights.dat file. To simplify the program, the user is requested to enter the number of layers and size of layers, although you could have the program figure this out from the weights file.

Compiling and Running the Backpropagation Simulator

Compiling the backprop.cpp file will compile the simulator since layer.cpp is included in backprop.cpp. To run the simulator, once you have created an executable (using 80X87 floating point hardware if available), you type in **backprop** and see the following screen (user input in italic):

```
C++ Neural Networks and Fuzzy Logic
        Backpropagation simulator
               version 1
Please enter 1 for TRAINING on, or 0 for off:

Use training to change weights according to your
expected outputs. Your training.dat file should contain
a set of inputs and expected outputs. The number of
inputs determines the size of the first (input) layer
while the number of outputs determines the size of the
last (output) layer :


1


-> Training mode is *ON*. weights will be saved
in the file weights.dat at the end of the
current set of input (training) data
 Please enter in the error_tolerance
 -- between 0.001 to 100.0, try 0.1 to start -

and the learning_parameter, beta
 -- between 0.01 to 1.0, try 0.5 to start -

 separate entries by a space
 example: 0.1 0.5 sets defaults mentioned :


0.2 0.25


Please enter the maximum cycles for the simulation
A cycle is one pass through the data set.
Try a value of 10 to start with
Please enter in the number of layers for your network.
You can have a minimum of three to a maximum of five.
three implies one hidden layer; five implies three hidden layers:
```

3

Enter in the layer sizes separated by spaces.
For a network with three neurons in the input layer,
two neurons in a hidden layer, and four neurons in the
output layer, you would enter: *3 2 4.*
You can have up to three hidden layers for five maximum entries :

2 2 1

1	0.353248
2	0.352684
3	0.352113
4	0.351536
5	0.350954
...	
299	0.0582381
300	0.0577085

done: results in file output.dat
        training: last vector only
        not training: full cycle

        weights saved in file weights.dat

—>average error per cycle = 0.20268 <—
—>error last cycle = 0.0577085 <—
->error last cycle per pattern= 0.0577085 <—
———>total cycles = 300 <—
———>total patterns = 300 <—

N O T E

The cycle number and the average error per pattern is displayed as the simulation progresses (not all values shown). You can monitor this to make sure the simulator is converging on a solution. If the error does not seem to decrease beyond a certain point, but instead drifts or blows up, then you should start the simulator again with a new starting point defined by the random weights initializer. Also, you could try decreasing the size of the learning rate parameter. Learning may be slower, but this may allow a better minimum to be found.

This example shows just one pattern in the training set with two inputs and one output. The results along with the (one) last pattern are shown as follows from the file output.dat:

```
for input vector:
0.400000   -0.400000
output vector is:
0.842291
expected output vector is:
0.900000
```

The match is pretty good, as can be expected, since the optimization is easy for the network; there is only one pattern to worry about. Let's look at the final set of weights for this simulation in weights.dat. These weights were obtained by updating the weights for 300 cycles with the learning law:

1 0.175039 0.435039

1 -1.319244 -0.559244

2 0.358281

2 2.421172

We'll leave the backpropagation simulator for now and return to it in a later chapter for further exploration. You can experiment a number of different ways with the simulator:

- Try a different number of layers and layer sizes for a given problem.
- Try different learning rate parameters and see its effect on convergence and training time.
- Try a very large learning rate parameter (should be between 0 and 1); try a number over 1 and note the result.

SUMMARY

In this chapter, you learned about one of the most powerful neural network algorithms called backpropagation. Without having feedback connections, propagating only errors appropriately to the hidden layer and input layer connections, the algorithm uses the so-called generalized delta rule and trains the network with exemplar pairs of patterns. It is difficult to determine how many hidden-layer neurons are to be provided for. The number of hidden layers could be more than one. In general, the size of the hidden layer(s) is related to the features or distinguishing characteristics that should be discerned from the data. Our example in this chapter relates to a simple case where there is a single hidden layer. The outputs of the output neurons, and therefore of the network, are vectors with components between 0 and 1, since the **thresholding** function is the **sigmoid** function. These values can be scaled, if necessary, to get values in another interval.

177

Our example does not relate to any particular function to be computed by the network, but inputs and outputs were randomly chosen. What this can tell you is that, if you do not know the functional equation between two sets of vectors, the feedback backpropagation network can find the mapping for any vector in the domain, even if the functional equation is not found. For all we know, that function could be nonlinear as well.

There is one important fact you need to remember about the backpropagation algorithm. Its steepest descent procedure in training does not guarantee finding a global or overall minimum, it can find only a local minimum of the energy surface.

CHAPTER 8

BAM: Bidirectional Associative Memory

Introduction

The *Bidirectional Associative Memory* (*BAM*) model has a neural network of two layers and is fully connected from each layer to the other. That is, there are feedback connections from the output layer to the input layer. However, the weights on the connections between any two given neurons from different layers are the same. You may even consider it to be a single bidirectional connection with a single weight. The matrix of weights for the connections from the output layer to the input layer is simply the transpose of the matrix of weights for the connections between the input and output layer. If we denote the matrix for forward connection weights by $\mathbf{W}$, then $\mathbf{W}^T$ is the matrix of weights for the output layer to input layer connections. As you recall, the transpose of a matrix is obtained simply by interchanging the rows and the columns of the matrix.

There are two layers of neurons, an input layer and an output layer. There are no lateral connections, that is, no two neurons within the same layer are connected. *Recurrent* connections, which are feedback connections to a neuron from itself, may or may not be present. The architecture is quite simple. Figure 8.1 shows the layout for this neural network model, using only three input neurons and two output neurons. There are feedback connections from Field A to Field B and vice-versa. This figure also indicates the presence of inputs and outputs at each of the two fields for the bidirectional associative memory network. Connection weights are also shown as labels on only a few connections in this figure, to avoid cluttering. The general case is analogous.

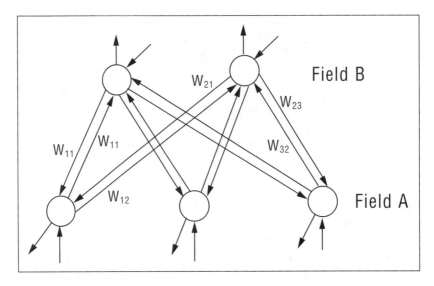

FIGURE 8.1 LAYOUT OF A BAM NETWORK

INPUTS AND OUTPUTS

The input to a BAM network is a vector of real numbers, usually in the set { −1, +1 }. The output is also a vector of real numbers, usually in the set { −1, +1 }, with the same or different dimension from that of the input. These vectors can be considered patterns, and the network makes heteroassociation of patterns. If the output is required to be the same as input, then you are asking the network to make autoassociation, which it does, and it becomes a special case of the general activity of this type of neural network.

For inputs and outputs that are not outside the set containing just −1 and +1, try following this next procedure. You can first make a mapping into binary numbers, and then a mapping of each binary digit into a bipolar digit. For example, if your inputs are first names of people, each character in the name can be replaced by its ASCII code, which in turn can be changed to a binary number, and then each binary digit 0 can be replaced by −1. For example, the ASCII code for the letter R is 82, which is 1010010, as a binary number. This is mapped onto the bipolar string 1 −1 1 −1 −1 1 −1. If a name consists of three characters, their ASCII codes in binary can be concatenated or juxtaposed and the corresponding bipolar string obtained. This bipolar string can also be looked upon as a vector of bipolar characters.

WEIGHTS AND TRAINING

BAM does not modify weights during its operation, and as mentioned in Chapter 6, like the Hopfield network, uses one-shot training. The adaptive variety of BAM, called the *Adaptive Bidirectional Associative Memory*, *(ABAM)* undergoes supervised iterative training. BAM needs some exemplar pairs of vectors. The pairs used as exemplars are those that require heteroassociation. The weight matrix, there are two, but one is just the transpose of the other as already mentioned, is constructed in terms of the exemplar vector pairs.

The use of exemplar vectors is a one-shot learning—to determine what the weights should be. Once weights are so determined, and an input vector is presented, a potentially associated vector is output. It is taken as input in the opposite direction, and its potentially associated vector is obtained back at the input layer. If the last vector found is the same as what is originally input, then there is resonance. Suppose the vector B is obtained at one end, as a result of C being input at the other end. If B in turn is input during the next cycle of operation at the end where it was obtained, and produces C at the opposite end, then you have a pair of heteroassociated vectors. This is what is basically happening in a BAM neural network.

The BAM and Hopfield memories are closely related. You can think of the Hopfield memory as a special case of the BAM.

N O T E

What follow are the equations for the determination of the weight matrix, when the k pairs of exemplar vectors are denoted by ($\mathbf{X_i}$, $\mathbf{Y_i}$), i ranging from 1 to k. Note that T in the superscript of a matrix stands for the transpose of the matrix. While you interchange the rows and columns to get the transpose of a matrix, you write a column vector as a row vector, and vice versa to get the transpose of a vector. The following equations refer to the vector pairs after their components are changed to bipolar values, only for obtaining the weight matrix $\mathbf{W}$. Once $\mathbf{W}$ is obtained, further use of these exemplar vectors is made in their original form.

$$\mathbf{W} = X_1{}^T Y_1 + \ldots + X_k{}^T Y_k$$

and

$$\mathbf{W}^T = Y_1{}^T X_1 + \ldots + Y_k{}^T X_k$$

Example

Suppose you choose two pairs of vectors as possible exemplars. Let them be:

$$X_1 = (1, 0, 0, 1), Y_1 = (0, 1, 1)$$

and

$$X_2 = (0, 1, 1, 0), Y_2 = (1, 0, 1)$$

These you change into bipolar components and get, respectively, $(1, -1, -1, 1)$, $(-1, 1, 1)$, $(-1, 1, 1, -1)$, and $(1, -1, 1)$.

$$
W =
\begin{array}{l}
1 \ [-1 \ 1 \ 1] \\
-1 \\
-1 \\
1
\end{array}
+
\begin{array}{l}
-1 \ [1 \ -1 \ 1] \\
1 \\
1 \\
-1
\end{array}
=
\begin{array}{ccc}
-1 & 1 & 1 \\
1 & -1 & -1 \\
1 & -1 & -1 \\
-1 & 1 & 1
\end{array}
+
\begin{array}{ccc}
-1 & 1 & -1 \\
1 & -1 & 1 \\
1 & -1 & 1 \\
-1 & 1 & -1
\end{array}
=
\begin{array}{ccc}
-2 & 2 & 0 \\
2 & -2 & 0 \\
2 & -2 & 0 \\
-2 & 2 & 0
\end{array}
$$

and

$$
W^T =
\begin{array}{cccc}
-2 & 2 & 2 & -2 \\
2 & -2 & -2 & 2 \\
0 & 0 & 0 & 0
\end{array}
$$

You may think that the last column of W being all zeros presents a problem in that when the input is X_1 and whatever the output, when this output is presented back in the backward direction, it does not produce X_1, which does not have a zero in the last component. There is a **thresholding** function needed here. The **thresholding** function is transparent when the activations are all either $+1$ or -1. It then just looks like you do an inverse mapping from bipolar values to binary, which is done by replacing each -1 by a 0. When the activation is zero, you simply leave the output of that neuron as it was in the previous cycle or iteration. We now present the **thresholding** function for BAM outputs.

$$
b_j|_{t+1} =
\begin{cases}
1 & \text{if } y_j > 0 \\
b_j|_t & \text{if } y_j = 0 \\
0 & \text{if } y_j < 0
\end{cases}
\quad \text{and} \quad
a_i|_{t+1} =
\begin{cases}
1 & \text{if } x_i > 0 \\
a_i|_t & \text{if } x_i = 0 \\
0 & \text{if } x_i < 0
\end{cases}
$$

where x_i and y_j are the activations of neurons i and j in the input layer and output layer, respectively, and $b_j|_t$ refers to the output of the jth neuron in the output layer in the cycle t, while $a_i|_t$ refers to the output of the ith neuron in the input layer in the cycle t. Note that at the start, the a_i and b_j values are the same as the corresponding components in the exemplar pair being used.

If $X_1 = (1, 0, 0, 1)$ is presented to the input neurons, their activations are given by the vector $(-4, 4, 0)$. The output vector, after using the **threshold** function just described is $(0, 1, 1)$. The last component here is supposed to be the same as the output in the previous cycle, since the corresponding activation value is 0. Since X_1 and Y_1 are one exemplar pair, the third component of Y_1 is what we need as the third component of the output in the current cycle of operation; therefore, we fed X_1 and received Y_1. If we feed Y_1 at the other end (B field) , the activations in the A field will be $(2, -2, -2, 2)$, and the output vector will be $(1, 0, 0, 1)$, which is X_1.

With A field input X_2 you get B field activations $(4, -4, 0)$, giving the output vector as $(1, 0, 1)$, which is Y_2. Thus X_2 and Y_2 are heteroassociated with each other.

Let us modify our X_1 to be $(1, 0, 1, 1)$. Then the weight matrix W becomes

$$W = \begin{matrix} 1 \\ -1 \\ 1 \\ 1 \end{matrix} \begin{bmatrix} -1 & 1 & 1 \end{bmatrix} + \begin{matrix} -1 \\ 1 \\ 1 \\ -1 \end{matrix} \begin{bmatrix} 1 & -1 & 1 \end{bmatrix} = \begin{matrix} -2 & 2 & 0 \\ 2 & -2 & 0 \\ 0 & 0 & 2 \\ -2 & 2 & 0 \end{matrix}$$

and

$$W_T = \begin{matrix} -2 & 2 & 0 & -2 \\ 2 & -2 & 0 & 2 \\ 0 & 0 & 2 & 0 \end{matrix}$$

Now this is a different set of two exemplar vector pairs. The pairs are $X_1 = (1, 0, 1, 1)$, $Y_1 = (0, 1, 1)$, and $X_2 = (0, 1, 1, 0)$, $Y_2 = (1, 0, 1)$. Naturally, the weight matrix is different, as is its transpose, correspondingly. As stated before, the weights do not change during the operation of the network with whatever inputs presented to it. The results are as shown in Table 8.1 below.

Table 8.1 Results for the Example

Input vector	activation	output vector
$X_1 = (1, 0, 1, 1)$	$(-4, 4, 2)$	$(0, 1, 1) = Y_1$
$X_2 = (0, 1, 1, 0)$	$(2, -2, 2)$	$(1, 0, 1) = Y_2$
$Y_1 = (0, 1, 1)$	$(2, -2, 2, 2)$	$(1, 0, 1, 1) = X_1$
$Y_2 = (1, 0, 1)$	$(-2, 2, 2, -2)$	$(0, 1, 1, 0) = X_2$

You may think that you will encounter a problem when you input a new vector and one of the neurons has activation 0. In the original example, you did find this situation when you got the third output neuron's activation as 0. The thresholding function asked you to use the same output for this neuron as existed in the earlier time cycle. So you took it to be 1, the third component in (0, 1, 1). But if your input vector is a new **X** vector for which you are trying to find an associated **Y** vector, then you do not have a **Y** component to fall back on when the activation turns out to be 0. How then can you use the **thresholding** function as stated? What guidance do you have in this situation? If you keep track of the inputs used and outputs received thus far, you realize that the **Field B** (where you get your **Y** vector) neurons are in some state, meaning that they had some outputs perhaps with some training vector. If you use that output component as the one existing in the previous cycle, you have no problem in using the **thresholding** function.

As an example, consider the input vector $X_3 = (1, 0, 0, 0)$, with which the activations of neurons in **Field B** would be (-2, 2, 0). The first component of the output vector is clearly 0, and the second clearly 1. The third component is what is in doubt. Considering the last row of the table where Y_2 gives the state of the neurons in **Field B**, you can accept 1, the last component of Y_2, as the value you get from the thresholding function corresponding to the activation value 0. So the output would be the vector (0, 1, 1), which is Y_1. But Y_1 is heteroassociated with X_1. Well, it means that $X_3 = (1, 0, 0, 0)$ is not heteroassociated with any **X** vector.

RECALL OF VECTORS

When X_1 is presented at the input layer, the activation at the output layer will give $(-4, 4, 2)$ to which we apply the **thresholding** function, which replaces a positive value by 1, and a negative value by 0.

This then gives us the vector (0, 1, 1) as the output, which is the same as our Y_1. Now Y_1 is passed back to the input layer through the feedback connections, and the activation of the input layer becomes the vector (2, −2, 2, 2), which after thresholding gives the output vector (1, 0, 1, 1), same as X_1. When X_2 is presented at the input layer, the activation at the output layer will give (2, −2, 2) to which the **thresholding** function, which replaces a positive value by 1 and a negative value by 0, is applied. This then gives the vector (1, 0, 1) as the output, which is the same as Y_2. Now Y_2 is passed back to the input layer through the feedback connections to get the activation of the input layer as (−2, 2, 2, −2), which after thresholding gives the output vector (0, 1, 1, 0), which is X_2.

NOTE

The two vector pairs chosen here for encoding worked out fine, and the BAM network with four neurons in Field A and three neurons in Field B is all set for finding a vector under heteroassociation with a given input vector.

Continuation of Example

Let us now use the vector X_3 = (1, 1, 1, 1). The vector Y_3 = (0, 1, 1) is obtained at the output layer. But the next step in which we present Y_3 in the backward direction does not produce X_3 , instead it gives an X_1 = (1, 0, 1, 1). We already have X_1 associated with Y_1. This means that X_3 is not associated with any vector in the output space. On the other hand, if instead of getting X_1 we obtained a different X_4 vector, and if this in the feed forward operation produced a different Y vector, then we repeat the operation of the network until no changes occur at either end. Then we will have possibly a new pair of vectors under the heteroassociation established by this BAM network.

Special Case—Complements

If a pair of (distinct) patterns X and Y are found to be heteroassociated by BAM, and if you input the complement of X, complement being obtained by interchanging the 0's and 1's in X, BAM will show that the complement of Y is the pattern associated with the complement of X. An example will be seen in the illustrative run of the program for C++ implementation of BAM, which follows.

C++ IMPLEMENTATION

In our C++ implementation of a discrete bidirectional associative memory network, we create classes for **neuron** and **network**. Other classes created are called **exemplar**, **assocpair**, **potlpair**, for the exemplar pair of vectors, associated pair of vectors, and potential pairs of vectors, respectively, for finding heteroassociation between them. We could have made one class of *pairvect* for a pair of vectors and derived the exemplar and so on from it. The **network** class is declared as a **friend** class in these other classes. Now we present the header and source files, called bamntwrk.h and bamntwrk.cpp. Since we reused our previous code from the Hopfield network of Chapter 4, there are a few data members of classes that we did not put to explicit use in the program. We call the **neuron** class **bmneuron** to remind us of BAM.

Program Details and Flow

A neuron in the first layer is referred to as **anrn**, and the number of neurons in this layer is referred to as **anmbr**. We give the name**bnrn** to the array of neurons in the second layer, and **bnmbr** denotes the size of that array. The sequence of operations in the program is as follows:

- We ask the user to input the exemplar vectors, and we transform them into their bipolar versions. The **trnsfrm ( )** function in the exemplar class is for this purpose.

- We give the network the **X** vector, in its bipolar version, in one exemplar pair. We find the activations of the elements of **bnrn** array and get corresponding output vector as a binary pattern. If this is the **Y** in the exemplar pair, the network has made a desired association in one direction, and we go on to the next.step. Otherwise we have a potential associated pair, one of which is **X** and the other is what we just got as the output vector in the opposite layer. We say potential associated pair because we have the next step to confirm the association.

- We run the **bnrn** array through the transpose of the weight matrix and calculate the outputs of the **anrn** array elements. If , as a result, we get the vector **X** as the **anrn** array, we found an associated pair, **(X, Y)**. Otherwise, we repeat the two steps just described until we find an associated pair.

- We now work with the next pair of exemplar vectors in the same manner as above, to find an associated pair.

- We assign serial numbers, denoted by the variable **idn**, to the associated pairs so we can print them all together at the end of the program. The pair is called **(X, Y)** where **X** produces **Y** through the weight matrix **W**, and **Y** produces **X** through the weight matrix which is the transpose of **W**.

- A flag is used to have value 0 until confirmation of association is obtained, when the value of the flag changes to 1.

- Functions **compr1** and **compr2** in the network class verify if the potential pair is indeed an associated pair and set the proper value of the flag mentioned above.

- Functions **comput1** and **comput2** in the network class carry out the calculations to get the activations and then find the output vector, in the proper directions of the bidirectional associative memory network.

PROGRAM EXAMPLE FOR BAM

For our illustration run, we provided for six neurons in the input layer and five in the output layer. We used three pairs of exemplars for encoding. We used two additional input vectors, one of which is the complement of the **X** of an exemplar pair, after the encoding is done, to see what association will be established in these cases, or what recall will be made by the BAM network.

Header File

As expected, the complement of the **Y** of the exemplar is found to be associated with the complement of the **X** of that pair. When the second input vector is presented, however, a new pair of associated vectors is found. After the code is presented, we list the computer output also.

LISTING 8.1 BAMNTWRK.H

```
//bamntwrk.h    V. Rao,  H. Rao

//Header file for BAM network program

#include <iostream.h>
#include <math.h>
#include <stdlib.h>
#define MXSIZ 10   // determines the maximum size of the network


class bmneuron
{
```

```
protected:
        int nnbr;
        int inn,outn;
        int output;
        int activation;
        int outwt[MXSIZ];
        char *name;
        friend class network;

public:
        bmneuron() { };
        void getnrn(int,int,int,char *);
};

class exemplar
{
protected:
        int xdim,ydim;
        int v1[MXSIZ],v2[MXSIZ];
        int u1[MXSIZ],u2[MXSIZ];
        friend class network;
        friend class mtrx;

public:
        exemplar() { };
        void getexmplr(int,int,int *,int *);
        void prexmplr();
        void trnsfrm();
        void prtrnsfrm();
};

class asscpair
{
protected:
        int xdim,ydim,idn;
        int v1[MXSIZ],v2[MXSIZ];
        friend class network;

public:
```

```
        asscpair() { };
        void getasscpair(int,int,int);
        void prasscpair();
};
```

```
class potlpair
{
protected:
        int xdim,ydim;
        int v1[MXSIZ],v2[MXSIZ];
        friend class network;

public:
        potlpair() { };
        void getpotlpair(int,int);
        void prpotlpair();
};


class network
{
public:
        int   anmbr,bnmbr,flag,nexmplr,nasspr,ninpt;
        bmneuron (anrn)[MXSIZ],(bnrn)[MXSIZ];
        exemplar (e)[MXSIZ];
        asscpair (as)[MXSIZ];
        potlpair (pp)[MXSIZ];
        int outs1[MXSIZ],outs2[MXSIZ];
        int mtrx1[MXSIZ][MXSIZ],mtrx2[MXSIZ][MXSIZ];

        network() { };
        void getnwk(int,int,int,int [][6],int [][5]);
        void compr1(int,int);
        void compr2(int,int);
        void prwts();
        void iterate();
        void findassc(int *);
        void asgninpt(int *);
        void asgnvect(int,int *,int *);
        void comput1();
```

```
        void comput2();
        void prstatus();
};
```

Source File

The program source file is presented as follows.

LISTING 8.2 BAMNTWRK.CPP

```
//bamntwrk.cpp    V. Rao, H. Rao

//Source file for BAM network program

#include "bamntwrk.h"

void bmneuron::getnrn(int m1,int m2,int m3,char *y)
{
int i;
name = y;
nnbr = m1;
outn = m2;
inn  = m3;

for(i=0;i<outn;++i){
        outwt[i] = 0 ;
        }

output = 0;
activation = 0;
}


void exemplar::getexmplr(int k,int l,int *b1,int *b2)
{
int i2;
xdim = k;
ydim = l;

for(i2=0;i2<xdim;++i2){
```

```
        v1[i2] = b1[i2]; }

for(i2=0;i2<ydim;++i2){
        v2[i2] = b2[i2]; }
}


void exemplar::prexmplr()
{
int i;
cout<<"\nX vector you gave is:\n";

for(i=0;i<xdim;++i){
        cout<<v1[i]<<"   ";}

cout<<"\nY vector you gave is:\n";

for(i=0;i<ydim;++i){
        cout<<v2[i]<<"   ";}

cout<<"\n";
}


void exemplar::trnsfrm()
{
int i;

for(i=0;i<xdim;++i){
        u1[i] = 2*v1[i] -1;}

for(i=0;i<ydim;++i){
        u2[i] = 2*v2[i] - 1;}


}

void exemplar::prtrnsfrm()
{
int i;
cout<<"\nbipolar version of X vector you gave is:\n";

for(i=0;i<xdim;++i){
```

```
            cout<<u1[i]<<"   ";}

cout<<"\nbipolar version of Y vector you gave is:\n";

for(i=0;i<ydim;++i){
        cout<<u2[i]<<"   ";}

cout<<"\n";
}

void asscpair::getasscpair(int i,int j,int k)
{
idn = i;
xdim = j;
ydim = k;
}

void asscpair::prasscpair()
{
int i;
cout<<"\nX vector in the associated pair no. "<<idn<<"        is:\n";

for(i=0;i<xdim;++i){
        cout<<v1[i]<<"   ";}

cout<<"\nY vector in the associated pair no. "<<idn<<"        is:\n";

for(i=0;i<ydim;++i){
        cout<<v2[i]<<"   ";}

cout<<"\n";
}

void potlpair::getpotlpair(int k,int j)
{

xdim = k;
ydim = j;
```

```
}

void potlpair::prpotlpair()
{
int i;
cout<<"\nX vector in possible associated pair is:\n";

for(i=0;i<xdim;++i){
        cout<<v1[i]<<"   ";}

cout<<"\nY vector in possible associated pair is:\n";

for(i=0;i<ydim;++i){
        cout<<v2[i]<<"   ";}

cout<<"\n";
}

void network::getnwk(int k,int l,int k1,int b1[][6],int
        b2[][5])
{
anmbr = k;
bnmbr = l;
nexmplr = k1;
nasspr = 0;
ninpt = 0;
int i,j,i2;
flag =0;
char *y1="ANEURON", *y2="BNEURON" ;

for(i=0;i<nexmplr;++i){
        e[i].getexmplr(anmbr,bnmbr,b1[i],b2[i]);
        e[i].prexmplr();
        e[i].trnsfrm();
        e[i].prtrnsfrm();
        }

for(i=0;i<anmbr;++i){
```

```
            anrn[i].bmneuron::getnrn(i,bnmbr,0,y1);}

for(i=0;i<bnmbr;++i){
        bnrn[i].bmneuron::getnrn(i,0,anmbr,y2);}

for(i=0;i<anmbr;++i){

        for(j=0;j<bnmbr;++j){
                mtrx1[i][j]  = 0;

                for(i2=0;i2<nexmplr;++i2){
                        mtrx1[i][j]  += e[i2].u1[i]*e[i2].u2[j];}

                mtrx2[j][i] = mtrx1[i][j];
                anrn[i].outwt[j] = mtrx1[i][j];
                bnrn[j].outwt[i] = mtrx2[j][i];
        }
}

prwts();
cout<<"\n";
}

void network::asgninpt(int *b)
{
int i;
cout<<"\n";

for(i=0;i<anmbr;++i){
        anrn[i].output = b[i];
        outs1[i] = b[i];
        }


}

void network::compr1(int j,int k)
{
int i;
```

```
for(i=0;i<anmbr;++i){

        if(pp[j].v1[i] != pp[k].v1[i]) flag = 1;
        break;
        }

}


void network::compr2(int j,int k)
{
int i;

for(i=0;i<anmbr;++i){

        if(pp[j].v2[i] != pp[k].v2[i]) flag = 1;
        break;}

}


void network::comput1()
{
int j;

for(j=0;j<bnmbr;++j){
        int ii1;
        int c1 =0,d1;
        cout<<"\n";

        for(ii1=0;ii1<anmbr;++ii1){
                d1 = outs1[ii1] * mtrx1[ii1][j];
                c1 += d1;
                }

        bnrn[j].activation = c1;
        cout<<"\n output layer neuron  "<<j<<" activation is"
                <<c1<<"\n";

        if(bnrn[j].activation <0) {
```

```
                    bnrn[j].output = 0;
                    outs2[j] = 0;}

        else

                if(bnrn[j].activation>0) {

                        bnrn[j].output = 1;
                        outs2[j] = 1;}

                else

                        {cout<<"\n A 0 is obtained, use previous output
value \n";

                        if(ninpt<=nexmplr){

                                bnrn[j].output = e[ninpt-1].v2[j];}

                        else

                                { bnrn[j].output = pp[0].v2[j];}
                                outs2[j] = bnrn[j].output; }

                cout<<"\n output layer neuron  "<<j<<" output is"
                        <<bnrn[j].output<<"\n";
                        }
        }


void network::comput2()
{
int i;

for(i=0;i<anmbr;++i){
        int ii1;
        int c1=0;

        for(ii1=0;ii1<bnmbr;++ii1){
```

```
                c1 += outs2[ii1] * mtrx2[ii1][i];   }


        anrn[i].activation = c1;
        cout<<"\ninput layer neuron "<<i<<"activation is "
            <<c1<<"\n";


        if(anrn[i].activation <0 ){

                anrn[i].output = 0;
                outs1[i] = 0;}


        else

                if(anrn[i].activation >0 ) {

                        anrn[i].output = 1;
                        outs1[i] = 1;
                        }


                else

                { cout<<"\n A 0 is obtained, use previous value if
available\n";

                if(ninpt<=nexmplr){

                        anrn[i].output = e[ninpt-1].v1[i];}

                else

                        {anrn[i].output = pp[0].v1[i];}

                outs1[i] = anrn[i].output;}
                cout<<"\n input layer neuron  "<<i<<" output is "
                        <<anrn[i].output<<"\n";
                }
}


void network::asgnvect(int j1,int *b1,int *b2)
```

```
{
int  j2;

for(j2=0;j2<j1;++j2){
        b2[j2] = b1[j2];}


}

void network::prwts()
{
int i3,i4;
cout<<"\n  weights-  input layer to output layer: \n\n";

for(i3=0;i3<anmbr;++i3){

        for(i4=0;i4<bnmbr;++i4){
                cout<<anrn[i3].outwt[i4]<<"  ";}

        cout<<"\n";  }

cout<<"\n";

cout<<"\nweights-  output layer to input layer: \n\n";

for(i3=0;i3<bnmbr;++i3){

        for(i4=0;i4<anmbr;++i4){
                cout<<bnrn[i3].outwt[i4]<<"  ";}

        cout<<"\n";   }

cout<<"\n";
}

void network::iterate()
{
int i1;

for(i1=0;i1<nexmplr;++i1){
```

```
            findassc(e[i1].v1);
            }


}
```

```
void network::findassc(int *b)
{
int j;
flag = 0;
        asgninpt(b);
        ninpt ++;
        cout<<"\nInput vector is:\n" ;

        for(j=0;j<6;++j){
                cout<<b[j]<<" ";}

        cout<<"\n";
        pp[0].getpotlpair(anmbr,bnmbr);
        asgnvect(anmbr,outs1,pp[0].v1);

        comput1();

        if(flag>=0){
                asgnvect(bnmbr,outs2,pp[0].v2);

                cout<<"\n";
                pp[0].prpotlpair();
                cout<<"\n";

                comput2(); }

        for(j=1;j<MXSIZ;++j){
                pp[j].getpotlpair(anmbr,bnmbr);
                asgnvect(anmbr,outs1,pp[j].v1);

                comput1();

                asgnvect(bnmbr,outs2,pp[j].v2);
```

```
                pp[j].prpotlpair();
                cout<<"\n";

                compr1(j,j-1);
                compr2(j,j-1);

                if(flag == 0) {

                        int j2;
                        nasspr += 1;
                        j2 = nasspr;

                        as[j2].getasscpair(j2,anmbr,bnmbr);
                        asgnvect(anmbr,pp[j].v1,as[j2].v1);
                        asgnvect(bnmbr,pp[j].v2,as[j2].v2);

                        cout<<"\nPATTERNS ASSOCIATED:\n";
                        as[j2].prasscpair();
                        j = MXSIZ ;
                }

                else

                        if(flag == 1)

                                {
                                flag = 0;
                                comput1();
                                }


        }
}


void network::prstatus()
{
int j;
cout<<"\nTHE FOLLOWING ASSOCIATED PAIRS WERE FOUND BY BAM\n\n";

for(j=1;j<=nasspr;++j){
```

```
        as[j].prasscpair();
        cout<<"\n";}


}
```

```
void main()
{
int ar = 6, br = 5, nex = 3;
int inptv[][6]={1,0,1,0,1,0,1,1,1,0,0,0,0,1,1,0,0,0,0,1,0,1,0,1,
        1,1,1,1,1,1};
int outv[][5]={1,1,0,0,1,0,1,0,1,1,1,0,0,1,0};

cout<<"\n\nTHIS PROGRAM IS FOR A BIDIRECTIONAL ASSOCIATIVE MEMORY NET-
WORK.\n";
cout<<" THE NETWORK ISSET UP FOR ILLUSTRATION WITH "<<ar<<
        " INPUT NEURONS, AND "<<br;
cout<<" OUTPUT NEURONS.\n"<<nex
        <<" exemplars are used to encode \n";

static network bamn;
bamn.getnwk(ar,br,nex,inptv,outv) ;
bamn.iterate();
bamn.findassc(inptv[3]);
bamn.findassc(inptv[4]);
bamn.prstatus();


}
```

Program Output

The output from an illustrative run of the program is listed next. You will notice that we provided for a lot of information to be output as the program is executed. The fourth vector we input is not from an exemplar, but it is the complement of the **X** of the first exemplar pair. The network found the complement of the **Y** of this exemplar to be associated with this input. The fifth vector we input is (1, 1, 1, 1, 1, 1). But BAM recalled in this case the complement pair for the third exemplar pair, which is **X** = (1, 0, 0, 1, 1, 1) and **Y** = (0, 1, 1, 0, 1). You notice that the Hamming distance of this input pattern from the X's of the three exemplars are 3, 3, and 4, respectively. The hamming distance from the complement of the **X** of the first exemplar pair is also 3. But the Hamming distance of (1, 1, 1, 1) from the **X** of the complement of the **X** of the third exemplar pair is only 2. It would be instructive if you would use the input pattern (1, 1, 1, 1, 1, 0) to see to what associated pair it leads. Now the output:

```
THIS PROGRAM IS FOR A BIDIRECTIONAL ASSOCIATIVE
     MEMORY NETWORK.
THE NETWORK IS SET UP FOR ILLUSTRATION WITH SIX INPUT
     NEURONS AND FIVE OUTPUT NEURONS.
Three exemplars are used to encode

X vector you gave is:
1  0  1  0  1  0
Y vector you gave is:
1  1  0  0  1

bipolar version of X vector you gave is:
1  -1  1  -1  1  -1
bipolar version of Y vector you gave is:
1  1  -1  -1  1

X vector you gave is:
1  1  1  0  0  0
Y vector you gave is:
0  1  0  1  1

bipolar version of X vector you gave is:
1  1  1  -1  -1  -1
bipolar version of Y vector you gave is:
-1  1  -1  1  1

X vector you gave is:
0  1  1  0  0  0
Y vector you gave is:
1  0  0  1  0

bipolar version of X vector you gave is:
-1  1  1  -1  -1  -1
bipolar version of Y vector you gave is:
```

```
1  -1  -1  1  -1

   weights—  input layer to output layer:

        -1       3      -1      -1       3
        -1       1      -1       3      -1
         1       1      -3       1       1
        -1      -1       3      -1      -1
         1       1       1      -3       1
        -1      -1       3      -1      -1


weights—  output layer to input layer:

        -1      -1       1      -1       1      -1
         3      -1       1      -1       1      -1
        -1      -1      -3       3       1       3
        -1       3       1      -1      -3      -1
         3      -1       1      -1       1      -1




Input vector is:
1 0 1 0 1 0


   output layer neuron  0 activation is 1

   output layer neuron  0 output is 1


   output layer neuron  1 activation is 5

   output layer neuron  1 output is 1


   output layer neuron  2 activation is -3
```

output layer neuron 2 output is 0

output layer neuron 3 activation is -3

output layer neuron 3 output is 0

output layer neuron 4 activation is 5

output layer neuron 4 output is 1

X vector in possible associated pair is:
1 0 1 0 1 0
Y vector in possible associated pair is:
1 1 0 0 1

input layer neuron 0 activation is 5

 input layer neuron 0 output is 1

input layer neuron 1 activation is -3

 input layer neuron 1 output is 0

input layer neuron 2 activation is 3

 input layer neuron 2 output is 1

input layer neuron 3 activation is -3

 input layer neuron 3 output is 0

input layer neuron 4 activation is 3

 input layer neuron 4 output is 1

```
input layer neuron 5 activation is -3

 input layer neuron  5 output is 0


 output layer neuron  0 activation is 1

 output layer neuron  0 output is 1


 output layer neuron  1 activation is 5

 output layer neuron  1 output is 1


 output layer neuron  2 activation is -3

 output layer neuron  2 output is 0


 output layer neuron  3 activation is -3

 output layer neuron  3 output is 0


 output layer neuron  4 activation is 5

 output layer neuron  4 output is 1

X vector in possible associated pair is:
1  0  1  0  1  0
Y vector in possible associated pair is:
1  1  0  0  1


PATTERNS ASSOCIATED:

X vector in the associated pair no. 1 is:
1  0  1  0  1  0
```

Y vector in the associated pair no. 1 is:
1 1 0 0 1

Input vector is:
1 1 1 0 0 0

// We get here more of the detailed output as in the previous case. We
will simply not present it here.

PATTERNS ASSOCIATED:

X vector in the associated pair no. 1 is:
1 1 1 0 0 0
Y vector in the associated pair no. 1 is:
0 1 0 1 1

Input vector is:
0 1 1 0 0 0

output layer neuron 0 activation is 0

A 0 is obtained, use previous output value

output layer neuron 0 output is 1

output layer neuron 1 activation is 0

A 0 is obtained, use previous output value

output layer neuron 1 output is 0

output layer neuron 2 activation is -4

output layer neuron 2 output is 0

output layer neuron 3 activation is 4

output layer neuron 3 output is 1

output layer neuron 4 activation is 0

A 0 is obtained, use previous output value

output layer neuron 4 output is 0

X vector in possible associated pair is:
0 1 1 0 0 0
Y vector in possible associated pair is:
1 0 0 1 0

// We get here more of the detailed output as in the previous case. We
will simply not present it here.

PATTERNS ASSOCIATED:

X vector in the associated pair no. 1 is:
0 1 1 0 0 0
Y vector in the associated pair no. 1 is:
1 0 0 1 0

Input vector is:
0 1 0 1 0 1

// We get here more of the detailed output as in the previous case. We
will simply not present it here.

X vector in possible associated pair is:

```
0  1  0  1  0  1
Y vector in possible associated pair is:
0  0  1  1  0
```

```
// We get here more of the detailed output as in the previous case. We
will simply not present it here.
```

```
X vector in possible associated pair is:
0  1  0  1  0  1
Y vector in possible associated pair is:
0  0  1  1  0
```

```
PATTERNS ASSOCIATED:
```

```
X vector in the associated pair no. 1 is:
0  1  0  1  0  1
Y vector in the associated pair no. 1 is:
0  0  1  1  0
```

```
Input vector is:
1 1 1 1 1 1
```

```
output layer neuron  0 activation is -2
```

```
output layer neuron  0 output is 0
```

```
output layer neuron  1 activation is 2
```

```
output layer neuron  1 output is 1
```

```
output layer neuron  2 activation is 2
```

```
output layer neuron  2 output is 1
```

```
output layer neuron  3 activation is -2

output layer neuron  3 output is 0


output layer neuron  4 activation is 2

output layer neuron .4 output is 1


X vector in possible associated pair is:
1  1  1  1  1  1
Y vector in possible associated pair is:
0  1  1  0  1


input layer neuron 0 activation is 5

input layer neuron  0 output is 1

input layer neuron 1 activation is -3

input layer neuron  1 output is 0

input layer neuron 2 activation is -1

input layer neuron  2 output is 0

input layer neuron 3 activation is 1

input layer neuron  3 output is 1

input layer neuron 4 activation is 3

input layer neuron  4 output is 1

input layer neuron 5 activation is 1

input layer neuron  5 output is 1
```

output layer neuron 0 activation is -2

output layer neuron 0 output is 0

output layer neuron 1 activation is 2

output layer neuron 1 output is 1

output layer neuron 2 activation is 6

output layer neuron 2 output is 1

output layer neuron 3 activation is -6

output layer neuron 3 output is 0

output layer neuron 4 activation is 2

output layer neuron 4 output is 1

X vector in possible associated pair is:
1 0 0 1 1 1
Y vector in possible associated pair is:
0 1 1 0 1

PATTERNS ASSOCIATED:

X vector in the associated pair no. 1 is:
1 0 0 1 1 1
Y vector in the associated pair no. 1 is:
0 1 1 0 1

THE FOLLOWING ASSOCIATED PAIRS WERE FOUND BY BAM

```
X vector in the associated pair no. 1 is:     //first exemplar pair
1  0  1  0  1  0
Y vector in the associated pair no. 1 is:
1  1  0  0  1
```

```
X vector in the associated pair no. 2 is:     //second exemplar pair
1  1  1  0  0  0
Y vector in the associated pair no. 2 is:
0  1  0  1  1
```

```
X vector in the associated pair no. 3 is:     //third exemplar pair
0  1  1  0  0  0
Y vector in the associated pair no. 3 is:
1  0  0  1  0
```

```
X vector in the associated pair no. 4 is:     //complement of X of the
0  1  0  1  0  1                                   first exemplar pair

Y vector in the associated pair no. 4 is:     //complement of Y of the
0  0  1  1  0                                      first exemplar pair
```

```
X vector in the associated pair no. 5 is:     //input was X = (1, 1, 1,
1  0  0  1  1  1                               1, 1) but result was complement
                                              of third exemplar pair
Y vector in the associated pair no. 5 is:     with X of which Hamming
0  1  1  0  1                                  distance is the  least.
```

ADDITIONAL ISSUES

If you desire to use vectors with real number components, instead of binary numbers, you can do so. Your model is then called a *Continuous Bidirectional Associative Memory*. A matrix **W** and its transpose are used for the connection weights. However, the matrix **W** is not formulated as we described so far. The matrix is arbitrarily chosen, and kept constant. The

thresholding function is also chosen as a continuous function, unlike the one used before. The changes in activations of neurons during training are according to extensions the Cohen-Grossberg paradigm.

Michael A. Cohen and Stephen Grossberg developed their model for autoassociation, with a symmetric matrix of weights. Stability is ensured by the Cohen-Grossberg theorem; there is no learning.

If, however, you desire the network to learn new pattern pairs, by modifying the weights so as to find association between new pairs, you are designing what is called an *Adaptive Bidirectional Associative Memory* (*ABAM*).

The law that governs the weight changes determines the type of ABAM you have, namely, the *Competitive* ABAM, or the *Differential Hebbian* ABAM, or the *Differential Competitive* ABAM. Unlike in the ABAM model, which is additive type, some products of outputs from the two layers, or the derivatives of the threshold functions are used in the other models.

Here we present a brief description of a model, which is a variation of the BAM. It is called *UBBAM* (*Unipolar Binary Bidirectional Associative Memory*).

UNIPOLAR BINARY BIDIRECTIONAL ASSOCIATIVE MEMORY

T. C. B. Yu and R. J. Mears describe a design of unipolar binary bidirectional associative memory, and its implementation with a Smart Advanced Spatial Light Modulator (SASLM). The SASLM device is a ferroelectric liquid crystal spatial light modulator. It is driven by a silicon CMOS backplane. We use their paper to present some features of a unipolar binary bidirectional associative memory, and ignore the hardware implementation of it.

Recall the procedure by which you determine the weight matrix **W** for a BAM network, as described in the previous pages. As a first step, you convert each vector in the exemplar pairs into its bipolar version. If **X** and **Y** are an exemplar pair (in bipolar versions), you take the product $\mathbf{X}^T \mathbf{Y}$ and add it to similar products from other exemplar pairs, to get a weight matrix **W**. Some of the elements of the matrix **W** may be negative numbers. In the unipolar context you do not have negative values and positive values at the same time. Only one of them is allowed. Suppose you do not want any negative numbers; then one way of remedying the situation is by adding a large enough constant to each element of the matrix. You cannot choose to add to only the negative numbers that show up in the matrix. Let us look at an example.

Suppose you choose two pairs of vectors as possible exemplars. Let them be,

$X_1 = (1, 0, 0, 1), Y_1 = (0, 1, 1)$

and

$X_2 = (0, 1, 1, 0), Y_2 = (1, 0, 1)$

These you change into bipolar components and get, respectively, $(1, -1, -1, 1)$, $(-1, 1, 1)$, $(-1, 1, 1, -1)$, and $(1, -1, 1)$. The calculation of W for the BAM was done as follows.

$$
W = \begin{matrix} 1 \\ -1 \\ -1 \\ 1 \end{matrix} \begin{bmatrix} -1 & 1 & 1 \end{bmatrix} + \begin{matrix} -1 \\ 1 \\ 1 \\ -1 \end{matrix} \begin{bmatrix} 1 & -1 & 1 \end{bmatrix} = \begin{matrix} -1 & 1 & 1 \\ 1 & -1 & -1 \\ 1 & -1 & -1 \\ -1 & 1 & 1 \end{matrix} + \begin{matrix} -1 & 1 & -1 \\ 1 & -1 & 1 \\ 1 & -1 & 1 \\ -1 & 1 & -1 \end{matrix} = \begin{matrix} -2 & 2 & 0 \\ 2 & -2 & 0 \\ 2 & -2 & 0 \\ -2 & 2 & 0 \end{matrix}
$$

and

$$
W^T = \begin{matrix} -2 & 2 & 2 & -2 \\ 2 & -2 & -2 & 2 \\ 0 & 0 & 0 & 0 \end{matrix}
$$

You see some negative numbers in the matrix W. If you add the constant **m**, $m = 2$, to all the elements in the matrix, you get the following matrix, denoted by **W~**.

$$
W^\sim = \begin{matrix} 0 & 4 & 2 \\ 4 & 0 & 2 \\ 4 & 0 & 2 \\ 0 & 4 & 2 \end{matrix}
$$

You need a modification to the thresholding function as well. Earlier you had the following function.

$$
b_j|_{t+1} = \begin{cases} 1 & \text{if } y_j > 0 \\ b_j|_t & \text{if } y_j = 0 \\ 0 & \text{if } y_j < 0 \end{cases} \quad \text{and} \quad a_i|_{t+1} = \begin{cases} 1 & \text{if } x_i > 0 \\ a_i|_t & \text{if } x_i = 0 \\ 0 & \text{if } x_i < 0 \end{cases}
$$

Now you need to change the right-hand sides of the conditions from 0 to the product of **m** and sum of the inputs of the neurons, that is to **m** times the sum of the inputs. For brevity, let us use $\mathbf{S_i}$ for the sum of the inputs. This is not a constant, and its value depends on what values you get for neuron inputs in any one direction. Then the **thresholding** function can be given as follows:

$$b_j|_{t+1} = \begin{cases} 1 & \text{if } y_j > m\,S_i \\ b_j|_t & \text{if } y_j = m\,S_i \\ 0 & \text{if } y_j < m\,S_i \end{cases} \quad \text{and} \quad a_i|_{t+1} = \begin{cases} 1 & \text{if } x_i > m\,S_i \\ a_i|_t & \text{if } x_i = m\,S_i \\ 0 & \text{if } x_i < m\,S_i \end{cases}$$

For example, the vector $X_1 = (1, 0, 0, 1)$ has the activity vector $(0, 8, 4)$ and the corresponding output vector is $(0, 1, 1)$, which is, $Y_1 = (0, 1, 1)$. The value of S_i is $2 = 1 + 0 + 0 + 1$. Therefore, $mS_i = 4$. The first component of the activation vector is 0, which is smaller than 4, and so the first component of the output vector is 0. The second component of the activation vector is 8, which is larger than 4, and so the second component of the output vector is 1. The last component of the activity vector is 4, which is equal to mS_i, and so you use the third component of the vector $Y_1 = (0, 1, 1)$, namely 1, as the third component of the output.

The example of Yu and Mears, consists of eight 8-component vectors in both directions of the Bidirectional Associative Memory model. You can take these vectors, as they do, to define the pixel values in a 8x8 grid of pixels. What is being accomplished is the association of two spatial patterns. In terms of the binary numbers that show the pixel values, the patterns are shown in Figure 8.2. Call them Pattern I and Pattern II.

```
1 1 1 1 1 1 1 1        1 1 1 1 1 1 1 1
1 1 0 1 1 0 1 1        1 1 1 1 0 0 0 1
1 0 1 0 0 1 0 1        1 1 1 0 1 1 0 1
1 1 0 1 1 0 1 1        1 1 0 1 1 1 0 1
1 1 0 1 1 0 1 1        1 0 1 1 1 0 1 1
1 0 1 0 0 1 0 1        1 0 1 1 0 1 1 1
1 1 0 1 1 0 1 1        1 0 0 0 1 1 1 1
1 1 1 1 1 1 1 1        1 1 1 1 1 1 1 1

    Pattern I              Pattern II
```

FIGURE 8.2 TWO PATTERNS, PATTERN I AND PATTERN II, GIVEN BY PIXEL VALUES: 1 FOR BLACK, 0 FOR WHITE.

Instead of Pattern I, they used a corrupted form of it as given in Figure 8.3. There was no problem in the network finding the associated pair (Pattern I, Pattern II).

```
1 0 1 1 1 1 1 1          1 1 1 1 1 1 1 1
1 1 0 1 1 0 1 0          1 1 0 1 1 0 1 1
1 0 1 0 0 0 0 1          1 0 1 0 0 1 0 1
0 1 0 1 0 0 1 1          1 1 0 1 1 0 1 1
1 0 1 1 1 0 0 1          1 1 0 1 1 0 1 1
1 0 0 0 0 1 0 1          1 0 1 0 0 1 0 1
1 1 0 1 0 0 1 1          1 1 0 1 1 0 1 1
1 1 1 1 1 1 1 1          1 1 1 1 1 1 1 1

Corrupted Pattern I          Pattern I
```

FIGURE 8.3 PATTERN I AND CORRUPTED PATTERN I.

In Figure 8.3, the corrupted version of Pattern I differs from Pattern I, in 10 of the 64 places, a corruption of 15.6%. This corruption percentage is cited as the limit below which, Yu and Mears state, heteroassociative recall is obtained. Thus noise elimination to a certain extent is possible with this model. As we have seen with the Hopfield memory, an application of associative memory is pattern completion, where you are presented a corrupted version of a pattern and you recall the true pattern. This is autoassociation. In the case of BAM, you have heteroassociative recall with a corrupted input.

SUMMARY

In this chapter, bidirectional associative memories are presented. The development of these memories is largely due to Kosko. They share with Adaptive Resonance Theory the feature of resonance between the two layers in the network. The bidirectional associative memories (BAM) network finds heteroassociation between binary patterns, and these are converted to bipolar values to determine the connection weight matrix. Even though there are connections in both directions between neurons in the two layers, essentially only one weight matrix is involved. You use the transpose of this weight matrix for the connections in the opposite direction. When one input at one end leads to some output at the other end, which in turn leads to output that is the same as the previous input, resonance is reached and an associated pair is found.

The continuous bidirectional associative memory extends the binary model to the continuous case. Adaptive bidirectional memories of different flavors are the result of incorporating different learning paradigms. A unipolar binary version of BAM is also presented as an application of the BAM for pattern completion.

CHAPTER 9

FAM: FUZZY ASSOCIATIVE MEMORY

INTRODUCTION

In this chapter, you we will learn about fuzzy sets and their elements, both for input and output of an associative neural network. Every element of a fuzzy set has a degree of membership in the set. Unless this degree of membership is 1, an element does not belong to the set (in the sense of elements of an ordinary set belonging to the set). In a neural network of a fuzzy system the inputs, the outputs, and the connection weights all belong fuzzily to the spaces that define them. The weight matrix will be a fuzzy matrix, and the activations of the neurons in such a network have to be determined by rules of fuzzy logic and fuzzy set operations.

An expert system uses what are called *crisp rules* and applies them sequentially. The advantage in casting the same problem in a fuzzy system is that the rules you work with do not have to be crisp, and the processing is done in parallel. What the fuzzy systems can determine is a fuzzy association. These associations can be modified, and the underlying phenomena better understood, as experience is gained. That is one of the reasons for their growing popularity in applications. When we try to relate two things through a process of trial and error, we will be implicitly and intuitively establishing an association that is gradually modified and perhaps bettered in some sense. Several fuzzy variables may be present in such an exercise. That we did not have full knowledge at the beginning is not a hindrance; there is some difference in using probabilities and using fuzzy logic as well. The degree of membership assigned for an element of a set does not have to be as firm as the assignment of a probability.

The degree of membership is, like a probability, a real number between 0 and 1. The closer it is to 1, the less ambiguous is the membership of the element in the set concerned. Suppose you have a set that may or may not contain three elements, say, a, b, and c. Then the fuzzy set representation of it would be by the ordered triple (m_a, m_b, m_c), which is called the *fit vector*, and its components are called *fit values*. For example, the triple (0.5, 0.5,

0.5) shows that each of a, b, and c, have a membership equal to only one-half. This triple itself will describe the fuzzy set. It can also be thought of as a point in the three-dimensional space. None of such points will be outside the unit cube. When the number of elements is higher, the corresponding points will be on or inside the unit hypercube.

It is interesting to note that this fuzzy set, given by the triple (0.5, 0.5, 0.5), is its own complement, something that does not happen with regular sets. The complement is the set that shows the degrees of nonmembership.

The *height* of a fuzzy set is the maximum of its fit values, and the fuzzy set is said to be *normal*, if its height is 1. The fuzzy set with fit vector (0.3, 0.7, 0.4) has height 0.7, and it is not a normal fuzzy set. However, by introducing an additional dummy component with fit value 1, we can extend it into a normal fuzzy set. The desirability of normalcy of a fuzzy set will become apparent when we talk about recall in fuzzy associative memories.

The subset relationship is also different for fuzzy sets from the way it is defined for regular sets. For example, if you have a fuzzy set given by the triple (0.3, 0.7, 0.4), then any fuzzy set with a triple (a, b, c) such that $a \leq 0.3$, $b \leq 0.7$, and $c \leq 0.4$, is its fuzzy subset. For example, the fuzzy set given by the triple (0.1, 0.7, 0) is a subset of the fuzzy set (0.3, 0.7, 0.4).

ASSOCIATION

Consider two fuzzy sets, one perhaps referring to the popularity of (or interest in) an exhibition and the other to the price of admission. The popularity could be very high, high, fair, low, or very low. Accordingly, the fit vector will have five components. The price of admission could be high, modest, or low. Its fit vector has three components. A fuzzy associative memory system then has the association (popularity, price), and the fuzzy set pair encodes this association.

We describe the encoding process and the recall in the following sections. Once encoding is completed, associations between subsets of the first fuzzy set and subsets of the second fuzzy set are also established by the same fuzzy associative memory system.

FAM NEURAL NETWORK

The neural network for a fuzzy associative memory has an input and an output layer, with full connections in both directions, just as in a BAM neural network. Figure 9.1 shows the layout. To avoid cluttering, the figure shows the network with three neurons in field A and two neurons in field B, and only some of the connection weights. The general case is analogous.

This figure is the same as Figure 8.1 in Chapter 8, since the model is the same, except that fuzzy vectors are used with the current model.

N O T E

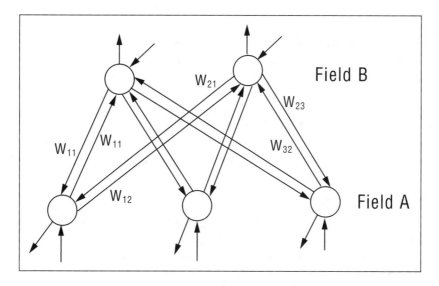

FIGURE 9.1 FUZZY ASSOCIATIVE MEMORY NEURAL NETWORK.

ENCODING

Encoding for fuzzy associative memory systems is similar in concept to the encoding process for bidirectional associative memories, with some differences in the operations involved. In the BAM encoding, bipolar versions of binary vectors are used just for encoding. Matrices of the type $X_i^T Y_i$ are added together to get the connection weight matrix. There are two basic operations with the elements of these vectors. They are multiplication and addition of products. There is no conversion of the fit values before the encoding by the fuzzy sets. The multiplication of elements is replaced by the operation of taking the **minimum**, and addition is replaced by the operation of taking the **maximum**.

There are two methods for encoding. The method just described is what is called max–min composition. It is used to get the connection weight matrix and also to get the outputs of neurons in the fuzzy associative memory neural network. The second method is called correlation–product encoding. It is obtained the same way as a BAM connection weight matrix is obtained. Max–min composition is what is frequently used in practice, and we will confine our attention to this method.

Example of Encoding

Suppose the two fuzzy sets we use to encode have the fit vectors (0.3, 0.7, 0.4, 0.2) and (0.4, 0.3, 0.9). Then the matrix W is obtained by using max–min composition as follows.

```
         0.3 [0.4 0.3 0.9]   min(0.3,0.4) min(0.3,0.3) min(0.3,0.9)   0.3 0.3 0.3
W  =   0.7                =  min(0.7,0.4) min(0.7,0.3) min(0.7,0.9) = 0.4 0.3 0.7
         0.4                   min(0.4,0.4) min(0.4,0.3) min(0.4,0.9)   0.4 0.3 0.4
         0.2                   min(0.2,0.4) min(0.2,0.3) min(0.2,0.9)   0.2 0.2 0.2
```

Recall for the Example

If we input the fit vector (0.3, 0.7, 0.4, 0.2), the output (b_1, b_2, b_3) is determined as follows, using $b_j = \max(\min(a_1, w_{1j}), ..., \min(a_m, w_{mj}))$, where m is the dimension of the 'a' fit vector, and w_{ij} is the ith row, jth column element of the matrix **W**.

```
b1  = max(min(0.3,  0.3), min(0.7,  0.4), min(0.4,  0.4),
      min(0.2, 0.2)) =  max(0.3, 0.4, 0.4, 0.2) = 0.4
b2  = max(min(0.3,  0.3), min(0.7,  0.3), min(0.4,  0.3),
      min(0.2, 0.2)) = max( 0.3, 0.3, 0.3, 0.2 ) = 0.3
b3  = max(min(0.3,  0.3), min(0.7,  0.7), min(0.4,  0.4),
      min(0.2, 0.2)) = max (0.3, 0.7, 0.4, 0.2) = 0.7
```

The output vector (0.4, 0.3, 0.7) is not the same as the second fit vector used, namely (0.4, 0.3, 0.9), but it is a subset of it, so the recall is not perfect. If you input the vector (0.4, 0.3, 0.7) in the opposite direction, using the transpose of the matrix W, the output is (0.3, 0.7, 0.4, 0.2), showing resonance. If on the other hand you input (0.4, 0.3, 0.9) at that end, the output vector is (0.3, 0.7, 0.4, 0.2), which in turn causes in the other direction an output of (0.4, 0.3, 0.7) at which time there is resonance. Can we foresee these results? The following section explains this further.

Recall

Let us use the operator o to denote max–min composition. Perfect recall occurs when the weight matrix is obtained using the max–min composition of fit vectors **U** and **V** as follows:

(i)   **U** o **W** = **V** if and only if height (U) ≥ height (V).
(ii) **V** o **W**^{T} = **U** if and only if height (V) ≥ height (U).

Also note that if **X** and **Y** are arbitrary fit vectors with the same dimensions as **U** and **V**, then:

(iii) **X** o **W** ⊂ **V**.
(iv) **Y** o **W**^{T} ⊂ **U**.

A ⊂ B is the notation to say A is a subset of B.

In the previous example, height of (0.3, 0.7, 0.4, 0.2) is 0.7, and height of (0.4, 0.3, 0.9) is 0.9. Therefore (0.4, 0.3, 0.9) as input, produced (0.3, 0.7, 0.4, 0.2) as output, but (0.3, 0.7, 0.4, 0.2) as input, produced only a subset of (0.4, 0.3, 0.9). That both (0.4, 0.3, 0.7) and (0.4, 0.3, 0.9) gave the same output, (0.3, 0.7, 0.4, 0.2) is in accordance with the corollary to the above, which states that if (X, Y) is a fuzzy associated memory, and if X is a subset of X', then (X', Y) is also a fuzzy associated memory.

C++ IMPLEMENTATION

We use the classes we created for BAM implementation in C++, except that we call the neuron class **fzneuron,** and we do not need some of the methods or functions in the **network** class. The header file, the source file, and the output from an illustrative run of the program are given in the following. The header file is called **fuzzyam.hpp**, and the source file is called **fuzzyam.cpp**.

222

Program details

The program details are analogous to the program details given in Chapter 8. The computations are done with fuzzy logic. Unlike in the nonfuzzy version, a single exemplar fuzzy vector pair is used here. There are no transformations to bipolar versions, since the vectors are fuzzy and not binary and crisp.

A neuron in the first layer is referred to as **anrn**, and the number of neurons in this layer is referred to as **anmbr**. **bnrn** is the name we give to the array of neurons in the second layer and **bnmbr** denotes the size of that array. The sequence of operations in the program are as follows:

- We ask the user to input the exemplar vector pair.
- We give the network the **X** vector, in the exemplar pair. We find the activations of the elements of **bnrn** array and get corresponding output vector as a binary pattern. If this is the **Y** in the exemplar pair, the network has made a desired association in one direction, and we go on to the next.step. Otherwise we have a potential associated pair, one of which is **X** and the other is what we just got as the output vector in the opposite layer. We say potential associated pair because we have the next step to confirm the association.
- We run the **bnrn** array through the transpose of the weight matrix and calculate the outputs of the **anrn** array elements. If, as a result, we get the vector **X** as the **anrn** array, we found an associated pair, **(X, Y)**. Otherwise, we repeat the two steps just described until we find an associated pair.
- We now work with the next pair of exemplar vectors in the same manner as above, to find an associated pair.
- We assign serial numbers, denoted by the variable **idn**, to the associated pairs so we can print them all together at the end of the program. The pair is called **(X, Y)** where **X** produces **Y** through the weight matrix **W**, and **Y** produces **X** through the weight matrix, which is the transpose of **W**.
- A flag is used to have value 0 until confirmation of association is obtained, when the value of the flag changes to 1.

- Functions **compr1** and **compr2** in the network class verify if the potential pair is indeed an associated pair and set the proper value of the flag mentioned above.

- Functions **comput1** and **comput2** in the network class carry out the calculations to get the activations and then find the output vector, in the respective directions of the fuzzy associative memory network.

223

A lot of the code from the bidirectional associative memory (BAM) is used for the FAM. Here are the listings, with comments added where there are differences between this code and the code for the BAM of Chapter 8.

Header File

LISTING 9.1 FUZZYAM.H

```
//fuzzyam.h     V. Rao, H. Rao

#include <iostream.h>
#define MXSIZ 10

class fzneuron
{

protected:
        int nnbr;
        int inn,outn;
        float output;
        float activation;
        float outwt[MXSIZ];
        char *name;
        friend class network;

public:
        fzneuron() { };
        void getnrn(int,int,int,char *);
};

class exemplar
```

```
{
protected:
        int xdim,ydim;
        float v1[MXSIZ],v2[MXSIZ];    // this is different from BAM
        friend class network;

public:
        exemplar() { };
        void getexmplr(int,int,float *,float *);
        void prexmplr();
};


class asscpair
{
protected:
        int xdim,ydim,idn;
        float v1[MXSIZ],v2[MXSIZ];
        friend class network;

public:
        asscpair() { };
        void getasscpair(int,int,int);
        void prasscpair();
};


class potlpair
{
protected:
        int xdim,ydim;
        float v1[MXSIZ],v2[MXSIZ];
        friend class network;

public:
        potlpair() { };
        void getpotlpair(int,int);
        void prpotlpair();
};


class network
```

```
{
public:
        int   anmbr,bnmbr,flag,nexmplr,nasspr,ninpt;
        fzneuron (anrn)[MXSIZ],(bnrn)[MXSIZ];
        exemplar (e)[MXSIZ];
        asscpair (as)[MXSIZ];
        potlpair (pp)[MXSIZ];
        float outs1[MXSIZ],outs2[MXSIZ];        // change from BAM to floats
        double mtrx1[MXSIZ][MXSIZ],mtrx2[MXSIZ][MXSIZ]; // change from
BAM to doubles

        network() { };
        void getnwk(int,int,int,float [][6],float [][4]);
        void compr1(int,int);
        void compr2(int,int);
        void prwts();
        void iterate();
        void findassc(float *);
        void asgninpt(float *);
        void asgnvect(int,float *,float *);
        void comput1();
        void comput2();
        void prstatus();
};
```

Source File

LISTING 9.2 FUZZYAM.CPP

```
//fuzzyam.cpp    V. Rao, H. Rao

#include "fuzzyam.h"

float max(float x,float y)    //new for FAM
{
float u;
u = ((x>y) ? x : y );
return u;
}
```

```
float min(float x,float y)              // new for FAM
{
float u;
u =( (x>y) ? y : x) ;
return u;
}



void fzneuron::getnrn(int m1,int m2,int m3,char *y)
{
int i;
name = y;
nnbr = m1;
outn = m2;
inn  = m3;

for(i=0;i<outn;++i){
        outwt[i] = 0 ;
        }

output = 0;
activation = 0;
}

void exemplar::getexmplr(int k,int l,float *b1,float *b2)    // changed
from BAM
{
int i2;
xdim = k;
ydim = l;

for(i2=0;i2<xdim;++i2){
        v1[i2] = b1[i2]; }

for(i2=0;i2<ydim;++i2){
        v2[i2] = b2[i2]; }


}
```

```
void exemplar::prexmplr()
{
int i;
cout<<"\nX vector you gave is:\n";

for(i=0;i<xdim;++i){
      cout<<v1[i]<<"   ";}

cout<<"\nY vector you gave is:\n";

for(i=0;i<ydim;++i){
      cout<<v2[i]<<"   ";}

cout<<"\n";
}


void asscpair::getasscpair(int i,int j,int k)
{
idn = i;
xdim = j;
ydim = k;
}


void asscpair::prasscpair()
{
int i;
cout<<"\nX vector in the associated pair no. "<<idn<<"is:\n";

for(i=0;i<xdim;++i){
      cout<<v1[i]<<"   ";}

cout<<"\nY vector in the associated pair no. "<<idn<<"is:\n";

for(i=0;i<ydim;++i){
      cout<<v2[i]<<"   ";}

cout<<"\n";
}
```

```
void potlpair::getpotlpair(int k,int j)
{
xdim = k;
ydim = j;

}

void potlpair::prpotlpair()
{
int i;
cout<<"\nX vector in possible associated pair is:\n";

for(i=0;i<xdim;++i){
        cout<<v1[i]<<"   ";}

cout<<"\nY vector in possible associated pair is:\n";

for(i=0;i<ydim;++i){
        cout<<v2[i]<<"   ";}

cout<<"\n";
}

void network::getnwk(int k,int l,int k1,float b1[][6],float
        b2[][4])
{
anmbr = k;
bnmbr = l;
nexmplr = k1;
nasspr = 0;
ninpt = 0;
int i,j,i2;
float tmp1,tmp2;
flag =0;
char *y1="ANEURON", *y2="BNEURON" ;

for(i=0;i<nexmplr;++i){
        e[i].getexmplr(anmbr,bnmbr,b1[i],b2[i]);
        e[i].prexmplr();
```

```
            cout<<"\n";
            }

for(i=0;i<anmbr;++i){
        anrn[i].fzneuron::getnrn(i,bnmbr,0,y1);}

for(i=0;i<bnmbr;++i){
        bnrn[i].fzneuron::getnrn(i,0,anmbr,y2);}

for(i=0;i<anmbr;++i){

        for(j=0;j<bnmbr;++j){
                tmp1 = 0.0;

                for(i2=0;i2<nexmplr;++i2){
                        tmp2 = min(e[i2].v1[i],e[i2].v2[j]);
                        tmp1 = max(tmp1,tmp2);
                }

                mtrx1[i][j] = tmp1;
                mtrx2[j][i] = mtrx1[i][j];
                anrn[i].outwt[j] = mtrx1[i][j];
                bnrn[j].outwt[i] = mtrx2[j][i];
                }

        }

prwts();
cout<<"\n";
}

void network::asgninpt(float *b)
{
int i,j;
cout<<"\n";

for(i=0;i<anmbr;++i){
        anrn[i].output = b[i];
        outs1[i] = b[i];
```

```
        }

}
```

```
void network::compr1(int j,int k)
{
int i;

for(i=0;i<anmbr;++i){

        if(pp[j].v1[i] != pp[k].v1[i]) flag = 1;

        break;
        }
}


void network::compr2(int j,int k)
{
int i;

for(i=0;i<anmbr;++i){

        if(pp[j].v2[i] != pp[k].v2[i]) flag = 1;

        break;}

}

void network::comput1()    //changed from BAM
{
int j;

for(j=0;j<bnmbr;++j){
        int ii1;
        float c1 =0.0,d1;
        cout<<"\n";

        for(ii1=0;ii1<anmbr;++ii1){
                d1 = min(outs1[ii1],mtrx1[ii1][j]);
```

```
                     c1 = max(c1,d1);
                     }

        bnrn[j].activation = c1;
        cout<<"\n output layer neuron  "<<j<<" activation is "
                <<c1<<"\n";
        bnrn[j].output = bnrn[j].activation;
        outs2[j] = bnrn[j].output;
        cout<<"\n output layer neuron  "<<j<<" output is "
                <<bnrn[j].output<<"\n";
        }
}


void network::comput2()                //changed from BAM
{
int i;

for(i=0;i<anmbr;++i){
        int ii1;
        float c1=0.0,d1;

        for(ii1=0;ii1<bnmbr;++ii1){
                d1 = min(outs2[ii1],mtrx2[ii1][i]);
                c1 = max(c1,d1);}

        anrn[i].activation = c1;
        cout<<"\ninput layer neuron "<<i<<"activation is "
                <<c1<<"\n";
        anrn[i].output = anrn[i].activation;
        outs1[i] = anrn[i].output;
        cout<<"\n input layer neuron  "<<i<<"output is "
                <<anrn[i].output<<"\n";
        }

}


void network::asgnvect(int j1,float *b1,float *b2)
{
int  j2;
```

```
for(j2=0;j2<j1;++j2){
        b2[j2] = b1[j2];}


}

void network::prwts()
{
int i3,i4;
cout<<"\n  weights-  input layer to output layer: \n\n";

for(i3=0;i3<anmbr;++i3){

        for(i4=0;i4<bnmbr;++i4){
                cout<<anrn[i3].outwt[i4]<<"   ";}

        cout<<"\n"; }

cout<<"\n";

cout<<"\nweights-  output layer to input layer: \n\n";

for(i3=0;i3<bnmbr;++i3){

        for(i4=0;i4<anmbr;++i4){
                cout<<bnrn[i3].outwt[i4]<<"   ";}

        cout<<"\n";   }

cout<<"\n";
}

void network::iterate()
{
int i1;

for(i1=0;i1<nexmplr;++i1){
        findassc(e[i1].v1);
        }
```

```
}

void network::findassc(float *b)
{
int j;
flag = 0;
asgninpt(b);
        ninpt ++;
        cout<<"\nInput vector is:\n" ;

        for(j=0;j<6;++j){
                cout<<b[j]<<" ";};

        cout<<"\n";
        pp[0].getpotlpair(anmbr,bnmbr);

asgnvect(anmbr,outs1,pp[0].v1);
comput1();

if(flag>=0){

        asgnvect(bnmbr,outs2,pp[0].v2);

        cout<<"\n";
        pp[0].prpotlpair();
        cout<<"\n";

        comput2(); }

for(j=1;j<MXSIZ;++j){
        pp[j].getpotlpair(anmbr,bnmbr);
        asgnvect(anmbr,outs1,pp[j].v1);
        comput1();

        asgnvect(bnmbr,outs2,pp[j].v2);

        pp[j].prpotlpair();
        cout<<"\n";
```

```
            compr1(j,j-1);
            compr2(j,j-1);

            if(flag == 0) {

                    int j2;
                    nasspr += 1;
                    j2 = nasspr;
                    as[j2].getasscpair(j2,anmbr,bnmbr);
                    asgnvect(anmbr,pp[j].v1,as[j2].v1);
                    asgnvect(bnmbr,pp[j].v2,as[j2].v2);

                    cout<<"\nPATTERNS ASSOCIATED:\n";
                    as[j2].prasscpair();
                    j = MXSIZ ;
                    }

            else

                    if(flag == 1)
                            {
                            flag = 0;
                            comput1();
                            }

                    }

    }

void network::prstatus()
{
int j;
cout<<"\nTHE FOLLOWING ASSOCIATED PAIRS WERE FOUND BY FUZZY AM\n\n";

for(j=1;j<=nasspr;++j){
        as[j].prasscpair();
        cout<<"\n";}

}
```

```
void main()
{
int ar = 6, br = 4, nex = 1;
float inptv[][6]={0.1,0.3,0.2,0.0,0.7,0.5,0.6,0.0,0.3,0.4,0.1,0.2};
float outv[][4]={0.4,0.2,0.1,0.0};

cout<<"\n\nTHIS PROGRAM IS FOR A FUZZY ASSOCIATIVE MEMORY NETWORK. THE
NETWORK \n";
cout<<"IS SET UP FOR ILLUSTRATION WITH "<<ar<<" INPUT NEURONS, AND "<<br;
cout<<" OUTPUT NEURONS.\n"<<nex<<" exemplars are used to encode \n";

static network famn;
famn.getnwk(ar,br,nex,inptv,outv);
famn.iterate();
famn.findassc(inptv[1]);
famn.prstatus();

}
```

Output

The illustrative run of the previous program uses the fuzzy sets with fit vectors (0.1, 0.3, 0.2, 0.0, 0.7, 0.5) and (0.4, 0.2, 0.1, 0.0). As you can expect according to the discussion earlier, recall is not perfect in the reverse direction and the fuzzy associated memory consists of the pairs (0.1, 0.3, 0.2, 0.0, 0.4, 0.4) with (0.4, 0.2, 0.1, 0.0) and (0.1, 0.2, 0.2, 0, 0.2, 0.2) with (0.2, 0.2, 0.1, 0). The computer output is in such detail as to be self-explanatory.

```
THIS PROGRAM IS FOR A FUZZY ASSOCIATIVE MEMORY NETWORK. THE NETWORK IS
SET UP FOR ILLUSTRATION WITH SIX INPUT NEURONS, AND FOUR OUTPUT NEURONS.
1 exemplars are used to encode

X vector you gave is:
0.1   0.3   0.2   0   0.7   0.5
Y vector you gave is:
0.4   0.2   0.1   0

    weights—input layer to output layer:
```

```
0.1   0.1   0.1   0
0.3   0.2   0.1   0
0.2   0.2   0.1   0
0     0     0     0
0.4   0.2   0.1   0
0.4   0.2   0.1   0


weights—output layer to input layer:

0.1   0.3   0.2   0   0.4   0.4
0.1   0.2   0.2   0   0.2   0.2
0.1   0.1   0.1   0   0.1   0.1
0     0     0     0   0     0

Input vector is:
0.1 0.3 0.2 0 0.7 0.5


output layer neuron  0 activation is 0.4

output layer neuron  0 output is 0.4


output layer neuron  1 activation is 0.2

output layer neuron  1 output is 0.2


output layer neuron  2 activation is 0.1

output layer neuron  2 output is 0.1


output layer neuron  3 activation is 0

 output layer neuron  3 output is 0
```

X vector in possible associated pair is:
0.1 0.3 0.2 0 0.7 0.5
Y vector in possible associated pair is:
0.4 0.2 0.1 0

input layer neuron 0 activation is 0.1

input layer neuron 0 output is 0.1

input layer neuron 1 activation is 0.3

input layer neuron 1 output is 0.3

input layer neuron 2 activation is 0.2

input layer neuron 2 output is 0.2

input layer neuron 3 activation is 0

input layer neuron 3 output is 0

input layer neuron 4 activation is 0.4

input layer neuron 4 output is 0.4

input layer neuron 5 activation is 0.4

input layer neuron 5 output is 0.4

output layer neuron 0 activation is 0.4

output layer neuron 0 output is 0.4

output layer neuron 1 activation is 0.2

output layer neuron   1 output is 0.2

output layer neuron   2 activation is 0.1

output layer neuron   2 output is 0.1

output layer neuron   3 activation is 0

output layer neuron   3 output is 0

X vector in possible associated pair is:
0.1   0.3   0.2   0   0.4   0.4
Y vector in possible associated pair is:
0.4   0.2   0.1   0

PATTERNS ASSOCIATED:

X vector in the associated pair no. 1 is:
0.1   0.3   0.2   0   0.4   0.4
Y vector in the associated pair no. 1 is:
0.4   0.2   0.1   0

Input vector is:
0.6 0 0.3 0.4 0.1 0.2

 output layer neuron   0 activation is 0.2

 output layer neuron   0 output is 0.2

 output layer neuron   1 activation is 0.2

 output layer neuron   1 output is 0.2

output layer neuron 2 activation is 0.1

output layer neuron 2 output is 0.1

output layer neuron 3 activation is 0

output layer neuron 3 output is 0

X vector in possible associated pair is:
0.6 0 0.3 0.4 0.1 0.2
Y vector in possible associated pair is:
0.2 0.2 0.1 0

input layer neuron 0 activation is 0.1

 input layer neuron 0 output is 0.1

input layer neuron 1 activation is 0.2

 input layer neuron 1 output is 0.2

input layer neuron 2 activation is 0.2

 input layer neuron 2 output is 0.2

input layer neuron 3 activation is 0

 input layer neuron 3 output is 0

input layer neuron 4 activation is 0.2

 input layer neuron 4 output is 0.2

input layer neuron 5 activation is 0.2

 input layer neuron 5 output is 0.2

```
output layer neuron  0 activation is 0.2

output layer neuron  0 output is 0.2


output layer neuron  1 activation is 0.2

output layer neuron  1 output is 0.2


output layer neuron  2 activation is 0.1

output layer neuron  2 output is 0.1


output layer neuron  3 activation is 0

output layer neuron  3 output is 0
X vector in possible associated pair is:
0.1  0.2  0.2  0  0.2  0.2
Y vector in possible associated pair is:
0.2  0.2  0.1  0

output layer neuron  0 activation is 0.2

output layer neuron  0 output is 0.2


output layer neuron  1 activation is 0.2

output layer neuron  1 output is 0.2


output layer neuron  2 activation is 0.1

output layer neuron  2 output is 0.1
```

```
output layer neuron  3 activation is 0

output layer neuron  3 output is 0


output layer neuron  0 activation is 0.2

output layer neuron  0 output is 0.2


output layer neuron  1 activation is 0.2

output layer neuron  1 output is 0.2


output layer neuron  2 activation is 0.1

output layer neuron  2 output is 0.1


output layer neuron  3 activation is 0

output layer neuron  3 output is 0

X vector in possible associated pair is:
0.1  0.2  0.2  0  0.2  0.2
Y vector in possible associated pair is:
0.2  0.2  0.1  0



PATTERNS ASSOCIATED:

X vector in the associated pair no. 2 is:
0.1  0.2  0.2  0  0.2  0.2
Y vector in the associated pair no. 2 is:
0.2  0.2  0.1  0


THE FOLLOWING ASSOCIATED PAIRS WERE FOUND BY FUZZY AM
```

```
X vector in the associated pair no. 1 is:
0.1  0.3  0.2  0  0.4  0.4
Y vector in the associated pair no. 1 is:
0.4  0.2  0.1  0


X vector in the associated pair no. 2 is:
0.1  0.2  0.2  0  0.2  0.2
Y vector in the associated pair no. 2 is:
0.2  0.2  0.1  0
```

SUMMARY

In this chapter, bidirectional associative memories are presented for fuzzy subsets. The development of these is largely due to Kosko. They share the feature of resonance between the two layers in the network with Adaptive Resonance theory. Even though there are connections in both directions between neurons in the two layers, only one weight matrix is involved. You use the transpose of this weight matrix for the connections in the opposite direction. When one input at one end leads to some output at the other, which in turn leads to output same as the previous input, resonance is reached and an associated pair is found. In the case of bidirectional fuzzy associative memories, one pair of fuzzy sets determines one fuzzy associative memory system. Fit vectors are used in max–min composition. Perfect recall in both directions is not the case unless the heights of both fit vectors are equal. Fuzzy associative memories can improve the performance of an expert system by allowing fuzzy rules.

CHAPTER 10

ADAPTIVE RESONANCE THEORY (ART)

INTRODUCTION

Grossberg's *Adaptive Resonance Theory*, developed further by Grossberg and Carpenter, is for the categorization of patterns using the *competitive learning* paradigm. It introduces a *gain control* and a *reset* to make certain that learned categories are retained even while new categories are learned and thereby addresses the plasticity–stability dilemma.

Adaptive Resonance Theory makes much use of a competitive learning paradigm. A criterion is developed to facilitate the occurrence of *winner-take-all* phenomenon. A single node with the largest value for the set criterion is declared the winner within its layer, and it is said to classify a pattern class. If there is a tie for the winning neuron in a layer, then an arbitrary rule, such as the first of them in a serial order, can be taken as the winner.

The neural network developed for this theory establishes a system that is made up of two subsystems, one being the attentional subsystem, and this contains the unit for gain control. The other is an orienting subsystem, and this contains the unit for reset. During the operation of the network modeled for this theory, patterns emerge in the attentional subsystem and are called traces of STM (*short-term memory*). Traces of LTM (*long-term memory*) are in the connection weights between the input layer and output layer.

The network uses processing with feedback between its two layers, until resonance occurs. Resonance occurs when the output in the first layer after feedback from the second layer matches the original pattern used as input for the first layer in that processing cycle. A match of this type does not have to be perfect. What is required is that the degree of match, measured suitably, exceeds a predetermined level, termed *vigilance parameter*. Just as a photograph matches the likeness of the subject to a greater degree when the granularity is higher, the pattern match gets finer when the vigilance parameter is closer to 1.

THE NETWORK FOR ART1

The neural network for the adaptive resonance theory or ART1 model consists of the following:

- A layer of neurons, called the F_1 layer (input layer or comparison layer)
- A node for each layer as a gain control unit
- A layer of neurons, called the F_2 layer (*output* layer or *recognition* layer)
- A node as a reset unit
- Bottom-up connections from F_1 layer to F_2 layer
- Top-down connections from F_2 layer to F_1 layer
- Inhibitory connection (negative weight) form F_2 layer to gain control
- Excitatory connection (positive weight) from gain control to a layer
- Inhibitory connection from F_1 layer to reset node
- Excitatory connection from reset node to F_2 layer

A SIMPLIFIED DIAGRAM OF NETWORK LAYOUT

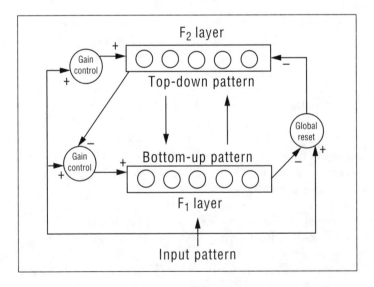

FIGURE 10.1 SIMPLIFIED DIAGRAM OF THE NEURAL NETWORK FOR AN ART1 MODEL.

PROCESSING IN ART1

The ART1 paradigm, just like the Kohonen Self-Organizing Map to be introduced in Chapter 11, performs data clustering on input data; like inputs are clustered together into a category. As an example, you can use a data clustering algorithm such as ART1 for *Optical Character Recognition* (OCR), where you try to match different samples of a letter to its ASCII equivalent. Particular attention is made in the ART1 paradigm to ensure that old information is not thrown away while new information is assimilated.

An input vector, when applied to an ART1 system, is first compared to existing patterns in the system. If there is a close enough match within a specified tolerance (as indicated by a vigilance parameter), then that stored pattern is made to resemble the input pattern further and the classification operation is complete. If the input pattern does not resemble any of the stored patterns in the system, then a new category is created with a new stored pattern that resembles the input pattern.

Special Features of the ART1 Model

One special feature of an ART1 model is that a *two-thirds rule* is necessary to determine the activity of neurons in the F_1 layer. There are three input sources to each neuron in layer F_1. They are the external input, the output of gain control, and the outputs of F_2 layer neurons. The F_1 neurons will not fire unless at least two of the three inputs are active. The gain control unit and the two-thirds rule together ensure proper response from the input layer neurons. A second feature is that a vigilance parameter is used to determine the activity of the reset unit, which is activated whenever there is no match found among existing patterns during classification.

Notation for ART1 Calculations

Let us list the various symbols we will use to describe the operation of a neural network for an ART1 model:

w_{ij}	Weight on the connection from the ith neuron in the F_1 layer to the jth neuron in the F_2 layer
v_{ji}	Weight on the connection from the jth neuron in the F_2 layer to the ith neuron on the F_1 layer
a_i	Activation of the ith neuron in the F_1 layer

b_j	Activation of the jth neuron in the F_2 layer
x_i	Output of the ith neuron in the F_1 layer
y_j	Output of the jth neuron in the F_2 layer
z_i	Input to the ith neuron in F_1 layer from F_2 layer
ρ	Vigilance parameter, positive and no greater than 1 ($0 < \rho \le 1$)
m	Number of neurons in the F_1 layer
n	Number of neurons in the F_2 layer
I	Input vector
S_i	Sum of the components of the input vector
S_x	Sum of the outputs of neurons in the F_1 layer
A, C, D	Parameters with positive values or zero
L	Parameter with value greater than 1
B	Parameter with value less than $D + 1$ but at least as large as either D or 1
r	Index of winner of competition in the F_2 layer

Algorithm for ART1 Calculations

The ART1 equations are not easy to follow. We follow the description of the algorithm found in James A. Freeman and David M. Skapura. The following equations, taken in the order given, describe the steps in the algorithm. Note that **binary** input patterns are used in ART1.

Initialization of Parameters

w_{ij}	should be positive and less than $L / (m - 1 + L)$
v_{ji}	should be greater than $(B - 1) / D$
$a_i = -B / (1 + C)$	

Equations for ART1 Computations

When you read below the equations for ART1 computations, keep in mind the following considerations. If a subscript i appears on the left-hand side of the equation, it means that there are m such equations, as the subscript i varies from 1 to m. Similarly, if instead a subscript j occurs, then there are n such equations as j ranges from 1 to n. The equations are used in the order they are given. They give a step-by-step description of the following algorithm. All the variables, you recall, are defined in the earlier section on notation. For example, I is the input vector.

F₁ layer calculations:

$$a_i = I_i / (1 + A (I_i + B) + C)$$
$$x_i = 1 \text{ if } a_i > 0$$
$$0 \text{ if } a_i \leq 0$$

F₂ layer calculations:

$$b_j = \Sigma w_{ij} x_i, \text{ the summation being on i from 1 to m}$$
$$y_j = 1 \text{ if jth neuron has the largest activation value in the F}_2 \text{ layer}$$
$$= 0 \text{ if jth neuron is not the winner in F}_2 \text{ layer}$$

Top-down inputs:

$$z_i = \Sigma v_{ji} y_j, \text{ the summation being on j from 1 to n (You will notice that exactly one term is nonzero)}$$

F₁ layer calculations:

$$a_i = (I_i + D z_i - B) / (1 + A (I_i + D z_i) + C)$$
$$x_i = 1 \text{ if } a_i > 0$$
$$= 0 \text{ if } a_i \leq 0$$

Checking with vigilance parameter:

If $(S_X / S_I) < \rho$, set $y_j = 0$ for all j, including the winner r in F_2 layer, and consider the jth neuron inactive (this step is reset, skip remaining steps).

If $(S_X / S_I) \geq \rho$, then continue.

Modifying top-down and bottom-up connection weight for winner r:

$$v_{ir} = (L / (S_x + L -1)) \text{ if } x_i = 1$$
$$= 0 \text{ if } x_i = 0$$
$$w_{ri} = 1 \text{ if } x_i = 1$$
$$= 0 \text{ if } x_i = 0$$

Having finished with the current input pattern, we repeat these steps with a new input pattern. We lose the index r given to one neuron as a winner and treat all neurons in the F_2 layer with their original indices (subscripts).

The above presentation of the algorithm is hoped to make all the steps as clear as possible. The process is rather involved. To recapitulate, first an input vector is presented to the F_1 layer neurons, their activations are determined, and then the **threshold** function is used. The outputs of the F_1 layer neurons constitute the inputs to the F_2 layer neurons, from which a winner is designated on the basis of the largest activation. The winner only is allowed to be active, meaning that the output is 1 for the winner and 0 for all the rest. The equations implicitly incorporate the use of the 2/3 rule that we mentioned earlier, and they also incorporate the way the gain control is used. The gain control is designed to have a value 1 in the phase of determining the activations of the neurons in the F_2 layer and 0 if either there is no input vector or output from the F_2 layer is propagated to the F_1 layer.

Other Models

Extensions of an ART1 model, which is for binary patterns, are ART2 and ART3. Of these, ART2 model categorizes and stores analog-valued patterns, as well as binary patterns, while ART3 addresses computational problems of hierarchies.

C++ Implementation

Again, the algorithm for ART1 processing as given in Freeman and Skapura is followed for our C++ implementation. Our objective in programming ART1 is to provide a feel for the workings of this paradigm with a very simple program implementation. For more details on the inner workings of ART1, you are encouraged to consult Freeman and Skapura, or other references listed at the back of the book.

A Header File for the C++ Program for the ART1 Model Network

The header file for the C++ program for the ART1 model network is art1net.hpp. It contains the declarations for two classes, an **artneuron** class for neurons in the ART1 model, and a **network** class, which is declared as a **friend** class in the **artneuron** class. Functions declared in the **network** class include one to do the iterations for the network operation, finding the winner in a given iteration, and one to inquire if reset is needed.

```
//art1net.h    V. Rao,  H. Rao
//Header file for ART1 model network program

#include <iostream.h>
#define MXSIZ 10


class artneuron
{

protected:
        int nnbr;
        int inn,outn;
        int output;
        double activation;
        double outwt[MXSIZ];
        char *name;
        friend class network;

public:
        artneuron() { };
        void getnrn(int,int,int,char *);
};

class network
{
public:
        int   anmbr,bnmbr,flag,ninpt,sj,so,winr;
        float ai,be,ci,di,el,rho;
```

```
artneuron (anrn)[MXSIZ],(bnrn)[MXSIZ];
int outs1[MXSIZ],outs2[MXSIZ];
int lrndptrn[MXSIZ][MXSIZ];
double acts1[MXSIZ],acts2[MXSIZ];
double mtrx1[MXSIZ][MXSIZ],mtrx2[MXSIZ][MXSIZ];

network() { };
void getnwk(int,int,float,float,float,float,float);
void prwts1();
void prwts2();
int winner(int k,double *v,int);
void practs1();
void practs2();
void prouts1();
void prouts2();
void iterate(int *,float,int);
void asgninpt(int *);
void comput1(int);
void comput2(int *);
void prlrndp();
void inqreset(int);
void adjwts1();
void adjwts2();
};
```

A Source File for C++ Program for an ART1 Model Network

The implementations of the functions declared in the header file are contained in the source file for the C++ program for an ART1 model network. It also has the **main** function, which contains specifications of the number of neurons in the two layers of the network, the values of the vigilance and other parameters, and the input vectors. Note that if there are n neurons in a layer, they are numbered serially from 0 to n–1, and not from 1 to n in the C++ program. The source file is called art1net.cpp. It is set up with six neurons in the F_1 layer and seven neurons in the F_2 layer. The **main** function also contains the parameters needed in the algorithm.

To initialize the bottom-up weights, we set each weight to be $-0.1 + L/(m - 1 + L)$ so that it is greater than 0 and less than $L/(m - 1 + L)$, as suggested before. Similarly, the top-down weights are initialized by setting each of them to $0.2 + (B - 1)/D$ so it would be greater than $(B - 1)/D$. Initial activations of the F_1 layer neurons are each set to $-B/(1 + C)$, as suggested earlier.

A **restrmax** function is defined to compute the maximum in an array when one of the array elements is not desired to be a candidate for the maximum. This facilitates the removal of the current winner from competition when reset is needed. Reset is needed when the degree of match is of a smaller magnitude than the vigilance parameter.

The function **iterate** is a **member** function of the **network** class and does the processing for the network. The **inqreset** function of the **network** class compares the **vigilance** parameter with the degree of match.

```cpp
//art1net.cpp  V. Rao, H. Rao
//Source file for ART1 network program

#include "art1net.h"

int restrmax(int j,double *b,int k)
      {
      int i,tmp;

      for(i=0;i<j;i++){
              if(i !=k)
              {tmp = i;
              i = j;}
              }

      for(i=0;i<j;i++){

      if( (i != tmp)&&(i != k))

        {if(b[i]>b[tmp]) tmp = i;}}

      return tmp;
      }

void artneuron::getnrn(int m1,int m2,int m3, char *y)
{
int i;
name = y;
```

```
nnbr = m1;
outn = m2;
inn  = m3;

for(i=0;i<outn;++i){

        outwt[i] = 0 ;
        }

output = 0;
activation = 0.0;
}

        void network::getnwk(int k,int l,float aa,float bb,float
        cc,float dd,float ll)
{
anmbr = k;
bnmbr = l;
ninpt = 0;
ai = aa;
be = bb;
ci = cc;
di = dd;
el = ll;
int i,j;
flag = 0;

char *y1="ANEURON", *y2="BNEURON" ;

for(i=0;i<anmbr;++i){

    anrn[i].artneuron::getnrn(i,bnmbr,0,y1);}

for(i=0;i<bnmbr;++i){

    bnrn[i].artneuron::getnrn(i,0,anmbr,y2);}

float tmp1,tmp2,tmp3;
tmp1 = 0.2 +(be - 1.0)/di;
```

```
tmp2 = -0.1 + el/(anmbr - 1.0 +el);
tmp3 = - be/(1.0 + ci);

for(i=0;i<anmbr;++i){

        anrn[i].activation = tmp3;
        acts1[i] = tmp3;

        for(j=0;j<bnmbr;++j){

                mtrx1[i][j]   = tmp1;
                mtrx2[j][i] = tmp2;
                anrn[i].outwt[j] = mtrx1[i][j];
                bnrn[j].outwt[i] = mtrx2[j][i];
                }
        }

prwts1();
prwts2();
practs1();
cout<<"\n";
}


int network::winner(int k,double *v,int kk){
int t1;

t1 = restrmax(k,v,kk);
return t1;
}


void network::prwts1()
{
int i3,i4;
cout<<"\nweights for F1 layer neurons: \n";

for(i3=0;i3<anmbr;++i3){

        for(i4=0;i4<bnmbr;++i4){
```

```
                    cout<<anrn[i3].outwt[i4]<<"   ";}

        cout<<"\n"; }
```

```
cout<<"\n";
}

void network::prwts2()
{
int i3,i4;
cout<<"\nweights for F2 layer neurons: \n";

for(i3=0;i3<bnmbr;++i3){

        for(i4=0;i4<anmbr;++i4){

                cout<<bnrn[i3].outwt[i4]<<"   ";};

        cout<<"\n";   }

cout<<"\n";
}

void network::practs1()
{
int j;
cout<<"\nactivations of F1 layer neurons: \n";

for(j=0;j<anmbr;++j){

        cout<<acts1[j]<<"    ";}

cout<<"\n";
}

void network::practs2()
{
int j;
cout<<"\nactivations of F2 layer neurons: \n";
```

```
for(j=0;j<bnmbr;++j){

        cout<<acts2[j]<<"    ";}

cout<<"\n";
}


void network::prouts1()
{
int j;
cout<<"\noutputs of F1 layer neurons: \n";

for(j=0;j<anmbr;++j){

        cout<<outs1[j]<<"    ";}

cout<<"\n";
}


void network::prouts2()
{
int j;
cout<<"\noutputs of F2 layer neurons: \n";

for(j=0;j<bnmbr;++j){

        cout<<outs2[j]<<"    ";}

cout<<"\n";
}


void network::asgninpt(int *b)
{
int j;
sj = so = 0;
cout<<"\nInput vector is:\n" ;

for(j=0;j<anmbr;++j){
```

```
            cout<<b[j]<<" ";}

cout<<"\n";

for(j=0;j<anmbr;++j){

        sj += b[j];
        anrn[j].activation = b[j]/(1.0 +ci +ai*(b[j]+be));
        acts1[j] = anrn[j].activation;

        if(anrn[j].activation > 0) anrn[j].output = 1;

        else
                anrn[j].output = 0;

        outs1[j] = anrn[j].output;
        so += anrn[j].output;
        }


practs1();
prouts1();
}

void network::inqreset(int t1)
{
int jj;
flag = 0;
jj = so/sj;
cout<<"\ndegree of match: "<<jj<<" vigilance:  "<<rho<<"\n";

if( jj > rho ) flag = 1;

        else
        {cout<<"winner is "<<t1;
        cout<<" reset required \n";}


}
```

```
void network::comput1(int k)
{
int j;

for(j=0;j<bnmbr;++j){

        int ii1;
        double c1 = 0.0;
        cout<<"\n";

        for(ii1=0;ii1<anmbr;++ii1){

                c1 += outs1[ii1] * mtrx2[j][ii1];
                }

        bnrn[j].activation = c1;
        acts2[j] = c1;};

winr = winner(bnmbr,acts2,k);
cout<<"winner is "<<winr;
for(j=0;j<bnmbr;++j){

        if(j == winr) bnrn[j].output = 1;

        else bnrn[j].output =  0;
        outs2[j] = bnrn[j].output;
        }

practs2();
prouts2();
}

void network::comput2(int *b)
{
double db[MXSIZ];
double tmp;
so = 0;
int i,j;
```

```
for(j=0;j<anmbr;++j){

        db[j] =0.0;

        for(i=0;i<bnmbr;++i){

                db[j] += mtrx1[j][i]*outs2[i];};

        tmp = b[j] + di*db[j];
        acts1[j] = (tmp - be)/(ci +1.0 +ai*tmp);
        anrn[j].activation = acts1[j];

        if(anrn[j].activation > 0) anrn[j].output = 1;

        else anrn[j].output = 0;

        outs1[j] = anrn[j].output;
        so += anrn[j].output;
        }

cout<<"\n";
practs1();
prouts1();
}

void network::adjwts1()
{
int i;

for(i=0;i<anmbr;++i){

        if(outs1[i] >0) {mtrx1[i][winr]  = 1.0;}

        else

                {mtrx1[i][winr] = 0.0;}

        anrn[i].outwt[winr] = mtrx1[i][winr];}
```

```
prwts1();
}

void network::adjwts2()
{
int i;
cout<<"\nwinner is "<<winr<<"\n";

for(i=0;i<anmbr;++i){

        if(outs1[i] > 0) {mtrx2[winr][i] = el/(so + el -1);}

        else

                {mtrx2[winr][i] = 0.0;}

        bnrn[winr].outwt[i]  = mtrx2[winr][i];}

prwts2();
}

void network::iterate(int *b,float rr,int kk)
{
int j;
rho = rr;
flag = 0;

asgninpt(b);
comput1(kk);
comput2(b);
inqreset(winr);

if(flag == 1){

        ninpt ++;
        adjwts1();
        adjwts2();
        int j3;
```

```
                for(j3=0;j3<anmbr;++j3){

                        lrndptrn[ninpt][j3] = b[j3];}

                prlrndp();
                }

        else

                {

                for(j=0;j<bnmbr;++j){

                        outs2[j] = 0;
                        bnrn[j].output = 0;}

                iterate(b,rr,winr);
                }
        }

void network::prlrndp()
{
int j;
cout<<"\nlearned vector # "<<ninpt<<"   :\n";

for(j=0;j<anmbr;++j){

        cout<<lrndptrn[ninpt][j]<<"   ";}

cout<<"\n";
}

void main()
{
int ar = 6, br = 7, rs = 8;
float aa = 2.0,bb = 2.5,cc = 6.0,dd = 0.85,ll = 4.0,rr =
        0.95;
int inptv[][6]={0,1,0,0,0,0,1,0,1,0,1,0,0,0,0,0,1,0,1,0,1,0,\
        1,0};
```

```
cout<<"\n\nTHIS PROGRAM IS FOR AN -ADAPTIVE RESONANCE THEORY\
        1 - NETWORK.\n";
cout<<"THE NETWORK IS SET UP FOR ILLUSTRATION WITH "<<ar<<" \
        INPUT NEURONS,\n";
cout<<" AND "<<br<<" OUTPUT NEURONS.\n";

static network bpn;
bpn.getnwk(ar,br,aa,bb,cc,dd,11) ;
bpn.iterate(inptv[0],rr,rs);
bpn.iterate(inptv[1],rr,rs);
bpn.iterate(inptv[2],rr,rs);
bpn.iterate(inptv[3],rr,rs);
}
```

PROGRAM OUTPUT

Four **input** vectors are used in the trial run of the program, and these are specified in the main function. The output is self-explanatory. We have included only in this text some comments regarding the output. These comments are enclosed within strings of asterisks. They are not actually part of the program output. Table 10.1 shows a summarization of the categorization of the inputs done by the network. Keep in mind that the numbering of the neurons in any layer, which has n neurons, is from 0 to $n - 1$, and not from 1 to n.

TABLE 10.1 CATEGORIZATION OF INPUTS

input	winner in F_2 layer
0 1 0 0 0 0	0, no reset
1 0 1 0 1 0	1, no reset
0 0 0 0 1 0	1, after reset 2
1 0 1 0 1 0	1, after reset 3

The input pattern 0 0 0 0 1 0 is considered a subset of the pattern 1 0 1 0 1 0 in the sense that in whatever position the first pattern has a 1, the second pattern also has a 1. Of course, the second pattern has 1's in other positions as well. At the same time, the pattern 1 0 1 0 1 0 is considered a superset of the pattern 0 0 0 0 1 0. The reason that the pattern 1 0 1 0 1 0 is repeated as input after the pattern 0 0 0 0 1 0 is processed, is to see what happens with this superset. In both cases, the degree of match falls short of the vigilance parameter, and a reset is needed.

Here's the output of the program:

```
THIS PROGRAM IS FOR AN ADAPTIVE RESONANCE THEORY
1-NETWORK. THE NETWORK IS SET UP FOR ILLUSTRATION WITH SIX INPUT NEURONS
AND SEVEN OUTPUT NEURONS.
**************************************************************
Initialization of connection weights and F1 layer activations. F1 layer
connection weights are all chosen to be equal to a random value subject
to the conditions given in the algorithm. Similarly, F2 layer connection
weights are all chosen to be equal to a random value subject to the con-
ditions given in the algorithm.
**************************************************************
weights for F1 layer neurons:
1.964706  1.964706  1.964706  1.964706  1.964706  1.964706  1.964706
1.964706  1.964706  1.964706  1.964706  1.964706  1.964706  1.964706
1.964706  1.964706  1.964706  1.964706  1.964706  1.964706  1.964706
1.964706  1.964706  1.964706  1.964706  1.964706  1.964706  1.964706
1.964706  1.964706  1.964706  1.964706  1.964706  1.964706  1.964706
1.964706  1.964706  1.964706  1.964706  1.964706  1.964706  1.964706


weights for F2 layer neurons:
0.344444  0.344444  0.344444  0.344444  0.344444  0.344444
0.344444  0.344444  0.344444  0.344444  0.344444  0.344444
0.344444  0.344444  0.344444  0.344444  0.344444  0.344444
0.344444  0.344444  0.344444  0.344444  0.344444  0.344444
0.344444  0.344444  0.344444  0.344444  0.344444  0.344444
0.344444  0.344444  0.344444  0.344444  0.344444  0.344444
0.344444  0.344444  0.344444  0.344444  0.344444  0.344444

activations of F1 layer neurons:
-0.357143 -0.357143 -0.357143 -0.357143 -0.357143 -0.357143
**************************************************************
A new input vector and a new iteration
**************************************************************
Input vector is:
0 1 0 0 0 0
```

```
activations of F1 layer neurons:
0   0.071429   0   0   0   0


outputs of F1 layer neurons:
0   1   0   0   0   0


winner is 0
activations of F2 layer neurons:
0.344444    0.344444    0.344444    0.344444    0.344444    0.344444
0.344444


outputs of F2 layer neurons:
1   0   0   0   0   0   0


activations of F1 layer neurons:
-0.080271    0.013776    -0.080271    -0.080271    -0.080271    -0.080271


outputs of F1 layer neurons:
0   1   0   0   0   0
****************************************************************
Top-down and bottom-up outputs at F1 layer match, showing resonance.
****************************************************************
degree of match: 1 vigilance:  0.95


weights for F1 layer neurons:
0   1.964706    1.964706    1.964706    1.964706    1.964706    1.964706
1   1.964706    1.964706    1.964706    1.964706    1.964706    1.964706
0   1.964706    1.964706    1.964706    1.964706    1.964706    1.964706
0   1.964706    1.964706    1.964706    1.964706    1.964706    1.964706
0   1.964706    1.964706    1.964706    1.964706    1.964706    1.964706
0   1.964706    1.964706    1.964706    1.964706    1.964706    1.964706


winner is 0


weights for F2 layer neurons:
0   1   0   0   0   0
0.344444    0.344444    0.344444    0.344444    0.344444    0.344444
0.344444    0.344444    0.344444    0.344444    0.344444    0.344444
0.344444    0.344444    0.344444    0.344444    0.344444    0.344444
```

```
0.344444   0.344444   0.344444   0.344444   0.344444   0.344444
0.344444   0.344444   0.344444   0.344444   0.344444   0.344444
0.344444   0.344444   0.344444   0.344444   0.344444   0.344444

learned vector # 1  :
0  1  0  0  0  0
****************************************************************
A new input vector and a new iteration
****************************************************************
Input vector is:
1 0 1 0 1 0

activations of F1 layer neurons:
0.071429   0   0.071429   0   0.071429   0

outputs of F1 layer neurons:
1   0   1   0   1   0

winner is 1
activations of F2 layer neurons:
0   1.033333   1.033333   1.033333   1.033333   1.033333   1.033333

outputs of F2 layer neurons:
0   1   0   0   0   0   0

activations of F1 layer neurons:
0.013776   -0.080271   0.013776   -0.080271   0.013776   -0.080271

outputs of F1 layer neurons:
1   0   1   0   1   0
****************************************************************
Top-down and bottom-up outputs at F1 layer match,
showing resonance.
****************************************************************
degree of match: 1 vigilance:  0.95

weights for F1 layer neurons:
0  1  1.964706   1.964706   1.964706   1.964706   1.964706
1  0  1.964706   1.964706   1.964706   1.964706   1.964706
```

```
0  1  1.964706   1.964706   1.964706   1.964706   1.964706
0  0  1.964706   1.964706   1.964706   1.964706   1.964706
0  1  1.964706   1.964706   1.964706   1.964706   1.964706
0  0  1.964706   1.964706   1.964706   1.964706   1.964706
```

```
winner is 1

weights for F2 layer neurons:
0   1   0   0   0   0
0.666667  0  0.666667  0  0.666667  0
0.344444   0.344444   0.344444   0.344444   0.344444   0.344444
0.344444   0.344444   0.344444   0.344444   0.344444   0.344444
0.344444   0.344444   0.344444   0.344444   0.344444   0.344444
0.344444   0.344444   0.344444   0.344444   0.344444   0.344444
0.344444   0.344444   0.344444   0.344444   0.344444   0.344444

learned vector # 2  :
1   0   1   0   1   0
*************************************************************
A new input vector and a new iteration
*************************************************************
Input vector is:
0 0 0 0 1 0

activations of F1 layer neurons:
0   0   0   0   0.071429   0

outputs of F1 layer neurons:
0   0   0   0   1   0

winner is 1
activations of F2 layer neurons:
0   0.666667   0.344444   0.344444   0.344444   0.344444   0.344444

outputs of F2 layer neurons:
0   1   0   0   0   0   0

activations of F1 layer neurons:
```

```
-0.189655    -0.357143    -0.189655    -0.357143    -0.060748    -0.357143
```

outputs of F1 layer neurons:
```
0   0   0   0   0   0
```

degree of match: 0 vigilance: 0.95
winner is 1 reset required
```
***************************************************************
```
Input vector repeated after reset, and a new iteration
```
***************************************************************
```
Input vector is:
```
0 0 0 0 1 0
```

activations of F1 layer neurons:
```
0    0    0    0    0.071429    0
```

outputs of F1 layer neurons:
```
0   0   0   0   1   0
```

winner is 2
activations of F2 layer neurons:
```
0   0.666667   0.344444   0.344444   0.344444   0.344444   0.344444
```
outputs of F2 layer neurons:
```
0   0   1   0   0   0   0
```

     activations of F1 layer neurons:
```
-0.080271    -0.080271    -0.080271    -0.080271    0.013776    -0.080271
```

outputs of F1 layer neurons:
```
0   0   0   0   1   0
```
```
***************************************************************
```
Top-down and bottom-up outputs at F1 layer match, showing resonance.
```
***************************************************************
```
degree of match: 1 vigilance: 0.95

weights for F1 layer neurons:
```
0   1   0   1.964706   1.964706   1.964706   1.964706
1   0   0   1.964706   1.964706   1.964706   1.964706
0   1   0   1.964706   1.964706   1.964706   1.964706
```

```
0   0   0   1.964706   1.964706   1.964706   1.964706
0   1   1   1.964706   1.964706   1.964706   1.964706
0   0   0   1.964706   1.964706   1.964706   1.964706
```

```
winner is 2

weights for F2 layer neurons:
0   1   0   0   0   0
0.666667   0   0.666667   0   0.666667   0
0   0   0   0   1   0
0.344444   0.344444   0.344444   0.344444   0.344444   0.344444
0.344444   0.344444   0.344444   0.344444   0.344444   0.344444
0.344444   0.344444   0.344444   0.344444   0.344444   0.344444
0.344444   0.344444   0.344444   0.344444   0.344444   0.344444

learned vector # 3  :
0   0   0   0   1   0
****************************************************************
An old (actually the second above) input vector is retried after trying a
subset vector, and a new iteration
****************************************************************
Input vector is:
1 0 1 0 1 0

activations of F1 layer neurons:
0.071429   0   0.071429   0   0.071429   0

outputs of F1 layer neurons:
1   0   1   0   1   0

winner is 1
activations of F2 layer neurons:
0   2   1   1.033333   1.033333   1.033333   1.03333

outputs of F2 layer neurons:
0   1   0   0   0   0   0

activations of F1 layer neurons:
-0.060748   -0.357143   -0.060748   -0.357143   -0.060748   -0.357143
```

```
outputs of F1 layer neurons:
0   0   0   0   0   0

degree of match: 0 vigilance:  0.95
winner is 1 reset required
****************************************************************
Input vector repeated after reset, and a new iteration
****************************************************************
Input vector is:
1 0 1 0 1 0

activations of F1 layer neurons:
0.071429   0   0.071429   0   0.071429   0

outputs of F1 layer neurons:
1   0   1   0   1   0

winner is 3
activations of F2 layer neurons:
0   2   1   1.033333   1.033333   1.033333   1.033333

outputs of F2 layer neurons:
0   0   0   1   0   0   0

activations of F1 layer neurons:
0.013776   -0.080271   0.013776   -0.080271   0.013776   -0.080271

outputs of F1 layer neurons:
1   0   1   0   1   0
****************************************************************
Top-down and Bottom-up outputs at F1layer match, showing resonance.
****************************************************************
degree of match: 1 vigilance:  0.95

weights for F1 layer neurons:
0   1   0   1   1.964706   1.964706   1.964706
1   0   0   0   1.964706   1.964706   1.964706
0   1   0   1   1.964706   1.964706   1.964706
0   0   0   0   1.964706   1.964706   1.964706
```

```
0  1  1  1  1.964706   1.964706   1.964706
0  0  0  0  1.964706   1.964706   1.964706
```

```
winner is 3
```

```
weights for F2 layer neurons:
0  1  0  0  0  0
0.666667  0  0.666667  0  0.666667  0
0  0  0  0  1  0
0.666667  0  0.666667  0  0.666667  0
0.344444   0.344444   0.344444   0.344444   0.344444   0.344444
0.344444   0.344444   0.344444   0.344444   0.344444   0.344444
0.344444   0.344444   0.344444   0.344444   0.344444   0.344444
```

```
learned vector # 4  :
1  0  1  0  1  0
```

SUMMARY

This chapter presented the basics of the Adaptive Resonance Theory of Grossberg and Carpenter and a C++ implementation of the neural network modeled for this theory. It is an elegant theory that addresses the stability–plasticity dilemma. The network relies on resonance. It is a self-organizing network and does categorization by associating individual neurons of the F_2 layer with individual patterns. By employing a so-called 2/3 rule, it ensures stability in learning patterns.

CHAPTER 11

THE KOHONEN SELF-ORGANIZING MAP

INTRODUCTION

This chapter discusses one type of unsupervised competitive learning, the *Kohonen feature map*, or *self-organizing map* (*SOM*). As you recall, in unsupervised learning there are no expected outputs presented to a neural network, as in a supervised training algorithm such as backpropagation. Instead, a network, by its self-organizing properties, is able to infer relationships and learn more as more inputs are presented to it. One advantage to this scheme is that you can expect the system to change with changing conditions and inputs. The system constantly learns. The Kohonen SOM is a neural network system developed by Teuvo Kohonen of Helsinki University of Technology and is often used to classify inputs into different categories. Applications for feature maps can be traced to many areas, including speech recognition and robot motor control.

COMPETITIVE LEARNING

A Kohonen feature map may be used by itself or as a layer of another neural network. A Kohonen layer is composed of neurons that compete with each other. Like in Adaptive Resonance Theory, the Kohonen SOM is another case of using a winner-take-all strategy. Inputs are fed into each of the neurons in the Kohonen layer (from the input layer). Each neuron determines its output according to a weighted sum formula:

$$\text{Output} = \sum w_{ij} x_i$$

The weights and the inputs are usually normalized, which means that the magnitude of the weight and input vectors are set equal to one. The neuron with the largest output is the winner. This neuron has a final output of 1.

All other neurons in the layer have an output of zero. Differing input patterns end up firing different winner neurons. Similar or identical input patterns classify to the same output neuron. You get like inputs clustered together. In Chapter 12, you will see the use of a Kohonen network in pattern classification.

Normalization of a Vector

Consider a vector, $\mathbf{A} = ax + by + cz$. The normalized vector $\mathbf{A'}$ is obtained by dividing each component of $\mathbf{A}$ by the square root of the sum of squares of all the components. In other words each component is multiplied by $1/\sqrt{(a^2 + b^2 + c^2)}$. Both the weight vector and the input vector are normalized during the operation of the Kohonen feature map. The reason for this is the training law uses subtraction of the weight vector from the input vector. Using normalization of the values in the subtraction reduces both vectors to a unitless status, and hence, makes the subtraction of like quantities possible. You will learn more about the training law shortly.

LATERAL INHIBITION

Lateral inhibition is a process that takes place in some biological neural networks. Lateral connections of neurons in a given layer are formed, and squash distant neighbors. The strength of connections is inversely related to distance. The positive, supportive connections are termed as *excitatory* while the negative, squashing connections are termed *inhibitory*.

A biological example of lateral inhibition occurs in the human vision system.

The Mexican Hat Function

Figure 11.1 shows a function, called the **mexican hat** function, which shows the relationship between the connection strength and the distance from the winning neuron. The effect of this function is to set up a competitive environment for learning. Only winning neurons and their neighbors participate in learning for a given input pattern.

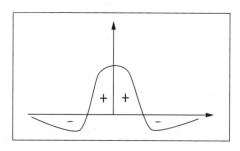

FIGURE 11.1 THE MEXICAN HAT FUNCTION SHOWING LATERAL INHIBITION.

TRAINING LAW FOR THE KOHONEN MAP

The training law for the Kohonen feature map is straightforward. The change in weight vector for a given output neuron is a gain constant, alpha, multiplied by the difference between the input vector and the old weight vector:

$$W_{new} = W_{old} + alpha * (Input - W_{old})$$

Both the old weight vector and the input vector are normalized to unit length. Alpha is a gain constant between 0 and 1.

Significance of the Training Law

Let us consider the case of a two-dimensional input vector. If you look at a unit circle, as shown in Figure 11.2, the effect of the training law is to try to align the weight vector and the input vector. Each pattern attempts to nudge the weight vector closer by a fraction determined by alpha. For three dimensions the surface becomes a unit sphere instead of a circle. For higher dimensions you term the surface a *hypersphere*. It is not necessarily ideal to have perfect alignment of the input and weight vectors. You use neural networks for their ability to recognize patterns, but also to generalize input data sets. By aligning all input vectors to the corresponding winner weight vectors, you are essentially *memorizing* the input data set classes. It may be more desirable to come close, so that noisy or incomplete inputs may still trigger the correct classification.

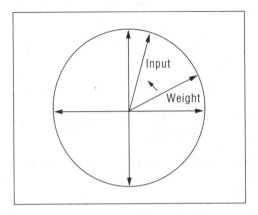

FIGURE 11.2 THE TRAINING LAW FOR THE KOHONEN MAP AS SHOWN ON A UNIT CIRCLE.

The Neighborhood Size and Alpha

In the Kohonen map, a parameter called the *neighborhood size* is used to model the effect of the **mexican hat** function. Those neurons that are within the distance specified by the neighborhood size participate in training and weight vector changes; those that are outside this distance do not participate in learning. The neighborhood size typically is started as an initial value and is decreased as the input pattern cycles continue. This process tends to support the winner-take-all strategy by eventually singling out a winner neuron for a given pattern.

Figure 11.3 shows a linear arrangement of neurons with a neighborhood size of 2. The hashed central neuron is the winner. The darkened adjacent neurons are those that will participate in training.

Besides the neighborhood size, **alpha** typically is also reduced during simulation. You will see these features when we develop a Kohonen map program.

FIGURE 11.3 WINNER NEURON WITH A NEIGHBORHOOD SIZE OF 2 FOR A KOHONEN MAP.

C++ CODE FOR IMPLEMENTING A KOHONEN MAP

The C++ code for the Kohonen map draws on much of the code developed for the backpropagation simulator. The Kohonen map is a much simpler program and may not rely on as large a data set for input. The Kohonen map program uses only two files, an input file and an output file. In order to use the program, you must create an input data set and save this in a file called **input.dat**. The output file is called **kohonen.dat** and is saved in your current working directory. You will get more details shortly on the formats of these files.

THE KOHONEN NETWORK

The Kohonen network has two layers, an *input* layer and a *Kohonen output* layer. (See Figure 11.4). The input layer is a size determined by the user and must match the size of each row (pattern) in the input data file.

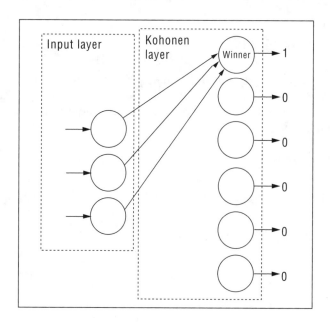

FIGURE 11.4 A KOHONEN NETWORK.

Modeling Lateral Inhibition and Excitation

The **mexican hat** function shows positive values for an immediate neighborhood around the neuron and negative values for distant neurons. A true method of modeling would incorporate mutual excitation or support for neurons that are within the neighborhood (with this excitation increasing for nearer neurons) and inhibition for distant neurons outside the neighborhood. For the sake of computational efficiency, we model lateral inhibition and excitation by looking at the maximum output for the output neurons and making that output belong to a winner neuron. Other outputs are inhibited by setting their outputs to zero. Training, or weight update, is performed on all outputs that are within a neighborhood size distance from the winner neuron. Neurons outside the neighborhood do not participate in training. The true way of modeling lateral inhibition would be too expensive since the number of lateral connections is quite large. You will find that this approximation will lead to a network with many if not all of the properties of a true modeling approach of a Kohonen network.

CLASSES TO BE USED

We use many of the classes from the backpropagation simulator. We require only two layers, the input layer and the Kohonen layer. We make a new layer class called the **Kohonen layer** class, and a new **network** class called the **Kohonen_network**.

Revisiting the Layer Class

The **layer** class needs to be slightly modified, as shown in Listing 11.1.

LISTING 11.1 MODIFICATION OF LAYER.H

```
// layer.h            V.Rao, H. Rao
// header file for the layer class hierarchy and
// the network class

#define MAX_LAYERS    5
#define MAX_VECTORS   100
```

```
class network;
class Kohonen_network;

class layer
{

protected:

        int num_inputs;
        int num_outputs;
        float *outputs;// pointer to array of outputs
        float *inputs; // pointer to array of inputs, which
                              // are outputs of some other layer

        friend network;
        friend Kohonen_network; // update for Kohonen model

public:

        virtual void calc_out()=0;
};
...
```

Here the changes are indicated in italic. You notice that the **Kohonen_network** is made a friend to the **layer** class, so that the **Kohonen_network** can have access to the data of a layer.

A New Layer Class for a Kohonen Layer

The next step to take is to create a **Kohonen_layer** class and a **Kohonen_network** class. This is shown in Listing 11.2.

LISTING 11.2 THE KOHONEN_LAYER CLASS AND KOHONEN_NETWORK CLASS IN LAYERK.H

```
// layerk.h           V.Rao, H. Rao
// header file for the Kohonen layer and
// the Kohonen network
```

```
class Kohonen_network;


class Kohonen_layer: public layer
{

protected:

        float * weights;
        int winner_index;
        float win_distance;
        int neighborhood_size;

        friend Kohonen_network;

public:

        Kohonen_layer(int, int, int);
        ~Kohonen_layer();
        virtual void calc_out();
        void randomize_weights();
        void update_neigh_size(int);
        void update_weights(const float);
        void list_weights();
        void list_outputs();
        float get_win_dist();



};


class Kohonen_network

{

private:

        layer *layer_ptr[2];
        int layer_size[2];
```

```
        int neighborhood_size;

public:
        Kohonen_network();
        ~Kohonen_network();
        void get_layer_info();
        void set_up_network(int);
        void randomize_weights();
        void update_neigh_size(int);
        void update_weights(const float);
        void list_weights();
        void list_outputs();
        void get_next_vector(FILE *);
        void process_next_pattern();
        float get_win_dist();
        int get_win_index();


};
```

The **Kohonen_layer** is derived from the **layer** class, so it has pointers inherited that point to a set of outputs and a set of inputs. Let's look at some of the functions and member variables.

Kohonen_layer:

- **float * weights** Pointer to the weights matrix.
- **int winner_index** Index value of the output, which is the winner.
- **float win_distance** The Euclidean distance of the winner weight vector from the input vector.
- **int neighborhood_size** The size of the neighborhood.
- **Kohonen_layer(int, int, int)** Constructor for the layer: inputs, outputs, and the neighborhood size.
- **~Kohonen_layer()** Destructor.
- **virtual void calc_out()** The function to calculate the outputs; for the Kohonen layer this models lateral competition.
- **void randomize_weights()** A function to initialize weights with random normal values.
- **void update_neigh_size(nt)** This function updates the neighborhood size with a new value.

- **void update_weights(const float)** This function updates the weights according to the training law using the passed parameter, alpha.
- **void list_weights()** This function can be used to list the weight matrix.
- **void list_outputs()** This function is used to write outputs to the output file.
- **float get_win_dist()** Returns the Euclidean distance between the winner weight vector and the input vector.

Kohonen_network:

- **layer *layer_ptr[2]** Pointer array; element 0 points to the input layer, element 1 points to the Kohonen layer.
- **int layer_size[2]** Array of layer sizes for the two layers.
- **int neighborhood_size** The current neighborhood size.
- **Kohonen_network()** Constructor.
- **~Kohonen_network()** Destructor.
- **void get_layer_info()** Gets information about the layer sizes.
- **void set_up_network(int)** Connects layers and sets up the Kohonen map.
- **void randomize_weights()** Creates random normalized weights.
- **void update_neigh_size(int)** Changes the neighborhood size.
- **void update_weights(const float)** Performs weight update according to the training law.
- **void list_weights()** Can be used to list the weight matrix.
- **void list_outputs()** Can be used to list outputs.
- **void get_next_vector(FILE *)** Function gets another input vector from the input file.
- **void process_next_pattern()** Applies pattern to the Kohonen map.
- **float get_win_dist()** Returns the winner's distance from the input vector.
- **int get_win_index()** Returns the index of the winner.

IMPLEMENTATION OF THE KOHONEN LAYER AND KOHONEN NETWORK

Listing 11.3 shows the *layerk.cpp* file, which has the implementation of the functions outlined.

LISTING 11.3 IMPLEMENTATION FILE FOR THE KOHONEN LAYER AND KOHONEN NETWORK :LAYERK.CPP

```
// layerk.cpp          V.Rao, H.Rao
// compile for floating point hardware if available
#include "layer.cpp"
#include "layerk.h"


// ————————————————-
//                       Kohonen layer
//————————————————



Kohonen_layer::Kohonen_layer(int i, int o, int
      init_neigh_size)
{
num_inputs=i;
num_outputs=o;
neighborhood_size=init_neigh_size;
weights = new float[num_inputs*num_outputs];
outputs = new float[num_outputs];
}


Kohonen_layer::~Kohonen_layer()
{
```

```
delete [num_outputs*num_inputs] weights;
delete [num_outputs] outputs;
}

void Kohonen_layer::calc_out()
{
// implement lateral competition
// choose the output with the largest
// value as the winner; neighboring
// outputs participate in next weight
// update. Winner's output is 1 while
// all other outputs are zero

int i,j,k;
float accumulator=0.0;
float maxval;
winner_index=0;
maxval=-1000000;

for (j=0; j<num_outputs; j++)
        {

        for (i=0; i<num_inputs; i++)

                {
                k=i*num_outputs;
                if (weights[k+j]*weights[k+j] > 1000000.0)
                        {
                        cout << "weights are blowing up\n";
                        cout << "try a smaller learning constant\n";
                        cout << "e.g. beta=0.02    aborting...\n";
                        exit(1);
                        }
                outputs[j]=weights[k+j]*(*(inputs+i));
                accumulator+=outputs[j];
                }
        // no squash function
        outputs[j]=accumulator;
        if (outputs[j] > maxval)
```

```
        {
        maxval=outputs[j];
        winner_index=j;
        }
    accumulator=0;
    }


// set winner output to 1
outputs[winner_index]=1.0;
// now zero out all other outputs
for (j=0; j< winner_index; j++)
        outputs[j]=0;
for (j=num_outputs-1; j>winner_index; j—)
        outputs[j]=0;



}



void Kohonen_layer::randomize_weights()
{
int i, j, k;
const unsigned first_time=1;

const unsigned not_first_time=0;
float discard;
float norm;

discard=randomweight(first_time);

for (i=0; i< num_inputs; i++)
        {
        k=i*num_outputs;
        for (j=0; j< num_outputs; j++)
                {
                weights[k+j]=randomweight(not_first_time);
                }
        }
```

```
// now need to normalize the weight vectors
// to unit length
// a weight vector is the set of weights for
// a given output

for (j=0; j< num_outputs; j++)
    {
    norm=0;
        for (i=0; i< num_inputs; i++)
            {
            k=i*num_outputs;
            norm+=weights[k+j]*weights[k+j];
}

    norm = 1/((float)sqrt((double)norm));

for (i=0; i< num_inputs; i++)
    {
    k=i*num_outputs;
    weights[k+j]*=norm;
    }
    }


}



void Kohonen_layer::update_neigh_size(int new_neigh_size)
{
neighborhood_size=new_neigh_size;
}



void Kohonen_layer::update_weights(const float alpha)
{
int i, j, k;
int start_index, stop_index;
// learning law: weight_change =
//              alpha*(input-weight)
//      zero change if input and weight
```

```
// vectors are aligned
// only update those outputs that
// are within a neighborhood's distance
// from the last winner
start_index = winner_index -
                       neighborhood_size;



if (start_index < 0)
        start_index =0;

stop_index = winner_index +
                       neighborhood_size;



if (stop_index > num_outputs-1)
        stop_index = num_outputs-1;



for (i=0; i< num_inputs; i++)
        {
        k=i*num_outputs;
        for (j=start_index; j<=stop_index; j++)
               weights[k+j] +=
                       alpha*((*(inputs+i))-weights[k+j]);
    }


}

void Kohonen_layer::list_weights()
{
int i, j, k;

for (i=0; i< num_inputs; i++)
        {
        k=i*num_outputs;
        for (j=0; j< num_outputs; j++)
```

```
                    cout << "weight["<<i<<","<<
                            j<<"] is: "<<weights[k+j];
        }
```

```
}

void Kohonen_layer::list_outputs()
{
int i;

for (i=0; i< num_outputs; i++)
        {
        cout << "outputs["<<i<<
                    "] is: "<<outputs[i];
        }


}



float Kohonen_layer::get_win_dist()
{
int i, j, k;
j=winner_index;
float accumulator=0;


float * win_dist_vec = new float [num_inputs];

for (i=0; i< num_inputs; i++)
        {
        k=i*num_outputs;
        win_dist_vec[i]=(*(inputs+i))-weights[k+j];
    accumulator+=win_dist_vec[i]*win_dist_vec[i];
        }

win_distance =(float)sqrt((double)accumulator);

delete [num_inputs]win_dist_vec;
```

```
return win_distance;

}
```

```
Kohonen_network::Kohonen_network()
{

}

Kohonen_network::~Kohonen_network()
{
}

void Kohonen_network::get_layer_info()
{
int i;

//————————————————
//
//      Get layer sizes for the Kohonen network
//
// ——————————————--



cout << " Enter in the layer sizes separated by spaces.\n";
cout << " A Kohonen network has an input layer \n";
cout << " followed by a Kohonen (output) layer \n";

for (i=0; i<2; i++)
        {
        cin >> layer_size[i];
        }

// ——————————————————
// size of layers:
//              input_layer             layer_size[0]
//              Kohonen_layer           layer_size[1]
```

```
//————————————————————————-



}

void Kohonen_network::set_up_network(int nsz)
{
int i;

// set up neighborhood size
neighborhood_size = nsz;

//————————————————————-
// Construct the layers
//
//————————————————————-



layer_ptr[0] = new input_layer(0,layer_size[0]);

layer_ptr[1] =
        new Kohonen_layer(layer_size[0],
                        layer_size[1],neighborhood_size);



for (i=0;i<2;i++)
        {
        if (layer_ptr[i] == 0)
                {
                cout << "insufficient memory\n";
                cout << "use a smaller architecture\n";
                exit(1);
                }
        }

//————————————————————-
// Connect the layers
```

```
//
//——————————————————-
// set inputs to previous layer outputs for the Kohonen layer

layer_ptr[1]->inputs = layer_ptr[0]->outputs;



}



void Kohonen_network::randomize_weights()
{

((Kohonen_layer *)layer_ptr[1])
            ->randomize_weights();
}



void Kohonen_network::update_neigh_size(int n)
{
((Kohonen_layer *)layer_ptr[1])
            ->update_neigh_size(n);
}


void Kohonen_network::update_weights(const float a)
{
((Kohonen_layer *)layer_ptr[1])
            ->update_weights(a);
}



void Kohonen_network::list_weights()
{
((Kohonen_layer *)layer_ptr[1])
            ->list_weights();
}


void Kohonen_network::list_outputs()
{
```

```
((Kohonen_layer *)layer_ptr[1])
            ->list_outputs();

}

void Kohonen_network::get_next_vector(FILE * ifile)
{
int i;
float normlength=0;
int num_inputs=layer_ptr[1]->num_inputs;
float *in = layer_ptr[1]->inputs;
// get a vector and normalize it
for (i=0; i<num_inputs; i++)
        {
        fscanf(ifile,"%f",(in+i));
        normlength += (*(in+i))*(*(in+i));
        }
fscanf(ifile,"\n");
normlength = 1/(float)sqrt((double)normlength);
for (i=0; i< num_inputs; i++)
        {
        (*(in+i)) *= normlength;
        }


}



void Kohonen_network::process_next_pattern()
{
        layer_ptr[1]->calc_out();
}

float Kohonen_network::get_win_dist()
{
float  retval;
retval=((Kohonen_layer *)layer_ptr[1])
            ->get_win_dist();

return retval;
}
```

```
int Kohonen_network::get_win_index()
{
return ((Kohonen_layer *)layer_ptr[1])
            ->winner_index;

}
```

FLOW OF THE PROGRAM AND THE MAIN() FUNCTION

The **main()** function is contained in a file called **kohonen.cpp**, which is shown in Listing 11.4. To compile this program, you need only compile and make this main file, kohonen.cpp. Other files are included in this.

LISTING 11.4 THE MAIN IMPLEMENTATION FILE, KOHONEN.CPP FOR THE KOHONEN MAP PROGRAM

```
// kohonen.cpp        V. Rao, H. Rao
// Program to simulate a Kohonen map

#include "layerk.cpp"

#define INPUT_FILE "input.dat"
#define OUTPUT_FILE "kohonen.dat"
#define dist_tol       0.05

void main()
{

int neighborhood_size, period;
float avg_dist_per_cycle=0.0;
float dist_last_cycle=0.0;
float avg_dist_per_pattern=100.0; // for the latest cycle
float dist_last_pattern=0.0;
float total_dist;
float alpha;
unsigned startup;
int max_cycles;
int patterns_per_cycle=0;
```

```
        int total_cycles, total_patterns;



// create a network object
Kohonen_network knet;



FILE * input_file_ptr, * output_file_ptr;

// open input file for reading
if ((input_file_ptr=fopen(INPUT_FILE,"r"))==NULL)
                {
                cout << "problem opening input file\n";
                exit(1);
                }

// open writing file for writing
if ((output_file_ptr=fopen(OUTPUT_FILE,"w"))==NULL)
                {
                cout << "problem opening output file\n";
                exit(1);
                }




//  ————————————-
//       Read in an initial values for alpha, and the
//   neighborhood size.
//   Both of these parameters are decreased with
//   time. The number of cycles to execute before
//   decreasing the value of these parameters is
//            called the period. Read in a value for the
//            period.
//  ————————————-
                cout << " Please enter initial values for:\n";
                cout << "alpha (0.01-1.0),\n";
                cout << "and the neighborhood size (integer between 0
                and
```

```
                       50)\n”;
                       cout << “separated by spaces, e.g. 0.3 5 \n “;

                       cin >> alpha >> neighborhood_size ;

                       cout << “\nNow enter the period, which is the\n”;
                       cout << “number of cycles after which the values\n”;
                       cout << “for alpha the neighborhood size are
                               decremented\n”;
                       cout << “choose an integer between 1 and 500 , e.g. 50
                               \n”;

                       cin >> period;


                       // Read in the maximum number of cycles
                       // each pass through the input data file is a cycle
                       cout << “\nPlease enter the maximum cycles for the
                               simulation\n”;
                       cout << “A cycle is one pass through the data set.\n”;
                       cout << “Try a value of 500 to start with\n\n”;

                       cin >> max_cycles;



// the main loop
//
//      continue looping until the average distance is less
//              than the tolerance specified at the top of this file
//              , or the maximum number of
//              cycles is exceeded;

// initialize counters
total_cycles=0; // a cycle is once through all the input data
total_patterns=0; // a pattern is one entry in the input data



// get layer information
```

```
knet.get_layer_info();

// set up the network connections
knet.set_up_network(neighborhood_size);

// initialize the weights

// randomize weights for the Kohonen layer
// note that the randomize function for the
// Kohonen simulator generates
// weights that are normalized to length = 1
knet.randomize_weights();

// write header to output file
fprintf(output_file_ptr,
        "cycle\tpattern\twin index\tneigh_size\tavg_dist_per_pa
            tern\n");

fprintf(output_file_ptr,
        "————————————————————————\n");

// main loop

startup=1;
total_dist=0;

while (

                    (avg_dist_per_pattern > dist_tol)
                    && (total_cycles < max_cycles)

                    || (startup==1)
                    )
{
startup=0;
dist_last_cycle=0; // reset for each cycle
patterns_per_cycle=0;
// process all the vectors in the datafile
```

```
while (!feof(input_file_ptr))
        {
        knet.get_next_vector(input_file_ptr);

        // now apply it to the Kohonen network
        knet.process_next_pattern();

    dist_last_pattern=knet.get_win_dist();

    // print result to output file
    fprintf(output_file_ptr,"%i\t%i\t%i\t\t%i\t\t%f\n",
        total_cycles,total_patterns,knet.get_win_index(),
        neighborhood_size,avg_dist_per_pattern);

        total_patterns++;

        // gradually reduce the neighborhood size
        // and the gain, alpha
        if (((total_cycles+1) % period) == 0)
                {
                if (neighborhood_size > 0)
                        neighborhood_size -;
                knet.update_neigh_size(neighborhood_size);
                if (alpha>0.1)
                        alpha -= (float)0.1;
                }


        patterns_per_cycle++;
        dist_last_cycle += dist_last_pattern;
        knet.update_weights(alpha);
        dist_last_pattern = 0;
  }

avg_dist_per_pattern= dist_last_cycle/patterns_per_cycle;
total_dist += dist_last_cycle;
total_cycles++;
```

```
        fseek(input_file_ptr, 0L, SEEK_SET); // reset the file
                                    pointer
                                    // to the beginning of
                                    // the file


} // end main loop

cout << "\n\n\n\n\n\n\n\n\n\n\n";
cout << "——————————————————-\n";
cout << "        done \n";


avg_dist_per_cycle= total_dist/total_cycles;



cout << "\n";
cout << "—>average dist per cycle = " << avg_dist_per_cycle
        << " <--\n";
cout << "—>dist last cycle = " << dist_last_cycle << " <--
        \n";
cout << "->dist last cycle per pattern= " <<
        avg_dist_per_pattern << " <--\n";
cout << "———>total cycles = " << total_cycles << " <--\n";
cout << "———>total patterns = " << total_patterns << " <--
        \n";
cout << "——————————————————-\n";
// close the input file
fclose(input_file_ptr);
}
```

Flow of the Program

The flow of the program is very similar to the backpropagation simulator. The criterion for ending the simulation in the Kohonen program is the *average winner distance*. This is a Euclidean distance measure between the input vector and the winner's weight vector. This distance is the square root of the sum of the squares of the differences between individual vector components between the two vectors.

RESULTS FROM RUNNING THE KOHONEN PROGRAM

Once you compile the program, you need to create an input file to try it. We will first use a very simple input file and examine the results.

A Simple First Example

Let us create an input file, input.dat, which contains only two arbitrary vectors:

0.4 0.98 0.1 0.2

0.5 0.22 0.8 0.9

NOTE The file contains two four-dimensional vectors. We expect to see output that contains a different winner neuron for each of these patterns. If this is the case, then the Kohonen map has assigned different categories for each of the input vectors, and, in the future, you can expect to get the same winner classification for vectors that are close to or equal to these vectors.

By running the Kohonen map program, you will see the following output (user input is italic):

```
Please enter initial values for:
alpha (0.01-1.0),
and the neighborhood size (integer between 0 and 50)
separated by spaces, e.g. 0.3 5
0.3 5
Now enter the period, which is the
number of cycles after which the values
for alpha the neighborhood size are decremented
choose an integer between 1 and 500 , e.g. 50
50
Please enter the maximum cycles for the simulation
A cycle is one pass through the data set.
Try a value of 500 to start with
500

Enter in the layer sizes separated by spaces.
```

```
A Kohonen network has an input layer
followed by a Kohonen (output) layer
4 10
```

```
              done
```

```
—>average dist per cycle = 0.544275 <—-
—>dist last cycle = 0.0827523 <—-
->dist last cycle per pattern= 0.0413762 <—-
——————>total cycles = 11 <—-
——————>total patterns = 22 <—-
```

The layer sizes are given as 4 for the input layer and 10 for the Kohonen layer. You should choose the size of the Kohonen layer to be larger than the number of distinct patterns that you think are in the input data set. One of the outputs reported on the screen is the distance for the last cycle per pattern. This value is listed as 0.04, which is less than the terminating value set at the top of the kohonen.cpp file of 0.05. The map converged on a solution. Let us look at the file, kohonen.dat, the output file, to see the mapping to winner indexes:

cycle	pattern	win index	neigh_size	avg_dist_per_pattern
0	0	1	5	100.000000
0	1	3	5	100.000000
1	2	1	5	0.304285
1	3	3	5	0.304285
2	4	1	5	0.568255
2	5	3	5	0.568255
3	6	1	5	0.542793
3	7	8	5	0.542793
4	8	1	5	0.502416
4	9	8	5	0.502416
5	10	1	5	0.351692
5	11	8	5	0.351692
6	12	1	5	0.246184

6	13	8	5	0.246184
7	14	1	5	0.172329
7	15	8	5	0.172329
8	16	1	5	0.120630
8	17	8	5	0.120630
9	18	1	5	0.084441
9	19	8	5	0.084441
10	20	1	5	0.059109
10	21	8	5	0.059109

In this example, the neighborhood size stays at its initial value of 5. In the first column you see the cycle number, and in the second the pattern number. Since there are two patterns per cycle, you see the cycle number repeated twice for each cycle.

N O T E

The Kohonen map was able to find two distinct winner neurons for each of the patterns. One has winner index 1 and the other index 8.

Orthogonal Input Vectors Example

For a second example, look at Figure 11.5, where we choose input vectors on a two-dimensional unit circle that are 90° apart. The input.dat file should look like the following:

```
1 0
0 1
-1 0
0 -1
```

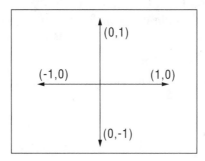

FIGURE 11.5 ORTHOGONAL INPUT VECTORS.

Using the same parameters for the Kohonen network, but with layer sizes of 2 and 10, what result would you expect? The output file, kohonen.dat, follows:

cycle	pattern	win index	neigh_size	avg_dist_per_pattern
0	0	4	5	100.000000
0	1	0	5	100.000000
0	2	9	5	100.000000
0	3	3	5	100.000000
1	4	4	5	0.444558
1	5	0	5	0.444558
497	1991	6	0	0.707107
498	1992	0	0	0.707107
498	1993	0	0	0.707107
498	1994	6	0	0.707107
498	1995	6	0	0.707107
499	1996	0	0	0.707107
499	1997	0	0	0.707107
499	1998	6	0	0.707107
499	1999	6	0	0.707107

You can see that this example doesn't quite work. Even though the neighborhood size gradually got reduced to zero, the four inputs did not get categorized to different outputs. The winner distance became stuck at the value of 0.707, which is the distance from a vector at 45°. In other words, the map generalizes a little too much, arriving at the middle value for all of the input vectors.

You can fix this problem by starting with a smaller neighborhood size, which provides for less generalization. By using the same parameters and a neighborhood size of 2, the following output is obtained.

cycle	pattern	win index	neigh_size	avg_dist_per_pattern
0	0	5	2	100.000000
0	1	6	2	100.000000
0	2	4	2	100.000000

0	3	9	2	100.000000
1	4	0	2	0.431695
1	5	6	2	0.431695
1	6	3	2	0.431695
1	7	9	2	0.431695
2	8	0	2	0.504728
2	9	6	2	0.504728
2	10	3	2	0.504728
2	11	9	2	0.504728
3	12	0	2	0.353309
3	13	6	2	0.353309
3	14	3	2	0.353309
3	15	9	2	0.353309
4	16	0	2	0.247317
4	17	6	2	0.247317
4	18	3	2	0.247317
4	19	9	2	0.247317
5	20	0	2	0.173122
5	21	6	2	0.173122
5	22	3	2	0.173122
5	23	9	2	0.173122
6	24	0	2	0.121185
6	25	6	2	0.121185
6	26	3	2	0.121185
6	27	9	2	0.121185
7	28	0	2	0.084830
7	29	6	2	0.084830
7	30	3	2	0.084830
7	31	9	2	0.084830
8	32	0	2	0.059381
8	33	6	2	0.059381
8	34	3	2	0.059381
8	35	9	2	0.059381

For this case, the network quickly converges on a unique winner for each of the four input patterns, and the distance criterion is below the set criterion within eight cycles. You can experiment with other input data sets and combinations of Kohonen network parameters.

VARIATIONS AND APPLICATIONS OF KOHONEN NETWORKS

There are many variations of the Kohonen network. Some of these will be briefly discussed in this section.

Using a Conscience

DeSieno has used a *conscience* factor in a Kohonen network. For a winning neuron, if the neuron is winning more than a fair share of the time (roughly more than $1/n$, where n is the number of neurons), then this neuron has a threshold that is applied temporarily to allow other neurons the chance to win. The purpose of this modification is to allow more uniform weight distribution while learning is taking place.

LVQ: Learning Vector Quantizer

You have read about LVQ (Learning Vector Quantizer) in previous chapters. In light of the Kohonen map, it should be pointed out that the LVQ is simply a supervised version of the Kohonen network. Inputs and expected output categories are presented to the network for training. You get data clustered, just as a Kohonen network, according to the similarity to other data inputs.

Counterpropagation Network

A neural network topology, called a *counterpropagation* network, is a combination of a Kohonen layer with a Grossberg layer. This network was developed by Robert Hecht-Nielsen and is useful for prototyping of systems, with a fairly rapid training time compared to backpropagation. The Kohonen layer provides for categorization, while the Grossberg layer allows for Hebbian conditioned learning. Counterpropagation has been used successfully in data compression applications for images. Compression ratios of 10:1 to 100:1 have been obtained, using a lossy compression scheme that codes the image with a technique called *vector quantization*, where the image is broken up into representative subimage vectors. The statistics of these vectors is such that you find that a large part of the image can be adequately represented by a subset of all the vectors. The vectors with the highest frequency of occurrence are coded with the shortest bit strings, hence you achieve data compression.

Application to Speech Recognition

Kohonen created a *phonetic typewriter* by classifying speech waveforms of different phonemes of Finnish speech into different categories using a Kohonen SOM. The Kohonen phoneme map used 50 samples of each phoneme for calibration. These samples caused excitation in a neighborhood of cells more strongly than in other cells. A neighborhood was labeled with the particular phoneme that caused excitation. For an utterance of speech made to the network, the exact neighborhoods that were active during the utterance were noted, and for how long, and in what sequence. Short excitations were taken as transitory sounds. The information obtained from the network was then pieced together to find out the words in the utterance made to the network.

SUMMARY

In this chapter, you have learned about one of the important types of competitive learning called Kohonen feature map. The most significant points of this discussion are outlined as follows:

- The Kohonen feature map is an example of an unsupervised neural network that is mainly used as a classifier system or data clustering system. As more inputs are presented to this network, the network improves its learning and is able to adapt to changing inputs.

- The training law for the Kohonen network tries to align the weight vectors along the same direction as input vectors.

- The Kohonen network models lateral competition as a form of self-organization. One winner neuron is derived for each input pattern to categorize that input.

- Only neurons within a certain distance (neighborhood) from the winner are allowed to participate in training for a given input pattern.

CHAPTER 12

APPLICATION TO PATTERN RECOGNITION

USING THE KOHONEN FEATURE MAP

In this chapter, you will use the Kohonen program developed in Chapter 11 to recognize patterns. You will modify the Kohonen program for the display of patterns.

An Example Problem: Character Recognition

The problem that is presented in this chapter is to recognize or categorize alphabetic characters. You will input various alphabetic characters to a Kohonen map and train the network to recognize these as separate categories. This program can be used to try other experiments that will be discussed at the end of this chapter.

Representing Characters

Each character is represented by a 5x7 grid of pixels. We use the graphical printing characters of the IBM extended ASCII character set to show a grayscale output for each pixel. To represent the letter A, for example, you could use the pattern shown in Figure 12.1. Here the blackened boxes represent value 1, while empty boxes represent a zero. You can represent all characters this way, with a binary map of 35 pixel values.

The letter A is represented by the values:

```
0 0 1 0 0
0 1 0 1 0
1 0 0 0 1
1 0 0 0 1
1 1 1 1 1
1 0 0 0 1
1 0 0 0 1
```

For use in the Kohonen program, we need to serialize the rows, so that all entries appear on one line.

For the characters A and X you would end up with the following entries in the input file, input.dat:

```
0 0 1 0 0   0 1 0 1 0  1 0 0 0 1  1 0 0 0 1  1 1 1 1 1  1 0 0 0 1  1 0 0 0 1  << the letter A
1 0 0 0 1   0 1 0 1 0  0 0 1 0 0  0 0 1 0 0  0 0 1 0 0  0 1 0 1 0  1 0 0 0 1  << the letter X
```

FIGURE 12.1 REPRESENTATION OF THE LETTER A WITH A 5X7 PATTERN.

Monitoring the Weights

We will present the Kohonen map with many such characters and find the response in output. You will be able to watch the Kohonen map as it goes through its cycles and learns the input patterns. At the same time, you should be able to watch the weight vectors for the winner neurons to see the pattern that is developing in the weights. Remember that for a Kohonen map the weight vectors tend to become aligned with the input vectors. So after a while, you will notice that the weight vector for the input will resemble the input pattern that you are categorizing.

Representing the Weight Vector

Although on and off values are fine for the input vectors mentioned, you need to see grayscale values for the weight vector. This can be accomplished by quantizing the weight vector into four bins, each represented by a different ASCII graphic character, as shown in Table 12.1.

TABLE 12.1 QUANTIZING THE WEIGHT VECTOR

<= 0	White rectangle (space)
0 < weight <= 0.25	Light-dotted rectangle
0.25 < weight <= 0.50	Medium-dotted rectangle
0.50 < weight <= 0.75	Dark-dotted rectangle
weight > 0.75	Black rectangle

The ASCII values for the graphics characters to be used are listed in Table 12.2.

TABLE 12.2 ASCII VALUES FOR RECTANGLE GRAPHIC CHARACTERS

White rectangle	255
Light-dotted rectangle	176
Medium-dotted rectangle	177
Dark-dotted rectangle	178
Black rectangle	219

C++ CODE DEVELOPMENT

The changes to the Kohonen program are relatively minor. The following listing indicates these changes.

Changes to the Kohonen Program

The first change to make is to the Kohonen_network class definition. This is in the file, layerk.h, shown in Listing 12.1.

LISTING 12.1 UPDATED LAYERK.H FILE

```
class Kohonen_network

{
```

```
private:

        layer *layer_ptr[2];
        int layer_size[2];
        int neighborhood_size;

public:
        Kohonen_network();
        ~Kohonen_network();
        void get_layer_info();
        void set_up_network(int);
        void randomize_weights();
        void update_neigh_size(int);
        void update_weights(const float);
        void list_weights();
        void list_outputs();
        void get_next_vector(FILE *);
        void process_next_pattern();
        float get_win_dist();
        int get_win_index();
        void display_input_char();
        void display_winner_weights();

};
```

The new member functions are shown in italic. The functions
display_input_char() and **display_winner_weights()** are used to display the
input and weight maps on the screen to watch weight character map converge to the input map.

The implementation of these functions is in the file, layerk.cpp. The
portion of this file containing these functions is shown in Listing 12.2.

LISTING 12.2 ADDITIONS TO THE LAYERK.CPP IMPLEMENTATION FILE

```
void Kohonen_network::display_input_char()
{
int i, num_inputs;
unsigned char ch;
float temp;
```

```
int col=0;
float * inputptr;

num_inputs=layer_ptr[1]->num_inputs;
inputptr = layer_ptr[1]->inputs;
// we've got a 5x7 character to display

for (i=0; i<num_inputs; i++)
        {
        temp = *(inputptr);
        if (temp <= 0)
                ch=255;// blank
        else if ((temp > 0) && (temp <= 0.25))
                ch=176; // dotted rectangle -light
        else if ((temp > 0.25) && (temp <= 0.50))
                ch=177; // dotted rectangle -medium
        else if ((temp >0.50) && (temp <= 0.75))
                ch=178; // dotted rectangle -dark
        else if (temp > 0.75)
                ch=219; // filled rectangle
        printf("%c",ch); //fill a row
        col++;
        if ((col % 5)==0)
                printf("\n"); // new row
        inputptr++;
        }
printf("\n\n\n");
}

void Kohonen_network::display_winner_weights()
{
int i, k;
unsigned char ch;
float temp;
float * wmat;
int col=0;
int win_index;
int num_inputs, num_outputs;
```

```
num_inputs= layer_ptr[1]->num_inputs;
wmat = ((Kohonen_layer*)layer_ptr[1])
            ->weights;
win_index=((Kohonen_layer*)layer_ptr[1])
            ->winner_index;

num_outputs=layer_ptr[1]->num_outputs;

// we've got a 5x7 character to display

for (i=0; i<num_inputs; i++)
    {
    k= i*num_outputs;
    temp = wmat[k+win_index];
    if (temp <= 0)
        ch=255;// blank
    else if ((temp > 0) && (temp <= 0.25))
        ch=176; // dotted rectangle -light
    else if ((temp > 0.25) && (temp <= 0.50))
        ch=177; // dotted rectangle -medium
    else if ((temp > 0.50) && (temp <= 0.75))
        ch=178; // dotted rectangle -dark
    else if (temp > 0.75)
        ch=219; // filled rectangle
    printf("%c",ch); //fill a row
    col++;
    if ((col % 5)==0)
        printf("\n"); // new row
    }
printf("\n\n");
printf("————-\n");

}
```

The final change to make is to the kohonen.cpp file. The new file is called pattern.cpp and is shown in Listing 12.3.

LISTING 12.3 THE IMPLEMENTATION FILE PATTERN.CPP

```
// pattern.cpp          V. Rao, H. Rao
// Kohonen map for pattern recognition
#include "layerk.cpp"
```

```
#define INPUT_FILE  "input.dat"
#define OUTPUT_FILE "kohonen.dat"
#define dist_tol       0.001
#define wait_cycles    10000 // creates a pause to
                             // view the character maps


void main()
{

int neighborhood_size, period;
float avg_dist_per_cycle=0.0;
float dist_last_cycle=0.0;
float avg_dist_per_pattern=100.0; // for the latest cycle
float dist_last_pattern=0.0;
float total_dist;
float alpha;
unsigned startup;
int max_cycles;
int patterns_per_cycle=0;

int total_cycles, total_patterns;
int i;


// create a network object
Kohonen_network knet;


FILE * input_file_ptr, * output_file_ptr;
```

```
// open input file for reading
if ((input_file_ptr=fopen(INPUT_FILE,"r"))==NULL)
                {
                cout << "problem opening input file\n";
                exit(1);
                }


// open writing file for writing
if ((output_file_ptr=fopen(OUTPUT_FILE,"w"))==NULL)
                {
                cout << "problem opening output file\n";
                exit(1);
                }




// ——————————————-
//      Read in an initial values for alpha, and the
//   neighborhood size.
//   Both of these parameters are decreased with
//   time. The number of cycles to execute before
//   decreasing the value of these parameters is
//                called the period. Read in a value for the
//                period.
// ——————————————-
                cout << " Please enter initial values for:\n";
                cout << "alpha (0.01-1.0),\n";
                cout << "and the neighborhood size (integer between 0\
                        and 50)\n";
                cout << "separated by spaces, e.g. 0.3 5 \n ";

                cin >> alpha >> neighborhood_size ;

                cout << "\nNow enter the period, which is the\n";
                cout << "number of cycles after which the values\n";
                cout << "for alpha the neighborhood size are \
                        decremented\n";
                cout << "choose an integer between 1 and 500 , e.g. \ 50
\n";
```

```
        cin >> period;



//      Read in the maximum number of cycles
//      each pass through the input data file is a cycle
        cout << "\nPlease enter the maximum cycles for the
        simulation\n";
        cout << "A cycle is one pass through the data set.\n";
        cout << "Try a value of 500 to start with\n\n";


        cin >> max_cycles;
```

313

```
// the main loop
//
//      continue looping until the average distance is less
than
//              the tolerance specified at the top of this file
//              , or the maximum number of
//              cycles is exceeded;

// initialize counters
total_cycles=0; // a cycle is once through all the input data
total_patterns=0; // a pattern is one entry in the input data



// get layer information
knet.get_layer_info();

// set up the network connections
knet.set_up_network(neighborhood_size);

// initialize the weights

// randomize weights for the Kohonen layer
// note that the randomize function for the
// Kohonen simulator generates
// weights that are normalized to length = 1
```

314

```
knet.randomize_weights();

// write header to output file
fprintf(output_file_ptr,
        "cycle\tpattern\twin index\tneigh_size\\
                tavg_dist_per_pattern\n");

fprintf(output_file_ptr,
        "————————————————\n");



startup=1;
total_dist=0;

while (

                    (avg_dist_per_pattern > dist_tol)
                    && (total_cycles < max_cycles)

                    || (startup==1)
                    )
{
startup=0;
dist_last_cycle=0; // reset for each cycle
patterns_per_cycle=0;
// process all the vectors in the datafile



while (!feof(input_file_ptr))
        {
        knet.get_next_vector(input_file_ptr);

        // now apply it to the Kohonen network
        knet.process_next_pattern();

    dist_last_pattern=knet.get_win_dist();

    // print result to output file
```

```
fprintf(output_file_ptr,"%i\t%i\t%i\t\t%i\t\t%f\n",
    total_cycles,total_patterns,knet.get_win_index(),
    neighborhood_size,avg_dist_per_pattern);


        // display the input character and the
        // weights for the winner to see match

        knet.display_input_char();
        knet.display_winner_weights();
        // pause for a while to view the
        // character maps
        for (i=0; i<wait_cycles; i++)
                {;}

        total_patterns++;

        // gradually reduce the neighborhood size
        // and the gain, alpha
        if (((total_cycles+1) % period) == 0)
                {
                if (neighborhood_size > 0)
                        neighborhood_size -;
                knet.update_neigh_size(neighborhood_size);
                if (alpha>0.1)
                        alpha -= (float)0.1;
                }


        patterns_per_cycle++;
        dist_last_cycle += dist_last_pattern;
        knet.update_weights(alpha);
        dist_last_pattern = 0;
    }

avg_dist_per_pattern= dist_last_cycle/patterns_per_cycle;
total_dist += dist_last_cycle;
total_cycles++;
```

```
fseek(input_file_ptr, 0L, SEEK_SET); // reset the file
        pointer
                                    // to the beginning of
                                    // the file


} // end main loop

cout << "\n\n\n\n\n\n\n\n\n\n\n";
cout << "————————————————\n";
cout << "      done \n";


avg_dist_per_cycle= total_dist/total_cycles;



cout << "\n";
cout << "—>average dist per cycle = " << avg_dist_per_cycle << " <—-\n";
cout << ">dist last cycle = " << dist_last_cycle << " <     \n";
cout << "->dist last cycle per pattern= " <<
        avg_dist_per_pattern << " <—-\n";
cout << "—->total cycles = " << total_cycles << " <—-\n";
cout << "———->total patterns = " <<
        total_patterns << " <—-\n";
cout << "————————————————\n";
// close the input file
fclose(input_file_ptr);
}
```

Changes to the program are indicated in italic. Compile this program by compiling and making the pattern.cpp file, after modifying the layerk.cpp and layerk.h files, as indicated previously.

Testing the Program

Let us run the example that we have created an input file for. We have an input.dat file with the characters A and X defined. A run of the program with these inputs is shown as follows:

```
Please enter initial values for:
alpha (0.01-1.0),
and the neighborhood size (integer between 0 and 50)
separated by spaces, e.g., 0.3 5
0.3 5
Now enter the period, which is the
number of cycles after which the values
for alpha the neighborhood size are decremented
choose an integer between 1 and 500, e.g., 50
50
Please enter the maximum cycles for the simulation
A cycle is one pass through the data set.
Try a value of 500 to start with
500
Enter in the layer sizes separated by spaces.
A Kohonen network has an input layer
followed by a Kohonen (output) layer
35 100
```

The output of the program is contained in file kohonen.dat as usual. This shows the following result.

cycle	pattern	win index	neigh_size	avg_dist_per_pattern
0	0	42	5	100.000000
0	1	47	5	100.000000
1	2	42	5	0.508321
1	3	47	5	0.508321
2	4	40	5	0.742254
2	5	47	5	0.742254
3	6	40	5	0.560121
3	7	47	5	0.560121
4	8	40	5	0.392084
4	9	47	5	0.392084
5	10	40	5	0.274459
5	11	47	5	0.274459
6	12	40	5	0.192121
6	13	47	5	0.192121

7	14	40	5	0.134485
7	15	47	5	0.134485
8	16	40	5	0.094139
8	17	47	5	0.094139
9	18	40	5	0.065898
9	19	47	5	0.065898
10	20	40	5	0.046128
10	21	47	5	0.046128
11	22	40	5	0.032290
11	23	47	5	0.032290
12	24	40	5	0.022603
12	25	47	5	0.022603
13	26	40	5	0.015822
13	27	47	5	0.015822
14	28	40	5	0.011075
14	29	47	5	0.011075
15	30	40	5	0.007753
15	31	47	5	0.007753
16	32	40	5	0.005427
16	33	47	5	0.005427
17	34	40	5	0.003799
17	35	47	5	0.003799
18	36	40	5	0.002659
18	37	47	5	0.002659
19	38	40	5	0.001861
19	39	47	5	0.001861
20	40	40	5	0.001303
20	41	47	5	0.001303

The tolerance for the distance was set to be 0.001 for this program, and the program was able to converge to this value. Both of the inputs were successfully classified into two different winning output neurons. In Figures 12.2 and 12.3 you see two snapshots of the input and weight vectors that you will find with this program. The weight vector resembles the input as you can see, but it is not an exact replication.

FIGURE 12.2 SAMPLE SCREEN OUTPUT OF THE LETTER A FROM THE INPUT AND WEIGHT VECTORS.

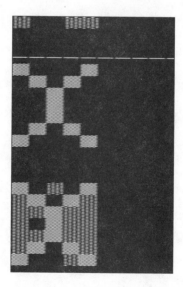

FIGURE 12.3 SAMPLE SCREEN OUTPUT OF THE LETTER X FROM THE INPUT AND WEIGHT VECTORS.

Generalization versus Memorization

As mentioned in Chapter 11, you actually don't desire the exact replication of the input pattern for the weight vector. This would amount to memorizing of the input patterns with no capacity for generalization.

For example, a typical use of this alphabet classifier system would be to use it to process noisy data, like handwritten characters. In such a case, you would need a great deal of latitude in scoping a class for a letter A.

Adding Characters

The next step of the program is to add characters and see what categories they end up in. There are many alphabetic characters that look alike, such as H and B for example. You can expect the Kohonen classifier to group these like characters into the same class.

We now modify the input.dat file to add the characters H, B, and I. The new input.dat file is shown as follows.

```
0 0 1 0 0    0 1 0 1 0   1 0 0 0 1   1 0 0 0 1   1 1 1 1 1   1 0 0 0 1   1 0 0 0 1
1 0 0 0 1    0 1 0 1 0   0 0 1 0 0   0 0 1 0 0   0 0 1 0 0   0 1 0 1 0   1 0 0 0 1
1 0 0 0 1    1 0 0 0 1   1 0 0 0 1   1 1 1 1 1   1 0 0 0 1   1 0 0 0 1   1 0 0 0 1
1 1 1 1 1    1 0 0 0 1   1 0 0 0 1   1 1 1 1 1   1 0 0 0 1   1 0 0 0 1   1 1 1 1 1
0 0 1 0 0    0 0 1 0 0   0 0 1 0 0   0 0 1 0 0   0 0 1 0 0   0 0 1 0 0   0 0 1 0 0
```

The output using this input file is shown as follows.

---

    done

—>average dist per cycle = 0.732607 <--
—>dist last cycle = 0.00360096 <--
->dist last cycle per pattern= 0.000720192 <--
———>total cycles = 37 <--
———>total patterns = 185 <--

The file kohonen.dat with the output values is now shown as follows.

cycle	pattern	win index	neigh_size	avg_dist_per_pattern
0	0	69	5	100.000000
0	1	93	5	100.000000
0	2	18	5	100.000000
0	3	18	5	100.000000
0	4	78	5	100.000000
1	5	69	5	0.806743
1	6	93	5	0.806743
1	7	18	5	0.806743
1	8	18	5	0.806743
1	9	78	5	0.806743
2	10	69	5	0.669678
2	11	93	5	0.669678
2	12	18	5	0.669678
2	13	18	5	0.669678
2	14	78	5	0.669678
3	15	69	5	0.469631
3	16	93	5	0.469631
3	17	18	5	0.469631
3	18	18	5	0.469631
3	19	78	5	0.469631
4	20	69	5	0.354791
4	21	93	5	0.354791
4	22	18	5	0.354791
4	23	18	5	0.354791
4	24	78	5	0.354791
5	25	69	5	0.282990
5	26	93	5	0.282990
5	27	18	5	0.282990
...				
35	179	78	5	0.001470
36	180	69	5	0.001029
36	181	93	5	0.001029
36	182	13	5	0.001029
36	183	19	5	0.001029
36	184	78	5	0.001029

Again, the network does not find a problem in classifying these vectors.

 Until cycle 21, both the **H** and the **B** were classified as output neuron 18. The ability to distinguish these vectors is largely due to the small tolerance we have assigned as a termination criterion.

N O T E

Other Experiments to Try

You can try other experiments with the program. For example, you can repeat the input file but with the order of the entries changed. In other words, you can present the same inputs a number of times in different order. This actually helps the Kohonen network train faster. You can try applying garbled versions of the characters to see if the network distinguishes them. Just as in the backpropagation program, you can save the weights in a weight file to freeze the state of training, and then apply new inputs. You can enter all of the characters from **A** to **Z** and see the classification that results. Do you need to train on all of the characters or a subset? You can change the size of the Kohonen layer. How many neurons do you need to recognize the complete alphabet?

There is no restriction on using digital inputs of 1 and 0 as we had used. You can apply grayscale analog values. The program will display the input pattern according to the quantization levels that were set. This set can be expanded, and you can use a graphics interface to display more levels. You can then try pattern recognition of arbitrary images, but remember that processing time will increase rapidly with the number of neurons used. The number of input neurons you choose is dictated by the image resolution, unless you filter and/or subsample the image before presenting it to the network. Filtering is the process of using a type of averaging function applied to groups of pixels. Subsampling is the process of choosing a lower-output image resolution by selecting fewer pixels than a source image. If you start with an image that is 100 x 100 pixels, you can subsample this image 2:1 in each direction to obtain an image that is one-fourth the size, or 50 x 50 pixels. Whether you throw away every other pixel to get this output resolution or apply a filter is up to you. You could average every two pixels to get one output pixel as an example of a very simple filter.

SUMMARY

The following list highlights the important Kohonen program features which you learned in this chapter.

- This chapter presented a simple character recognition program using a Kohonen feature map.
- The input vectors and the weight vectors were displayed to show convergence and note similarity between the two vectors.
- As training progresses, the weight vector for the winner neuron resembles the input character map.

CHAPTER 13

BACKPROPAGATION II

ENHANCING THE SIMULATOR

In Chapter 7, you developed a backpropagation simulator. In this chapter, you will put it to use with examples and also add some new features to the simulator: a term called *momentum*, and the capability of adding noise to the inputs during simulation. There are many variations of the algorithm that try to alleviate two problems with backpropagation. First, like other neural networks, there is a strong possibility that the solution found with backpropagation is not a global error minimum, but a local one. You may need to shake the weights a little by some means to get out of the local minimum, and possibly arrive at a lower minimum. The second problem with backpropagation is speed. The algorithm is very slow at learning. There are many proposals for speeding up the search process. Neural networks are inherently parallel processing architectures and are suited for simulation on parallel processing hardware. While there are a few plug-in neural net or digital signal processing boards available in the market, the low-cost simulation platform of choice remains the personal computer. Speed enhancements to the training algorithm are therefore very necessary.

Another Example of Using Backpropagation

Before modifying the simulator to add features, let's look at the same problem we used the Kohonen map to analyze in Chapter 12. As you recall, we would like to be able to distinguish alphabetic characters by assigning them to different bins. For backpropagation, we would apply the inputs and train the network with anticipated responses. Here is the input file that we used for distinguishing five different characters, **A**, **X**, **H**, **B**, and **I**:

```
0 0 1 0 0   0 1 0 1 0   1 0 0 0 1   1 0 0 0 1   1 1 1 1 1   1 0 0 0 1   1 0 0 0 1
1 0 0 0 1   0 1 0 1 0   0 0 1 0 0   0 0 1 0 0   0 0 1 0 0   0 1 0 1 0   1 0 0 0 1
1 0 0 0 1   1 0 0 0 1   1 0 0 0 1   1 1 1 1 1   1 0 0 0 1   1 0 0 0 1   1 0 0 0 1
1 1 1 1 1   1 0 0 0 1   1 0 0 0 1   1 1 1 1 1   1 0 0 0 1   1 0 0 0 1   1 1 1 1 1
0 0 1 0 0   0 0 1 0 0   0 0 1 0 0   0 0 1 0 0   0 0 1 0 0   0 0 1 0 0   0 0 1 0 0
```

Each line has a 5x7 dot representation of each character. Now we need to name each of the output categories. We can assign a simple 3-bit representation as follows:

A	000
X	010
H	100
B	101
I	111

Let's train the network to recognize these characters. The **training.dat** file looks like the following.

```
0 0 1 0 0   0 1 0 1 0   1 0 0 0 1   1 0 0 0 1   1 1 1 1 1   1 0 0 0 1   1 0 0 0 1   0 0 0
1 0 0 0 1   0 1 0 1 0   0 0 1 0 0   0 0 1 0 0   0 0 1 0 0   0 1 0 1 0   1 0 0 0 1   0 1 0
1 0 0 0 1   1 0 0 0 1   1 0 0 0 1   1 1 1 1 1   1 0 0 0 1   1 0 0 0 1   1 0 0 0 1   1 0 0
1 1 1 1 1   1 0 0 0 1   1 0 0 0 1   1 1 1 1 1   1 0 0 0 1   1 0 0 0 1   1 1 1 1 1   1 0 1
0 0 1 0 0   0 0 1 0 0   0 0 1 0 0   0 0 1 0 0   0 0 1 0 0   0 0 1 0 0   0 0 1 0 0   1 1 1
```

Now you can start the simulator. Using the parameters (beta = 0.1, tolerance = 0.001, and max_cycles = 1000) and with three layers of size 35 (input), 5 (middle), and 3 (output), you will get a typical result like the following.

```
                         --
        done:   results in file output.dat
                        training: last vector only
                        not training: full cycle

                        weights saved in file weights.dat

—>average error per cycle = 0.035713<--
—>error last cycle = .0.008223 <--
->error last cycle per pattern= 0.00164455 <--
```

```
———>total cycles = 1000 <—-
———>total patterns = 5000 <—-
```

The simulator stopped at the 1000 maximum cycles specified in this case. Your results will be different since the weights start at a random point. Note that the tolerance specified was nearly met. Let us see how close the output came to what we wanted. Look at the **output.dat** file. You can see the match for the last pattern as follows:

```
for input vector:
0.000000   0.000000   1.000000   0.000000   0.000000   0.000000   0.000000
1.000000   0.000000   0.000000   0.000000   0.000000   1.000000   0.000000
0.000000   0.000000   0.000000   1.000000   0.000000   0.000000   0.000000
0.000000   1.000000   0.000000   0.000000   0.000000   0.000000   1.000000
0.000000   0.000000   0.000000   0.000000   1.000000   0.000000   0.000000
output vector is:
0.999637   0.998721   0.999330
expected output vector is:
1.000000   1.000000   1.000000
```

To see the outputs of all the patterns, we need to copy the **training.dat** file to the **test.dat** file and rerun the simulator in **Test** mode. Remember to delete the expected output field once you copy the file.

Running the simulator in **Test** mode (0) shows the following result in the output.dat file:

```
for input vector:
0.000000   0.000000   1.000000   0.000000   0.000000   0.000000   1.000000
0.000000   1.000000   0.000000   1.000000   0.000000   0.000000   0.000000
1.000000   1.000000   0.000000   0.000000   0.000000   1.000000   1.000000
1.000000   1.000000   1.000000   1.000000   1.000000   0.000000   0.000000
0.000000   1.000000   1.000000   0.000000   0.000000   0.000000   1.000000
output vector is:
0.005010   0.002405   0.000141
```

```
for input vector:
1.000000   0.000000   0.000000   0.000000   1.000000   0.000000   1.000000
0.000000   1.000000   0.000000   0.000000   0.000000   1.000000   0.000000
0.000000   0.000000   0.000000   1.000000   0.000000   0.000000   0.000000
0.000000   1.000000   0.000000   0.000000   0.000000   1.000000   0.000000
1.000000   0.000000   1.000000   0.000000   0.000000   0.000000   1.000000
```

```
output vector is:
0.001230  0.997844  0.000663
```

```
for input vector:
1.000000  0.000000  0.000000  0.000000  1.000000  1.000000  0.000000
0.000000  0.000000  1.000000  1.000000  0.000000  0.000000  0.000000
1.000000  1.000000  1.000000  1.000000  1.000000  1.000000  1.000000
0.000000  0.000000  0.000000  1.000000  1.000000  0.000000  0.000000
0.000000  1.000000  1.000000  0.000000  0.000000  0.000000  1.000000
output vector is:
0.995348  0.000253  0.002677
```

```
for input vector:
1.000000  1.000000  1.000000  1.000000  1.000000  1.000000  0.000000
0.000000  0.000000  1.000000  1.000000  0.000000  0.000000  0.000000
1.000000  1.000000  1.000000  1.000000  1.000000  1.000000  1.000000
0.000000  0.000000  0.000000  1.000000  1.000000  0.000000  0.000000
0.000000  1.000000  1.000000  1.000000  1.000000  1.000000  1.000000
output vector is:
0.999966  0.000982  0.997594
```

```
for input vector:
0.000000  0.000000  1.000000  0.000000  0.000000  0.000000  0.000000
1.000000  0.000000  0.000000  0.000000  0.000000  1.000000  0.000000
0.000000  0.000000  0.000000  1.000000  0.000000  0.000000  0.000000
0.000000  1.000000  0.000000  0.000000  0.000000  0.000000  1.000000
0.000000  0.000000  0.000000  0.000000  1.000000  0.000000  0.000000
output vector is:
0.999637  0.998721  0.999330
```

The training patterns are learned very well. If a smaller tolerance is used, it would be possible to complete the learning in fewer cycles. What happens if we present a foreign character to the network? Let us create a new test.dat file with two entries for the letters **M** and **J**, as follows:

```
1 0 0 0 1  1 1 0 1 1  1 0 1 0 1  1 0 0 0 1  1 0 0 0 1  1 0 0 0 1  1 0 0 0 1
0 0 1 0 0  0 0 1 0 0  0 0 1 0 0  0 0 1 0 0  0 0 1 0 0  0 0 1 0 0  0 1 1 1 1
```

The results should show each foreign character in the category closest to it. The middle layer of the network acts as a feature detector. Since we specified five neurons, we have given the network the freedom to define five features in the input training set to use to categorize inputs. The results in the output.dat file are shown as follows.

```
for input vector:
1.000000  0.000000  0.000000  0.000000  1.000000  1.000000  1.000000
0.000000  1.000000  1.000000  1.000000  0.000000  1.000000  0.000000
1.000000  1.000000  0.000000  0.000000  0.000000  1.000000  1.000000
0.000000  0.000000  0.000000  1.000000  1.000000  0.000000  0.000000
0.000000  1.000000  1.000000  0.000000  0.000000  0.000000  1.000000
output vector is:
0.963513  0.000800  0.001231
```

```
for input vector:
0.000000  0.000000  1.000000  0.000000  0.000000  0.000000  0.000000
1.000000  0.000000  0.000000  0.000000  0.000000  1.000000  0.000000
0.000000  0.000000  0.000000  1.000000  0.000000  0.000000  0.000000
0.000000  1.000000  0.000000  0.000000  0.000000  0.000000  1.000000
0.000000  0.000000  0.000000  1.000000  1.000000  1.000000  1.000000
output vector is:
0.999469  0.996339  0.999157
```

In the first pattern, an **M** is categorized as an **H**, whereas in the second pattern, a **J** is categorized as an **I**, as expected. The case of the first pattern seems reasonable since the **H** and **M** share many pixels in common.

Other Experiments to Try

There are many other experiments you could try in order to get a better feel for how to train and use a backpropagation neural network.

- You could use the ASCII 8-bit code to represent each character, and try to train the network. You could also code all of the alphabetic characters and see if it's possible to distinguish all of them.

- You can garble a character, to see if you still get the correct output.

- You could try changing the size of the middle layer, and see the effect on training time and generalization ability.

- You could change the tolerance setting to see the difference between an overtrained and undertrained network in generalization capability. That is, given a foreign pattern, is the network able to find the closest match and use that particular category, or does it arrive at a new category altogether?

We will return to the same example after enhancing the simulator with momentum and noise addition capability.

Adding the Momentum Term

A simple change to the training law that sometimes results in much faster training is the addition of a *momentum* term. The training law for backpropagation as implemented in the simulator is:

```
Weight change = Beta * output_error * input
```

Now we add a term to the weight change equation as follows:

```
Weight change = Beta * output_error * input +
                Alpha*previous_weight_change
```

The second term in this equation is the momentum term. The weight change, in the absence of error, would be a constant multiple by the previous weight change. In other words, the weight change continues in the direction it was heading. The momentum term is an attempt to try to keep the weight change process moving, and thereby not get stuck in local minimas.

Code Changes

The effected files to implement this change are the layer.cpp file, to modify the **update_weights()** member function of the **output_layer** class, and the main backprop.cpp file to read in the value for **alpha** and pass it to the **member** function. There is some additional storage needed for storing previous weight changes, and this affects the layer.h file. The momentum term could be implemented in two ways:

1. Using the weight change for the previous pattern.
2. Using the weight change accumulated over the previous cycle.

Although both of these implementations are valid, the second is particularly useful, since it adds a term that is significant for all patterns, and hence would contribute to global error reduction. We implement the second choice by accumulating the value of the current cycle weight changes in a vector called **cum_deltas**. The past cycle weight changes are stored in a vector called **past_deltas**. These are shown as follows in a portion of the layer.h file.

```
class output_layer:    public layer
{
protected:

        float * weights;
        float * output_errors; // array of errors at output
        float * back_errors; // array of errors back-propagated
        float * expected_values;       // to inputs
        float * cum_deltas;    // for momentum
        float * past_deltas;   // for momentum

    friend network;
    ...
```

Changes to the layer.cpp File

The implementation file for the **layer** class changes in the **output_layer::update_weights()** routine and the constructor and destructor for **output_layer**. First, here is the constructor for **output_layer**. Changes are highlighted in italic.

```
output_layer::output_layer(int ins, int outs)
{
int i, j, k;
num_inputs=ins;
num_outputs=outs;
weights = new float[num_inputs*num_outputs];
output_errors = new float[num_outputs];
back_errors = new float[num_inputs];
outputs = new float[num_outputs];
expected_values = new float[num_outputs];
cum_deltas = new float[num_inputs*num_outputs];
past_deltas = new float[num_inputs*num_outputs];
if ((weights==0)||(output_errors==0)||(back_errors==0)
        ||(outputs==0)||(expected_values==0)
        ||(past_deltas==0)||(cum_deltas==0))
```

```
        {
        cout << "not enough memory\n";
        cout << "choose a smaller architecture\n";
        exit(1);
        }
// zero cum_deltas and past_deltas matrix
for (i=0; i< num_inputs; i++)
        {
        k=i*num_outputs;
        for (j=0; j< num_outputs; j++)
                {
                cum_deltas[k+j]=0;
                past_deltas[k+j]=0;
                }
        }


}
```

The destructor simply deletes the new vectors:

```
output_layer::~output_layer()
{
// some compilers may require the array
// size in the delete statement; those
// conforming to Ansi C++ will not
delete [num_outputs*num_inputs] weights;
delete [num_outputs] output_errors;
delete [num_inputs] back_errors;
delete [num_outputs] outputs;
delete [num_outputs*num_inputs] past_deltas;
delete [num_outputs*num_inputs] cum_deltas;
}
```

Now let's look at the **update_weights()** routine changes:

```
void output_layer::update_weights(const float beta,
                                  const float alpha)
{
int i, j, k;
```

```
float delta;
// learning law: weight_change =
//              beta*output_error*input + alpha*past_delta
for (i=0; i< num_inputs; i++)
        {
        k=i*num_outputs;
        for (j=0; j< num_outputs; j++)
                {
                delta=beta*output_errors[j]*(*(inputs+i))
                        +alpha*past_deltas[k+j];
                weights[k+j] += delta;
                cum_deltas[k+j]+=delta; // current cycle
                }


        }


}
```

The change to the training law amounts to calculating a delta and adding it to the cumulative total of weight changes in **cum_deltas**. At some point (at the start of a new cycle) you need to set the **past_deltas** vector to the **cum_delta** vector. Where does this occur? Since the layer has no concept of cycle, this must be done at the network level. There is a network level function called **update_momentum** at the beginning of each cycle that in turns calls a layer level function of the same name. The layer level function swaps the **past_deltas** vector and the **cum_deltas** vector, and reinitializes the **cum_deltas** vector to zero. We need to return to the layer.h file to see changes that are needed to define the two functions mentioned.

```
class output_layer:    public layer
{
protected:

        float * weights;
        float * output_errors; // array of errors at output
        float * back_errors; // array of errors back-propagated
        float * expected_values;    // to inputs
        float * cum_deltas;    // for momentum
        float * past_deltas;   // for momentum
```

```
    friend network;

public:


        output_layer(int, int);
        ~output_layer();
        virtual void calc_out();
        void calc_error(float &);
        void randomize_weights();
        void update_weights(const float, const float);
        void update_momentum();
        void list_weights();
        void write_weights(int, FILE *);
        void read_weights(int, FILE *);
        void list_errors();
        void list_outputs();
};



class network

{

private:

layer *layer_ptr[MAX_LAYERS];
    int number_of_layers;
    int layer_size[MAX_LAYERS];
    float *buffer;
    fpos_t position;
    unsigned training;

public:
  network();
    ~network();
                void set_training(const unsigned &);
                unsigned get_training_value();
                void get_layer_info();
```

```
void set_up_network();
void randomize_weights();
void update_weights(const float, const float);
void update_momentum();
...
```

At both the **network** and **output_layer** class levels the function prototype
for the **update_momentum** member functions are highlighted. The implementation for these functions are shown as follows from the **layer.cpp** class.

```
void output_layer::update_momentum()
{
// This function is called when a
// new cycle begins; the past_deltas
// pointer is swapped with the
// cum_deltas pointer. Then the contents
// pointed to by the cum_deltas pointer
// is zeroed out.
int i, j, k;
float * temp;

// swap
temp = past_deltas;
past_deltas=cum_deltas;
cum_deltas=temp;

// zero cum_deltas matrix
// for new cycle
for (i=0; i< num_inputs; i++)
        {
        k=i*num_outputs;
        for (j=0; j< num_outputs; j++)
                cum_deltas[k+j]=0;
        }
}

void network::update_momentum()
{
int i;
```

```
for (i=1; i<number_of_layers; i++)
        ((output_layer *)layer_ptr[i])
                ->update_momentum();
}
```

Adding Noise During Training

Another approach to breaking out of local minima as well as to enhance generalization ability is to introduce some noise in the inputs during training. A random number is added to each input component of the input vector as it is applied to the network. This is scaled by an overall noise factor, **NF**, which has a 0 to 1 range. You can add as much noise to the simulation as you want, or not any at all, by choosing **NF** = 0. When you are close to a solution and have reached a satisfactory minimum, you don't want noise at that time to interfere with convergence to the minimum. We implement a noise factor that decreases with the number of cycles, as shown in the following excerpt from the backprop.cpp file.

```
// update NF
// gradually reduce noise to zero
if (total_cycles>0.7*max_cycles)
                new_NF = 0;
else if (total_cycles>0.5*max_cycles)
                new_NF = 0.25*NF;
else if (total_cycles>0.3*max_cycles)
                new_NF = 0.50*NF;
else if (total_cycles>0.1*max_cycles)
                new_NF = 0.75*NF;

backp.set_NF(new_NF);
```

The noise factor is reduced at regular intervals. The new noise factor is updated with the network class function called **set_NF(float)**. There is a member variable in the network class called **NF** that holds the current value for the noise factor. The noise is added to the inputs in the **input_layer** member function **calc_out()**.

Another reason for using noise is to prevent memorization by the network. You are effectively presenting a different input pattern with each cycle so it becomes hard for the network to memorize patterns.

ONE OTHER CHANGE—STARTING TRAINING FROM A SAVED WEIGHT FILE

Shortly, we will look at the complete listings for the backpropagation simulator. There is one other enhancement to discuss. It is often useful in long simulations to be able to start from a known point, which is from an already saved set of weights. This is a simple change in the backprop.cpp program, which is well worth the effort. As a side benefit, this feature will allow you to run a simulation with a large beta value for, say, 500 cycles, save the weights, and then start a new simulation with a smaller beta value for another 500 or more cycles. You can take preset breaks in long simulations, which you will encounter in Chapter 14. At this point, let's look at the complete listings for the updated layer.h and layer.cpp files in Listings 13.1 and 13.2:

LISTING 13.1 LAYER.H FILE UPDATED TO INCLUDE NOISE AND MOMENTUM

```
// layer.h              V.Rao, H. Rao
// header file for the layer class hierarchy and
// the network class
 // added noise and momentum


#define MAX_LAYERS    5
#define MAX_VECTORS   100


class network;
class Kohonen_network;

class layer
{

protected:

        int num_inputs;
        int num_outputs;
        float *outputs;// pointer to array of outputs
        float *inputs; // pointer to array of inputs, which
                            // are outputs of some other layer
```

```
        friend network;
        friend Kohonen_network; // update for Kohonen model

public:

        virtual void calc_out()=0;
};



class input_layer: public layer
{

private:

float noise_factor;
float * orig_outputs;

public:

        input_layer(int, int);
        ~input_layer();
        virtual void calc_out();
        void set_NF(float);

        friend network;
};

class middle_layer;

class output_layer:    public layer
{
protected:

        float * weights;
        float * output_errors; // array of errors at output
        float * back_errors; // array of errors back-propagated
        float * expected_values;      // to inputs
        float * cum_deltas;    // for momentum
```

```
        float * past_deltas;   // for momentum

    friend network;

public:


        output_layer(int, int);
        ~output_layer();
        virtual void calc_out();
        void calc_error(float &);
        void randomize_weights();
        void update_weights(const float, const float);
        void update_momentum();
        void list_weights();
        void write_weights(int, FILE *);
        void read_weights(int, FILE *);
        void list_errors();
        void list_outputs();
};




class middle_layer:    public output_layer
{

private:

public:
    middle_layer(int, int);
    ~middle_layer();
        void calc_error();
};


class network

{
```

```
    private:

layer *layer_ptr[MAX_LAYERS];
    int number_of_layers;
    int layer_size[MAX_LAYERS];
    float *buffer;
    fpos_t position;
    unsigned training;

public:
    network();
    ~network();
                void set_training(const unsigned &);
                unsigned get_training_value();
                void get_layer_info();
                void set_up_network();
                void randomize_weights();
                void update_weights(const float, const float);
                void update_momentum();
                void write_weights(FILE *);
                void read_weights(FILE *);
                void list_weights();
                void write_outputs(FILE *);
                void list_outputs();
                void list_errors();
                void forward_prop();
                void backward_prop(float &);
                int fill_IObuffer(FILE *);
                void set_up_pattern(int);
                void set_NF(float);

    };
```

LISTING 13.2 LAYER.CPP FILE UPDATED TO INCLUDE NOISE AND MOMENTUM

```
// layer.cpp          V.Rao, H.Rao
// added momentum and noise


// compile for floating point hardware if available
```

```
#include <stdio.h>
#include <iostream.h>
#include <stdlib.h>
#include <math.h>
#include <time.h>
#include "layer.h"



inline float squash(float input)
// squashing function
// use sigmoid — can customize to something
// else if desired; can add a bias term too
//
{
if (input < -50)
        return 0.0;
else    if (input > 50)
                return 1.0;
        else return (float)(1/(1+exp(-(double)input)));

}



inline float randomweight(unsigned init)
{
int num;
// random number generator
// will return a floating point
// value between -1 and 1

if (init==1)    // seed the generator
        srand ((unsigned)time(NULL));

num=rand() % 100;

return 2*(float(num/100.00))-1;
}


// the next function is needed for Turbo C++
```

```
// and Borland C++ to link in the appropriate
// functions for fscanf floating point formats:
static void force_fpf()
{
        float x, *y;
        y=&x;
        x=*y;
}


// ——————————————--
//                              input layer
//——————————————
input_layer::input_layer(int i, int o)
{



num_inputs=i;
num_outputs=o;

outputs = new float[num_outputs];
orig_outputs = new float[num_outputs];
if ((outputs==0)||(orig_outputs==0))
        {
        cout << "not enough memory\n";
        cout << "choose a smaller architecture\n";
        exit(1);
        }

noise_factor=0;

}

input_layer::~input_layer()
{
delete [num_outputs] outputs;
delete [num_outputs] orig_outputs;
}
```

```
void input_layer::calc_out()
{
//add noise to inputs
// randomweight returns a random number
// between -1 and 1

int i;
for (i=0; i<num_outputs; i++)
        outputs[i] =orig_outputs[i]*
                (1+noise_factor*randomweight(0));

}


void input_layer::set_NF(float noise_fact)
{
noise_factor=noise_fact;
}




// ————————————-
//                                   output layer
//————————————




output_layer::output_layer(int ins, int outs)
{
int i, j, k;
num_inputs=ins;
num_outputs=outs;
weights = new float[num_inputs*num_outputs];
output_errors = new float[num_outputs];
back_errors = new float[num_inputs];
outputs = new float[num_outputs];
expected_values = new float[num_outputs];
```

```
cum_deltas = new float[num_inputs*num_outputs];
past_deltas = new float[num_inputs*num_outputs];

if ((weights==0)||(output_errors==0)||(back_errors==0)
        ||(outputs==0)||(expected_values==0)
        ||(past_deltas==0)||(cum_deltas==0))
        {
        cout << "not enough memory\n";
        cout << "choose a smaller architecture\n";
        exit(1);
        }

// zero cum_deltas and past_deltas matrix
for (i=0; i< num_inputs; i++)
        {
        k=i*num_outputs;
        for (j=0; j< num_outputs; j++)
                {
                cum_deltas[k+j]=0;
                past_deltas[k+j]=0;
                }
        }
}


output_layer::~output_layer()
{
// some compilers may require the array
// size in the delete statement; those
// conforming to Ansi C++ will not
delete [num_outputs*num_inputs] weights;
delete [num_outputs] output_errors;
delete [num_inputs] back_errors;
delete [num_outputs] outputs;
delete [num_outputs*num_inputs] past_deltas;
delete [num_outputs*num_inputs] cum_deltas;

}
```

```
void output_layer::calc_out()
{

int i,j,k;
float accumulator=0.0;



for (j=0; j<num_outputs; j++)
        {

        for (i=0; i<num_inputs; i++)

                {
                k=i*num_outputs;
                if (weights[k+j]*weights[k+j] > 1000000.0)
                        {
                        cout << "weights are blowing up\n";
                        cout << "try a smaller learning constant\n";
                        cout << "e.g. beta=0.02    aborting...\n";
                        exit(1);
                        }
                outputs[j]=weights[k+j]*(*(inputs+i));
                accumulator+=outputs[j];
                }
        // use the sigmoid squash function
        outputs[j]=squash(accumulator);
        accumulator=0;
        }

}



void output_layer::calc_error(float & error)
{
int i, j, k;
float accumulator=0;
```

```
float total_error=0;

for (j=0; j<num_outputs; j++)
    {
            output_errors[j] = expected_values[j]-outputs[j];
            total_error+=output_errors[j];
            }

error=total_error;

for (i=0; i<num_inputs; i++)
{
k=i*num_outputs;
for (j=0; j<num_outputs; j++)
        {
            back_errors[i]=
                    weights[k+j]*output_errors[j];
            accumulator+=back_errors[i];
            }
        back_errors[i]=accumulator;
        accumulator=0;
        // now multiply by derivative of
        // sigmoid squashing function, which is
        // just the input*(1-input)
        back_errors[i]*=(*(inputs+i))*(1-(*(inputs+i)));
        }

}


void output_layer::randomize_weights()
{
int i, j, k;
const unsigned first_time=1;

const unsigned not_first_time=0;
float discard;

discard=randomweight(first_time);
```

```
for (i=0; i< num_inputs; i++)
        {
        k=i*num_outputs;
        for (j=0; j< num_outputs; j++)
                weights[k+j]=randomweight(not_first_time);
        }
}


void output_layer::update_weights(const float beta,
                                  const float alpha)
{
int i, j, k;
float delta;

// learning law: weight_change =
//              beta*output_error*input + alpha*past_delta

for (i=0; i< num_inputs; i++)
        {
        k=i*num_outputs;
        for (j=0; j< num_outputs; j++)
                {
                delta=beta*output_errors[j]*(*(inputs+i))
                        +alpha*past_deltas[k+j];
                weights[k+j] += delta;
                cum_deltas[k+j]+=delta; // current cycle
                }

        }

}


void output_layer::update_momentum()
{
// This function is called when a
// new cycle begins; the past_deltas
// pointer is swapped with the
// cum_deltas pointer. Then the contents
```

```
// pointed to by the cum_deltas pointer
// is zeroed out.
int i, j, k;
float * temp;

// swap
temp = past_deltas;
past_deltas=cum_deltas;
cum_deltas=temp;

// zero cum_deltas matrix
// for new cycle
for (i=0; i< num_inputs; i++)
        {
        k=i*num_outputs;
        for (j=0; j< num_outputs; j++)
                cum_deltas[k+j]=0;
        }
}

void output_layer::list_weights()
{
int i, j, k;

for (i=0; i< num_inputs; i++)
        {
        k=i*num_outputs;
        for (j=0; j< num_outputs; j++)
                cout << "weight["<<i<<","<<
                        j<<"] is: "<<weights[k+j];
        }

}

void output_layer::list_errors()
{
int i, j;

for (i=0; i< num_inputs; i++)
```

```
              cout << "backerror["<<i<<
                  "] is : "<<back_errors[i]<<"\n";

for (j=0; j< num_outputs; j++)
        cout << "outputerrors["<<j<<
                     "] is: "<<output_errors[j]<<"\n";


}



void output_layer::write_weights(int layer_no,
              FILE * weights_file_ptr)
{
int i, j, k;

// assume file is already open and ready for
// writing

// prepend the layer_no to all lines of data
// format:
//           layer_no       weight[0,0] weight[0,1] ...
//           layer_no       weight[1,0] weight[1,1] ...
//           ...

for (i=0; i< num_inputs; i++)
        {
        fprintf(weights_file_ptr,"%i ",layer_no);
        k=i*num_outputs;
     for (j=0; j< num_outputs; j++)
        {
        fprintf(weights_file_ptr,"%f ",
                    weights[k+j]);
        }
     fprintf(weights_file_ptr,"\n");
     }



}
```

```
void output_layer::read_weights(int layer_no,
            FILE * weights_file_ptr)
{
int i, j, k;


// assume file is already open and ready for
// reading

// look for the prepended layer_no
// format:
//          layer_no      weight[0,0] weight[0,1] ...
//          layer_no      weight[1,0] weight[1,1] ...
//          ...
while (1)

        {

        fscanf(weights_file_ptr,"%i",&j);
        if ((j==layer_no)|| (feof(weights_file_ptr)))
                break;
        else
                {
                while (fgetc(weights_file_ptr) != '\n')
                        {;}// get rest of line
                }
        }

if (!(feof(weights_file_ptr)))
        {
        // continue getting first line
        i=0;
        for (j=0; j< num_outputs; j++)
                        {

                        fscanf(weights_file_ptr,"%f",
                                &weights[j]); // i*num_outputs = 0
        }
        fscanf(weights_file_ptr,"\n");
```

```
        // now get the other lines
        for (i=1; i< num_inputs; i++)
                {
                fscanf(weights_file_ptr,"%i",&layer_no);
        k=i*num_outputs;
        for (j=0; j< num_outputs; j++)
        {
        fscanf(weights_file_ptr,"%f",
                &weights[k+j]);
            }

    }
    fscanf(weights_file_ptr,"\n");
    }


else cout << "end of file reached\n";

}

void output_layer::list_outputs()
{
int j;

for (j=0; j< num_outputs; j++)
        {
        cout << "outputs["<<j
                <<"] is: "<<outputs[j]<<"\n";
        }


}



// ——————————————-
//                              middle layer
//——————————————


middle_layer::middle_layer(int i, int o):
```

```
        output_layer(i,o)
{

}
```

```
middle_layer::~middle_layer()
{
delete [num_outputs*num_inputs] weights;
delete [num_outputs] output_errors;
delete [num_inputs] back_errors;
delete [num_outputs] outputs;
}


void middle_layer::calc_error()
{
int i, j, k;
float accumulator=0;

for (i=0; i<num_inputs; i++)
        {
        k=i*num_outputs;
        for (j=0; j<num_outputs; j++)
                {
                back_errors[i]=
                        weights[k+j]*(*(output_errors+j));
                accumulator+=back_errors[i];
                }
        back_errors[i]=accumulator;
        accumulator=0;
        // now multiply by derivative of
        // sigmoid squashing function, which is
        // just the input*(1-input)
        back_errors[i]*=(*(inputs+i))*(1-(*(inputs+i)));
        }

}


network::network()
```

```
{
position=0L;
}


network::~network()
{
int i,j,k;
i=layer_ptr[0]->num_outputs;// inputs
j=layer_ptr[number_of_layers-1]->num_outputs; //outputs
k=MAX_VECTORS;


delete [(i+j)*k]buffer;
}


void network::set_training(const unsigned & value)
{
training=value;
}


unsigned network::get_training_value()
{
return training;
}



void network::get_layer_info()
{
int i;


//————————————
//
//       Get layer sizes for the network
//
// ————————————-



cout << " Please enter in the number of layers for your net work.\n";
```

```
cout << " You can have a minimum of 3 to a maximum of 5. \n";
cout << " 3 implies 1 hidden layer; 5 implies 3 hidden layers : \n\n";

cin >> number_of_layers;

cout << " Enter in the layer sizes separated by spaces.\n";
cout << " For a network with 3 neurons in the input layer,\n";
cout << " 2 neurons in a hidden layer, and 4 neurons in the\n";
cout << " output layer, you would enter: 3 2 4 .\n";
cout << " You can have up to 3 hidden layers,for five maximum entries
:\n\n";

for (i=0; i<number_of_layers; i++)
     {
     cin >> layer_size[i];
     }

// ─────────────────────────────
// size of layers:
//     input_layer            layer_size[0]
//     output_layer           layer_size[number_of_layers-1]
//     middle_layers          layer_size[1]
//     optional: layer_size[number_of_layers-3]
//     optional: layer_size[number_of_layers-2]
//─────────────────────────────--


}

void network::set_up_network()
{
int i,j,k;
//─────────────────────────────-
// Construct the layers
//
//─────────────────────────────-
```

```
layer_ptr[0] = new input_layer(0,layer_size[0]);

for (i=0;i<(number_of_layers-1);i++)
        {
        layer_ptr[i+1] =
        new middle_layer(layer_size[i],layer_size[i+1]);
        }

layer_ptr[number_of_layers-1] = new
output_layer(layer_size[number_of_layers-2],
layer_size[number_of_layers-1]);

for (i=0;i<(number_of_layers-1);i++)
        {
        if (layer_ptr[i] == 0)
                {
                cout << "insufficient memory\n";
                cout << "use a smaller architecture\n";
                exit(1);
                }
        }


//——————————————————-
// Connect the layers
//
//——————————————————-
// set inputs to previous layer outputs for all layers,
//           except the input layer

for (i=1; i< number_of_layers; i++)
        layer_ptr[i]->inputs = layer_ptr[i-1]->outputs;



// for back_propagation, set output_errors to next layer
//           back_errors for all layers except the output
//           layer and input layer
```

```
for (i=1; i< number_of_layers -1; i++)
        ((output_layer *)layer_ptr[i])->output_errors =
                ((output_layer *)layer_ptr[i+1])->back_errors;

// define the IObuffer that caches data from
// the datafile
i=layer_ptr[0]->num_outputs;// inputs
j=layer_ptr[number_of_layers-1]->num_outputs; //outputs
k=MAX_VECTORS;

buffer=new
        float[(i+j)*k];
if (buffer==0)
        {
        cout << "insufficient memory for buffer\n";
        exit(1);
        }
}

void network::randomize_weights()
{
int i;

for (i=1; i<number_of_layers; i++)
        ((output_layer *)layer_ptr[i])
                ->randomize_weights();
}


void network::update_weights(const float beta, const float alpha)
{
int i;

for (i=1; i<number_of_layers; i++)
        ((output_layer *)layer_ptr[i])
                ->update_weights(beta,alpha);
}


void network::update_momentum()
```

```
{
int i;

for (i=1; i<number_of_layers; i++)
        ((output_layer *)layer_ptr[i])
                ->update_momentum();
}



void network::write_weights(FILE * weights_file_ptr)
{
int i;

for (i=1; i<number_of_layers; i++)
        ((output_layer *)layer_ptr[i])
                ->write_weights(i,weights_file_ptr);
}



void network::read_weights(FILE * weights_file_ptr)
{
int i;

for (i=1; i<number_of_layers; i++)
        ((output_layer *)layer_ptr[i])
                ->read_weights(i,weights_file_ptr);
}



void network::list_weights()
{
int i;

for (i=1; i<number_of_layers; i++)
        {
        cout << "layer number : " <<i<< "\n";
        ((output_layer *)layer_ptr[i])
                ->list_weights();
        }
```

```
}

void network::list_outputs()
{
int i;

for (i=1; i<number_of_layers; i++)
        {
        cout << "layer number : " <<i<< "\n";
        ((output_layer *)layer_ptr[i])
                ->list_outputs();
        }
}

void network::write_outputs(FILE *outfile)
{
int i, ins, outs;
ins=layer_ptr[0]->num_outputs;
outs=layer_ptr[number_of_layers-1]->num_outputs;
float temp;

fprintf(outfile,"for input vector:\n");

for (i=0; i<ins; i++)
        {
        temp=layer_ptr[0]->outputs[i];
        fprintf(outfile,"%f  ",temp);
        }


fprintf(outfile,"\noutput vector is:\n");

for (i=0; i<outs; i++)
        {
        temp=layer_ptr[number_of_layers-1]->
        outputs[i];
        fprintf(outfile,"%f  ",temp);

        }
```

```
if (training==1)
{
fprintf(outfile,"\nexpected output vector is:\n");

for (i=0; i<outs; i++)
        {
        temp=((output_layer *)(layer_ptr[number_of_layers-1]))->
        expected_values[i];
        fprintf(outfile,"%f  ",temp);


        }
}

fprintf(outfile,"\n—————————\n");

}




void network::list_errors()
{
int i;

for (i=1; i<number_of_layers; i++)
        {
        cout << "layer number : " <<i<< "\n";
        ((output_layer *)layer_ptr[i])
                ->list_errors();
        }
}




int network::fill_IObuffer(FILE * inputfile)
{
// this routine fills memory with
// an array of input, output vectors
// up to a maximum capacity of
// MAX_INPUT_VECTORS_IN_ARRAY
```

```
// the return value is the number of read
// vectors


int i, k, count, veclength;

int ins, outs;

ins=layer_ptr[0]->num_outputs;

outs=layer_ptr[number_of_layers-1]->num_outputs;

if (training==1)
        veclength=ins+outs;
else
        veclength=ins;

count=0;
while   ((count<MAX_VECTORS)&&
                (!feof(inputfile)))
        {
        k=count*(veclength);
        for (i=0; i<veclength; i++)
                {
                fscanf(inputfile,"%f",&buffer[k+i]);
                }
        fscanf(inputfile,"\n");
        count++;
        }

if (!(ferror(inputfile)))
        return count;
else return -1; // error condition

}



void network::set_up_pattern(int buffer_index)
{
```

```
// read one vector into the network
int i, k;
int ins, outs;

ins=layer_ptr[0]->num_outputs;
outs=layer_ptr[number_of_layers-1]->num_outputs;
if (training==1)
        k=buffer_index*(ins+outs);
else
        k=buffer_index*ins;

for (i=0; i<ins; i++)
        ((input_layer*)layer_ptr[0])
                        ->orig_outputs[i]=buffer[k+i];

if (training==1)
{
        for (i=0; i<outs; i++)

                ((output_layer *)layer_ptr[number_of_layers-1])->
                        expected_values[i]=buffer[k+i+ins];
}


}



void network::forward_prop()
{
int i;
for (i=0; i<number_of_layers; i++)
        {
        layer_ptr[i]->calc_out(); //polymorphic
                                // function
        }
}


void network::backward_prop(float & toterror)
{
```

```
int i;

// error for the output layer
((output_layer*)layer_ptr[number_of_layers-1])->
                    calc_error(toterror);

// error for the middle layer(s)
for (i=number_of_layers-2; i>0; i—)
        {
        ((middle_layer*)layer_ptr[i])->
                    calc_error();


        }

}


void network::set_NF(float noise_fact)
{
((input_layer*)layer_ptr[0])->set_NF(noise_fact);
}
```

The New and Final backprop.cpp File

The last file to present is the backprop.cpp file. This is shown in Listing 13.3.

LISTING 13.3 IMPLEMENTATION FILE FOR THE BACKPROPAGATION SIMULATOR, WITH NOISE AND MOMENTUM BACKPROP.CPP

```
// backprop.cpp        V. Rao, H. Rao
#include "layer.cpp"

#define TRAINING_FILE  "training.dat"
#define WEIGHTS_FILE "weights.dat"
#define OUTPUT_FILE    "output.dat"
#define TEST_FILE      "test.dat"

void main()
{
```

```
float error_tolerance=0.1;
float total_error=0.0;
float avg_error_per_cycle=0.0;
float error_last_cycle=0.0;
float avgerr_per_pattern=0.0; // for the latest cycle
float error_last_pattern=0.0;
float learning_parameter=0.02;
float alpha; // momentum parameter
float NF; // noise factor
float new_NF;

unsigned temp, startup, start_weights;
long int vectors_in_buffer;
long int max_cycles;
long int patterns_per_cycle=0;

long int total_cycles, total_patterns;
int i;



// create a network object
network backp;


FILE * training_file_ptr, * weights_file_ptr, * output_file_ptr;
FILE * test_file_ptr, * data_file_ptr;

// open output file for writing
if ((output_file_ptr=fopen(OUTPUT_FILE,"w"))==NULL)
                {
                cout << "problem opening output file\n";
                exit(1);
                }

// enter the training mode : 1=training on      0=training off
cout << "————————————————————-\n";
cout << " C++ Neural Networks and Fuzzy Logic \n";
cout << "        Backpropagation simulator \n";
cout << "                version 2 \n";
```

```
cout << "————————————————————-\n";
cout << "Please enter 1 for TRAINING on, or 0 for off: \n\n";
cout << "Use training to change weights according to your\n";
cout << "expected outputs. Your training.dat file should contain\n";
cout << "a set of inputs and expected outputs. The number of\n";
cout << "inputs determines the size of the first (input) layer\n";
cout << "while the number of outputs determines the size of the\n";
cout << "last (output) layer :\n\n";

cin >> temp;
backp.set_training(temp);

if (backp.get_training_value() == 1)
        {
        cout << "–> Training mode is *ON*. weights will be saved\n";
        cout << "in the file weights.dat at the end of the\n";
        cout << "current set of input (training) data\n";
        }
else
        {
        cout << "–> Training mode is *OFF*. weights will be loaded\n";
        cout << "from the file weights.dat and the current\n";
        cout << "(test) data set will be used. For the test\n";
        cout << "data set, the test.dat file should contain\n";
        cout << "only inputs, and no expected outputs.\n";
        }

if (backp.get_training_value()==1)
        {
        // ———————————————-
        //      Read in values for the error_tolerance,
        //      and the learning_parameter
        // ———————————————-
        cout << " Please enter in the error_tolerance\n";
        cout << " –- between 0.001 to 100.0, try 0.1 to start - \n";
        cout << "\n";
        cout << "and the learning_parameter, beta\n";
        cout << " –- between 0.01 to 1.0, try 0.5 to start - \n\n";
        cout << " separate entries by a space\n";
```

```
cout << " example: 0.1 0.5 sets defaults mentioned :\n\n";

cin >> error_tolerance >> learning_parameter;

// —————————————————-
//      Read in values for the momentum
//      parameter, alpha (0-1.0)
//      and the noise factor, NF (0-1.0)
// —————————————————-
cout << "Enter values now for the momentum \n";
cout << "parameter, alpha(0-1.0)\n";
cout << " and the noise factor, NF (0-1.0)\n";
cout << "You may enter zero for either of these\n";
cout << "parameters, to turn off the momentum or\n";
cout << "noise features.\n";
cout << "If the noise feature is used, a random\n";
cout << "component of noise is added to the inputs\n";
cout << "This is decreased to 0 over the maximum\n";
cout << "number of cycles specified.\n";
cout << "enter alpha followed by NF, e.g., 0.3 0.5\n";

cin >> alpha >> NF;

//—————————————————-
// open training file for reading
//—————————————————-
if ((training_file_ptr=fopen(TRAINING_FILE,"r"))==NULL)
        {
        cout << "problem opening training file\n";
        exit(1);
        }
data_file_ptr=training_file_ptr; // training on

// Read in the maximum number of cycles
// each pass through the input data file is a cycle
cout << "Please enter the maximum cycles for the simulation\n";
cout << "A cycle is one pass through the data set.\n";
cout << "Try a value of 10 to start with\n";
```

```
        cin >> max_cycles;

        cout << "Do you want to read weights from weights.dat to
start?\n";
        cout << "Type 1 to read from file, 0 to randomize starting
weights\n";
        cin >> start_weights;

        }
else
        {
        if ((test_file_ptr=fopen(TEST_FILE,"r"))==NULL)
                {
                cout << "problem opening test file\n";
                exit(1);
                }

        data_file_ptr=test_file_ptr; // training off
        }




// training: continue looping until the total error is less than
//              the tolerance specified, or the maximum number of
//              cycles is exceeded; use both the forward signal propaga-
tion
//              and the backward error propagation phases. If the error
//              tolerance criteria is satisfied, save the weights in a
file.
// no training: just proceed through the input data set once in the
//              forward signal propagation phase only. Read the starting
//              weights from a file.
// in both cases report the outputs on the screen


// initialize counters
total_cycles=0; // a cycle is once through all the input data
total_patterns=0; // a pattern is one entry in the input data
new_NF=NF;
```

```
// get layer information
backp.get_layer_info();

// set up the network connections
backp.set_up_network();

// initialize the weights
if ((backp.get_training_value()==1)&&(start_weights!=1))
        {
        // randomize weights for all layers; there is no
        // weight matrix associated with the input layer
        // weight file will be written after processing

        backp.randomize_weights();
        // set up the noise factor value
        backp.set_NF(new_NF);
        }
else
        {
        // read in the weight matrix defined by a
        // prior run of the backpropagation simulator
        // with training on
        if ((weights_file_ptr=fopen(WEIGHTS_FILE,"r"))
                   ==NULL)
                {
                cout << "problem opening weights file\n";
                exit(1);
                }
        backp.read_weights(weights_file_ptr);
        fclose(weights_file_ptr);
        }




// main loop
// if training is on, keep going through the input data
//             until the error is acceptable or the maximum number of
cycles
```

```
//              is exceeded.
// if training is off, go through the input data once. report outputs
// with inputs to file output.dat

startup=1;
vectors_in_buffer = MAX_VECTORS; // startup condition
total_error = 0;


while ( ((backp.get_training_value()==1)
                    && (avgerr_per_pattern
                                    > error_tolerance)
                    && (total_cycles < max_cycles)
                    && (vectors_in_buffer !=0))
                    || ((backp.get_training_value()==0)
                    && (total_cycles < 1))
                    || ((backp.get_training_value()==1)
                    && (startup==1))
                    )
    {
startup=0;
error_last_cycle=0; // reset for each cycle
patterns_per_cycle=0;

backp.update_momentum(); // added to reset
                    // momentum matrices
                    // each cycle

// process all the vectors in the datafile
// going through one buffer at a time
// pattern by pattern



while ((vectors_in_buffer==MAX_VECTORS))
        {

        vectors_in_buffer=
```

```
backp.fill_IObuffer(data_file_ptr); // fill buffer
if (vectors_in_buffer < 0)
        {
        cout << "error in reading in vectors, aborting\n";
        cout << "check that there are no extra
linefeeds\n";

        cout << "in your data file, and that the
number\n";

        cout << "of layers and size of layers match
the\n";

        cout << "the parameters provided.\n";
        exit(1);
        }

// process vectors
for (i=0; i<vectors_in_buffer; i++)
        {
        // get next pattern
        backp.set_up_pattern(i);

        total_patterns++;
        patterns_per_cycle++;
        // forward propagate

        backp.forward_prop();

        if (backp.get_training_value()==0)
                backp.write_outputs(output_file_ptr);

        // back_propagate, if appropriate
        if (backp.get_training_value()==1)
                {

                backp.backward_prop(error_last_pattern);
                error_last_cycle +=
                        error_last_pattern*error_last_pat-
tern;

                avgerr_per_pattern=
((float)sqrt((double)error_last_cycle/patterns_per_cycle));
```

```
                            // if it's not the last cycle, update
weights
                            if ((avgerr_per_pattern
                                    > error_tolerance)
                                    && (total_cycles+1 < max_cycles))

                                    backp.update_weights(learning_para-
meter,
                                        alpha);
                            // backp.list_weights(); // can
                            // see change in weights by
                            // using list_weights before and
                            // after back_propagation
                            }


                }

        error_last_pattern = 0;
        }


total_error += error_last_cycle;
total_cycles++;

// update NF
// gradually reduce noise to zero
if (total_cycles>0.7*max_cycles)
                new_NF = 0;
else    if (total_cycles>0.5*max_cycles)
                    new_NF = 0.25*NF;
            else    if (total_cycles>0.3*max_cycles)
                        new_NF = 0.50*NF;
                    else    if (total_cycles>0.1*max_cycles)
                                new_NF = 0.75*NF;


backp.set_NF(new_NF);

// most character displays are 25 lines
// user will see a corner display of the cycle count
// as it changes
```

```
cout << "\n\n\n\n\n\n\n\n\n\n\n\n\n\n\n\n\n\n\n\n\n\n\n\n";
cout << total_cycles << "\t" << avgerr_per_pattern << "\n";


fseek(data_file_ptr, 0L, SEEK_SET); // reset the file pointer
                              // to the beginning of
                              // the file
vectors_in_buffer = MAX_VECTORS; // reset


} // end main loop


if (backp.get_training_value()==1)
      {
      if ((weights_file_ptr=fopen(WEIGHTS_FILE,"w"))
                  ==NULL)
            {
            cout << "problem opening weights file\n";
            exit(1);
            }
      }


cout << "\n\n\n\n\n\n\n\n\n\n\n";
cout << "—————————————————————\n";
cout << "      done:   results in file output.dat\n";
cout << "              training: last vector only\n";
cout << "              not training: full cycle\n\n";
if (backp.get_training_value()==1)
      {
      backp.write_weights(weights_file_ptr);
      backp.write_outputs(output_file_ptr);
      avg_error_per_cycle=(float)sqrt((double)total_error/
      total_cycles);
      error_last_cycle=(float)sqrt((double)error_last_cycle);
      fclose(weights_file_ptr);


cout << "              weights saved in file weights.dat\n";
cout << "\n";
cout << "—>average error per cycle = " << avg_error_per_cycle << " <--
\n";
cout << "—>error last cycle = " << error_last_cycle << " <--\n";
```

```
cout << "->error last cycle per pattern= " <<
avgerr_per_pattern << " <--\n";


        }


cout << "——————>total cycles = " << total_cycles <<
" <--\n";
cout << "——————>total patterns = " << total_patterns << " <--\n";
cout << "——————————————————\n";
// close all files
fclose(data_file_ptr);
fclose(output_file_ptr);


}
```

Trying the Noise and Momentum Features

You can test out the version 2 simulator, which you just compiled with the example that you saw at the beginning of the chapter. You will find that there is a lot of trial and error in finding optimum values for **alpha**, the noise factor, and **beta**. This is true also for the middle layer size and the number of middle layers. For some problems, the addition of momentum makes convergence much faster. For other problems, you may not find any noticeable difference. An example run of the five-character recognition problem discussed at the beginning of this chapter resulted in the following results with **beta** = 0.1, **tolerance** = 0.001, **alpha** = 0.25, **NF** = 0.1, and the layer sizes kept at 35 5 3.

```
        done:   results in file output.dat
                training: last vector only
                not training: full cycle


                weights saved in file weights.dat

—->average error per cycle = 0.02993<--
—->error last cycle = 0.00498<--
->error last cycle per pattern= 0.000996 <--
```

```
———>total cycles = 242 <--
———>total patterns = 1210 <--
```

N O T E The network was able to converge on a better solution (in terms of error measurement) in one-fourth the number of cycles. You can try varying **alpha** and **NF** to see the effect on overall simulation time. You can now start from the same initial starting weights by specifying a value of 1 for the starting weights question. For large values of **alpha** and **beta**, the network usually will not converge, and the weights will get unacceptably large (you will receive a message to that effect).

VARIATIONS OF THE BACKPROPAGATION ALGORITHM

Backpropagation is a versatile neural network algorithm that very often leads to success. Its Achilles heel is the slowness at which it converges for certain problems. Many variations of the algorithm exist in the literature to try to improve convergence speed and robustness. Variations have been proposed in the following portions of the algorithm:

- **Adaptive parameters**. You can set rules that modify **alpha**, the momentum parameter, and **beta**, the learning parameter, as the simulation progresses. For example, you can reduce beta whenever a weight change does not reduce the error. You can consider undoing the particular weight change, setting **alpha** to **zero** and redoing the weight change with the new value of **beta**.

- **Use other minimum search routines besides steepest descent**. For example, you could use Newton's method for finding a minimum, although this would be a fairly slow process. Other examples include the use of conjugate gradient methods or Levenberg-Marquardt optimization, both of which would result in very rapid training.

- **Use different cost functions**. Instead of calculating the error (as expected—actual output), you could determine another cost function that you want to minimize.

- **Modify the architecture**. You could use partially connected layers instead of fully connected layers. Also, you can use a recurrent network, that is, one in which some outputs feed back as inputs.

APPLICATIONS

Backpropagation remains the king of neural network architectures because of its ease of use and wide applicability. A few of the notable applications in the literature will be cited as examples.

- **NETTalk.** In 1987, Sejnowski and Rosenberg developed a network connected to a speech synthesizer that was able to utter English words, being trained to produce phonemes from English text. The architecture consisted of an input layer window of seven characters. The characters were part of English text that was scrolled by. The network was trained to pronounce the letter at the center of the window. The middle layer had 80 neurons, while the output layer consisted of 26 neurons. With 1024 training patterns and 10 cycles, the network started making intelligible speech, similar to the process of a child learning to talk. After 50 cycles, the network was about 95% accurate. You could purposely damage the network with the removal of neurons, but this did not cause performance to drop off a cliff; instead, the performance degraded gracefully. There was rapid recovery with retraining using fewer neurons also. This shows the fault tolerance of neural networks.

- **Sonar target recognition**. Neural nets using backpropagation have been used to identify different types of targets using the frequency signature (with a Fast Fourier transform) of the reflected signal.

- **Car navigation**. Pomerleau developed a neural network that is able to navigate a car based on images obtained from a camera mounted on the car's roof, and a range finder that coded distances in grayscale. The 30x32 pixel image and the 8x32 range finder image were fed into a hidden layer of size 29 feeding an output layer of 45 neurons. The output neurons were arranged in a straight line with each side representing a turn to a particular direction (right or left), while the center neurons represented "drive straight ahead." After 1200 road images were trained on the network, the neural network driver was able to negotiate a part of the Carnegie-Mellon campus at a speed of about 3 miles per hour, limited only by the speed of the real-time calculations done on a trained network in the Sun-3 computer in the car.

- **Image compression**. G.W. Cottrell, P. Munro, and D. Zipser used backpropagation to compress images with the result of an 8:1 compression ratio. They used standard backpropagation with 64 input neurons

(8x8 pixels), 16 hidden neurons, and 64 output neurons equal to the inputs. This is called *self-supervised backpropagation* and represents an autoassociative network. The compressed signal is taken from the hidden layer. The input to hidden layer comprised the compressor, while the hidden to output layer forms a decompressor.

- **Image recognition**. Le Cun reported a backpropagation network with three hidden layers that could recognize handwritten postal zip codes. He used a 16x16 array of pixel to represent each handwritten digit and needed to encode 10 outputs, each of which represented a digit from 0 to 9. One interesting aspect of this work is that the hidden layers were not fully connected. The network was set up with blocks of neurons in the first two hidden layers set up as feature detectors for different parts of the previous layer. All the neurons in the block were set up to have the same weights as those from the previous layer. This is called *weight sharing*. Each block would sample a different part of the previous layer's image. The first hidden layer had 12 blocks of 8x8 neurons, whereas the second hidden layer had 12 blocks of 4x4 neurons. The third hidden layer was fully connected and consisted of 30 neurons. There were 1256 neurons. The network was trained on 7300 examples and tested on 2000 cases with error rates of 1% on training set and 5% on the test set.

SUMMARY

You explored further the backpropagation algorithm in this chapter, continuing the discussion in Chapter 7.

- A momentum term was added to the training law and was shown to result in much faster convergence in some cases.

- A noise term was added to inputs to allow training to take place with random noise applied. This noise was made to decrease with the number of cycles, so that final stage learning could be done in a noise-free environment.

- The final version of the backpropagation simulator was constructed and used on the example from Chapter 12. Further application of the simulator will be made in Chapter 14.

- Several applications with the backpropagation algorithm were outlined, showing the wide applicability of this algorithm.

APPLICATION TO FINANCIAL FORECASTING

INTRODUCTION

In Chapters 7 and 13, the backpropagation simulator was developed. In this chapter, you will use the simulator to tackle a complex problem in financial forecasting. The application of neural networks to financial forecasting and modeling has been very popular over the last few years. Financial journals and magazines frequently mention the use of neural networks, and commercial tools and simulators are quite widespread.

This chapter gives you an overview of typical steps used in creating financial forecasting models. Many of the steps will be simplified, and so the results will not, unfortunately, be good enough for real life application. However, this chapter will hopefully serve as an introduction to the field with some added pointers for further reading and resources for those who want more detailed information.

WHO TRADES WITH NEURAL NETWORKS?

There has been a great amount of interest on Wall Street for neural networks. Bradford Lewis runs two Fidelity funds in part with the use of neural networks. Also, LBS Capital Management (Peoria, Illinois) manages part of its portfolio with neural networks. According to Barron's (February 27, 1995), LBS's $150 million fund beat the averages by three percentage points a year since 1992. Each weekend, neural networks are retrained with the latest technical and fundamental data including P/E ratios, earnings results and interest rates. Another of LBS's models has done worse than the S&P 500 for the past five years however. In the book *Virtual Trading*, Jeffrey Katz states that most of the successful neural network systems are proprietary and not publicly heard of. Clients who use neural networks usually don't want anyone else to know what they are doing, for fear of losing their competitive edge. Firms put in many person-years of engineering design with a lot of CPU cycles to achieve practical and profitable results. Let's look at the process:

DEVELOPING A FORECASTING MODEL

There are many steps in building a forecasting model, as listed below.

1. Decide on what your target is and develop a neural network (following these steps) for each target.
2. Determine the time frame that you wish to forecast.
3. Gather information about the problem domain.
4. Gather the needed data and get a feel for each inputs relationship to the target.
5. Process the data to highlight features for the network to discern.
6. Transform the data as appropriate.
7. Scale and bias the data for the network, as needed.
8. Reduce the dimensionality of the input data as much as possible.
9. Design a network architecture (topology, # layers, size of layers, parameters, learning paradigm).
10. Go through the train/test/redesign loop for a network.
11. Eliminate correlated inputs as much as possible, while in step 10.
12. Deploy your network on new data and test it and refine it as necessary.

Once you develop a forecasting model, you then must integrate this into your *trading system*. A neural network can be designed to predict direction, or magnitude, or maybe just turning points in a particular market or something else. Avner Mandelman of Cereus Investments (Los Altos Hills, California) uses a long-range trained neural network to tell him when the market is making a top or bottom (*Barron's*, December 14, 1992).

Now let's expand on the twelve aspects of model building:

The Target and the Timeframe

What should the output of your neural network forecast? Let's say you want to predict the stock market. Do you want to predict the S&P 500? Or, do you want to predict the direction of the S&P 500 perhaps? You could predict the volatility of the S&P 500 too (maybe if you're an options player). Further, like Mr. Mandelman, you could only want to predict tops and bottoms, say, for the Dow Jones Industrial Average. You need to decide on the market or markets and also on your specific objectives.

Another crucial decision is the timeframe you want to predict forward. It is easier to create neural network models for longer term predictions than it is for shorter term predictions. You can see a lot of market noise, or seemingly random, chaotic variations at smaller and smaller timescale resolutions that might explain this. Another reason is that the macroeconomic forces that *fundamentally* move market over long periods, move slowly. The U.S. dollar makes multiyear trends, shaped by economic policy of governments around the world. For a given error tolerance, a one-year forecast, or one-month forecast will take less effort with a neural network than a one-day forecast will.

Domain Expertise

So far we've talked about the target and the timeframe. Now one other important aspect of model building is knowledge of the domain. If you want to create an effective predictive model of the weather, then you need to know or be able to guess about the factors that influence weather. The same holds true for the stock market or other financial market. In order to create a real tradable Treasury bond trading system, you need to have a good idea of what really drives the market and works— i.e., talk to a Tbond trader and encapsulate his domain expertise!

Gather the Data

Once you know the factors that influence the target output, you can gather raw data. If you are predicting the S&P 500, then you may consider Treasury yields, 3-month T-bill yields, and earnings as some of the factors. Once you have the data, you can do scatter plots to see if there is some correlation between the input and the target output. If you are not satisfied with the plot, you may consider a different input in its place.

Pre processing the Data for the Network

Surprising as it may sound, you are most likely going to spend about 90% of your time, as a neural network developer, in massaging and transforming data into meaningful form for training your network. We actually defined three substeps in this area of preprocessing in our master list:

- Highlight features
- Transform
- Scale and bias

Highlighting Features in the Input Data

You should present the neural network, as much as possible, with an easy way to find patterns in your data. For time series data, like stock market prices over time, you may consider presenting quantities like rate of change and acceleration (the first and second derivatives of your input) as examples. Other ways to highlight data is to magnify certain occurrences. For example, if you consider Central bank intervention as an important qualifier to foreign exchange rates, then you may include as an input to your network, a value of 1 or 0, to indicate the presence or lack of presence of Central bank intervention. Now if you further consider the activity of the U.S. Federal Reserve bank to be important by itself, then you may wish to highlight that, by separating it out as another 1/0 input. Using 1/0 coding to separate composite effects is called *thermometer encoding*.

There is a whole body of study of market behavior called *Technical Analysis* from which you may also wish to present *technical studies* on your data. There is a wide assortment of mathematical technical studies that you perform on your data (see references), such as moving averages to smooth data as an example. There are also pattern recognition studies you can use, like the "double-top" formation, which purportedly results in a high probability of significant decline. To be able to recognize such a pattern, you may wish to present a mathematical function that aids in the identification of the double-top.

You may want to de-emphasize unwanted noise in your input data. If you see a spike in your data, you can lessen its effect, by passing it through a moving average filter for example. You should be careful about introducing excessive lag in the resulting data though.

Transform the Data If Appropriate

For time series data, you may consider using a *Fourier transform* to move to the frequency-phase plane. This will uncover periodic cyclic information if it exists. The Fourier transform will decompose the input discrete data series into a series of frequency spikes that measure the relevance of each frequency component. If the stock market indeed follows the so-called January effect, where prices typically make a run up, then you would expect a strong yearly component in the frequency spectrum. Mark Jurik suggests sampling data with intervals that catch different cycle periods, in his paper on neural network data preparation (see references).

You can use other signal processing techniques such as filtering. Besides the frequency domain, you can also consider moving to other spaces, such as with using the wavelet transform. You may also analyze the chaotic component of the data with chaos measures. It's beyond the scope of this book to discuss these techniques. (Refer to the Resources section of this chapter for more information.) If you are developing short-term trading neural network systems, these techniques may play a significant role in your preprocessing effort. All of these techniques will provide new ways of looking at your data, for possible features to detect in other domains.

Scale Your Data

Neurons like to see data in a particular input range to be most effective. If you present data like the S&P 500 that varies from 200 to 550 (as the S&P 500 has over the years) will not be useful, since the middle layer of neurons have a Sigmoid Activation function that squashes large inputs to either 0 or +1. In other words, you should choose data that fit a range that does not saturate, or overwhelm the network neurons. Choosing inputs from –1 to 1 or 0 to 1 is a good idea. By the same token, you should normalize the expected values for the outputs to the 0 to 1 sigmoidal range.

It is important to pay attention to the number of input values in the data set that are close to zero. Since the weight change law is proportional to the input value, then a close to zero input will mean that that weight will not participate in learning! To avoid such situations, you can add a constant bias to your data to move the data closer to 0.5, where the neurons respond very well.

Reduce Dimensionality

You should try to eliminate inputs wherever possible. This will reduce the *dimensionality* of the problem and make it easier for your neural network to generalize. Suppose that you have three inputs, x, y and z and one output, o. Now suppose that you find that all of your inputs are restricted only to one plane. You could redefine axes such that you have x' and y' for the new plane and map your inputs to the new coordinates. This changes the number of inputs to your problem to 2 instead of 3, without any loss of information. This is illustrated in Figure 14.1.

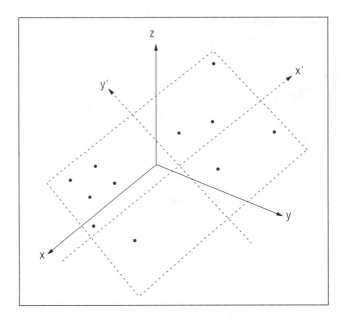

Figure 14.1 Reducing dimensionality from three to two dimensions.

Generalization versus Memorization

If your overall goal is beyond pattern classification, you need to track your network's ability to generalize. Not only should you look at the overall error with the training set that you define, but you should set aside some training examples as part of a test set (and do not train with them), with which you can see whether or not the network is able to correctly predict. If the network responds poorly to your test set, you know that you have overtrained, or you can say the network "memorized" the training patterns. If you look at the arbitrary curve-fitting analogy in Figure 14.2, you see curves for a generalized fit, labeled G, and an overfit, labeled O. In the case of the overfit, any data point outside of the training data results in highly erroneous prediction. Your test data will certainly show you large error in the case of an overfitted model.

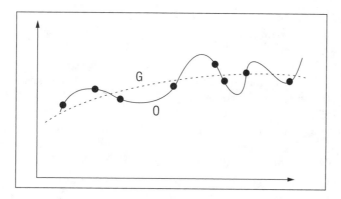

FIGURE 14.2 GENERAL (G) VERSUS OVER FITTING (O) OF DATA.

Another way to consider this issue is in terms of *Degrees Of Freedom (DOF)*. For the polynomial:

$$y = a0 + a1x + a2x2 + anxn...$$

the DOF equals the number of coefficients $a0$, $a1$ ... an, which is $N + 1$. So for the equation of a line ($y=a0 + a1x$), the DOF would be 2. For a parabola, this would be 3 and so on. The objective to not overfit data can be restated as an objective to obtain the function with the least DOF that fits the data adequately. For neural network models, the larger the number of trainable weights (which is a function of the number of inputs and the architecture), the larger the DOF. Be careful with having too many (unimportant) inputs. You may find terrific results with your training data, but extremely poor results with your test data.

Eliminate Correlated Inputs Where Possible

You have seen that getting to the minimum number of inputs for a given problem is important in terms of minimizing DOF and simplifying your model. Another way to reduce dimensionality is to look for correlated inputs and to *carefully* eliminate redundancy. For example, you may find that the

Swiss franc and German mark are highly correlated over a certain time period of interest. You may wish to eliminate one of these inputs to reduce dimensionality. You have to be careful in this process though. You may find that a seemingly redundant piece of information is actually very important. Mark Jurik, of Jurik Consulting, in his paper on data preprocessing, suggests that one of the best ways to determine if an input is really needed is to construct neural network models with and without the input and choose the model with the best error on training and test data. Although very iterative, you can try eliminating as many inputs as possible this way and be assured that you haven't eliminated a variable that really made a difference.

Another approach is *sensitivity analysis*, where you vary one input a little, while holding all others constant and note the effect on the output. If the effect is small you eliminate that input. This approach is flawed because in the real world, all the inputs *are not constant.* Jurik's approach is more time consuming but will lead to a better model.

The process of *decorrelation,* or eliminating correlated inputs, can also utilize a linear algebra technique called *principal component analysis.* The result of principal component analysis is a minimum set of variables that contain the maximum information. For further information on principal component analysis, you should consult a statistics reference or research two methods of principal component analysis: the *Karhunen-Loev transform* and the *Hotelling transform.*

Design a Network Architecture

Now it's time to actually design the neural network. For the backpropagation feed-forward neural network we have designed, this means making the following choices:

1. The number of hidden layers.
2. The size of hidden layers.
3. The learning constant, beta(β).
4. The momentum parameter, alpha(α).
5. The form of the squashing function (does not have to be the sigmoid).
6. The starting point, that is, initial weight matrix.
7. The addition of noise.

Some of the parameters listed can be made to vary with the number of cycles executed, similar to the current implementation of noise. For example, you can start with a learning constant β that is large and reduce this constant as learning progresses. This allows rapid initial learning in the beginning of the process and may speed up the overall simulation time.

385

The Train/Test/Redesign Loop

Much of the process of determining the best parameters for a given application is trial and error. You need to spend a great deal of time evaluating different options to find the best fit for your problem. You may literally create hundreds if not thousands of networks either manually or automatically to search for the best solution. Many commercial neural network programs use *genetic algorithms* to help to automatically arrive at an optimum network. A genetic algorithm makes up possible solutions to a problem from a set of starting *genes*. Analogous to biological evolution, the algorithm combines genetic solutions with a predefined set of operators to create new generations of solutions, who survive or perish depending on their ability to solve the problem. The key benefit of genetic algorithms (GA) is the ability to traverse an enormous search space for a possibly optimum solution. You would program a GA to search for the number of hidden layers and other network parameters, and gradually evolve a neural network solution. Some vendors use a GA only to assign a starting set of weights to the network, instead of randomizing the weights to start you off near a good solution.

Now let's review the steps:

1. **Split your data**. First, divide you data set into three pieces, a training set, a test set and a blind test set. Use about 80% of your data records for your training set, 10% for your test set and 10% for your blind test set.

2. **Train and test**. Next, start with a network topology and train your network on your training set data. When you have reached a satisfactory minimum error, save your weights and apply your trained network to the test data and note the error. Now restart the process with the same network topology for a different set of initial weights and see if you can achieve a better error on training and test sets. Reasoning: you may have found a local minimum on your first attempt and randomizing the initial weights will start you off to a different, maybe better solution.

3. **Eliminate correlated inputs**. You may optionally try at this point to see if you can eliminate correlated inputs, as mentioned before, by iteratively removing each input and noting the best error you can achieve on the training and test sets for each of these cases. Choose the case that leads to the best error and eliminate the input (if any) that achieved it. You can repeat this whole process again to try to eliminate another input variable.

4. **Iteratively train and test**. Now you can try other network parameters and repeat the train and test process to achieve a better result.

5. **Deploy your network**. You now can use the blind test data set to see how your optimized network performs. If the error is not satisfactory, then you need to re-enter the design phase or the train and test phase.

6. **Revisit your network design when conditions change**. You need to retrain your network when you have reason to think that you have new information relevant to the problem you are modeling. If you have a neural network that tries to predict the weekly change in the S&P 500, then you likely will need to retrain your network at least once a month, if not once a week. If you find that the network no longer generalizes well with the new information, you need to re-enter the design phase.

If this sounds like a lot of work, it is! Now, let's try our luck at forecasting by going through a subset of the steps outlined:

FORECASTING THE S&P 500

The S&P 500 index is a widely followed stock average, like the Dow Jones Industrial Average (DJIA). It has a broader representation of the stock market since this average is based on 500 stocks, whereas the DJIA is based on only 30. The problem to be approached in this chapter is to predict the S&P 500 index, given a variety of indicators and data for prior weeks.

Choosing the Right Outputs and Objective

Our objective is to forecast the S&P 500 ten weeks from now. Whereas the objective may be to predict the level of the S&P 500, it is important to simplify the job of the network by asking for a change in the level rather than for the absolute level of the index. What you want to do is give the network the ability to fit the problem at hand conveniently in the output space of the output layer. Practically speaking, you know that the output from the network cannot be outside of the 0 to 1 range, since we have used a sigmoid activation function. You could take the S&P 500 index and scale this absolute price level to this range, for example. However, you will likely end up with very small numbers that have a small range of variability. The difference from week to week, on the other hand, has a much smaller overall range, and when these differences are scaled to the 0 to 1 range, you have much more variability.

The output we choose is the change in the S&P 500 from the current week to 10 weeks from now as a percentage of the current week's value.

Choosing the Right Inputs

The inputs to the network need to be weekly changes of indicators that have some relevance to the S&P 500 index. This is a complex forecasting problem, and we can only guess at some of the relationships. This is one of the inherent strengths of using neural nets for forecasting; if a relationship is weak, the network will learn to ignore it automatically. Be cognizant that you do want to minimize the DOF as mentioned before though. In this example, we choose a data set that represents the state of the financial markets and the economy. The inputs chosen are listed as:

- Previous price action in the S&P 500 index, including the close or final value of the index

- Breadth indicators for the stock market, including the number of advancing issues and declining issues for the stocks in the New York Stock Exchange (NYSE)

- Other technical indicators, including the number of new highs and new lows achieved in the week for the NYSE market. This gives some indication about the strength of an uptrend or downtrend.

- Interest rates, including short-term interest rates in the Three-Month Treasury Bill Yield, and long-term rates in the 30-Year Treasury Bond Yield.

Other possible inputs could have been government statistics like the Consumer Price Index, Housing starts, and the Unemployment Rate. These were not chosen because long- and short-term interest rates tend to encompass this data already.

You are encouraged to experiment with other inputs and ideas. All of the data mentioned can be obtained in the public domain, such as from financial publications (*Barron's, Investor's Business Daily, Wall Street Journal*) and from ftp sites on the Internet for the Department of Commerce and the Federal Reserve, as well as from commercial vendors (see the Resource Guide at the end of the chapter). There are new sources cropping up on the Internet all the time. A sampling of World Wide Web addresses for commercial and noncommercial sources include:

- FINWeb, http://riskweb.bus.utexas.edu/finweb.html
- Chicago Mercantile Exchange, http://www.cme.com/cme
- SEC Edgar Database, http://town.hall.org/edgar/edgar.html
- CTSNET Business & Finance Center, http://www.cts.com/cts/biz/
- QuoteCom, http://www.quote.com
- Philadelphia Fed, http://compstat.wharton.upenn.edu:8001/~siler/fedpage.html
- Ohio state Financial Data Finder, http://cob.ohio-state.edu/dept/fin/osudata.html

Choosing a Network Architecture

The input and output layers are fixed by the number of inputs and outputs we are using. In our case, the output is a single number, the expected change in the S&P 500 index 10 weeks from now. The input layer size will be dictated by the number of inputs we have after preprocessing. You will see more on this soon. The middle layers can be either 1 or 2. It is best to choose the smallest number of neurons possible for a given problem to allow for generalization. If there are too many neurons, you will tend to get memorization of patterns. We will use one hidden layer. The size of the first hidden layer is generally recommended as between one-half to three times the size of the input layer. If a second hidden layer is present, you may have between three and ten times the number of output neurons. The best way to determine optimum size is by trial and error.

NOTE You should try to make sure that there are enough training examples for your trainable weights. In other words, your architecture may be dictated by the number of input training examples, or *facts,* you have. In an ideal world, you would want to have about 10 or more facts for each weight. For a 10-10-1 architecture, there are (10X10 + 10X1 = 110 weights), so you should aim for about 1100 facts. The smaller the ratio of facts to weights, the more likely you will be undertraining your network, which will lead to very poor generalization capability.

PREPROCESSING DATA

We now begin the preprocessing effort. As mentioned before, this will likely be where you, the neural network designer, will spend most of your time.

A View of the Raw Data

Let's look at the raw data for the problem we want to solve. There are a couple of ways we can start preprocessing the data to reduce the number of inputs and enhance the variability of the data:

- Use Advances/Declines ratio instead of each value separately.
- Use New Highs/New Lows ratio instead of each value separately.

We are left with the following indicators:

1. Three-Month Treasury Bill Yield
2. 30-Year Treasury Bond Yield
3. NYSE Advancing/Declining issues
4. NYSE New Highs/New Lows
5. S&P 500 closing price

Raw data for the period from January 4, 1980 to October 28, 1983 is taken as the training period, for a total of 200 weeks of data. The following 50 weeks are kept on reserve for a test period to see if the predictions are valid outside of the training interval. The last date of this period is October 19, 1984. Let's look at the raw data now. (You get data on the disk available with this book that covers the period from January, 1980 to December, 1992.) In Figures 14.3 through 14.5, you will see a number of these indicators plotted over the training plus test intervals:

- Figure 14.3 shows you the S&P 500 stock index.
- Figure 14.4 shows long-term bonds and short-term 3-month T-bill interest rates.
- Figure 14.5 shows some breadth indicators on the NYSE, the number of advancing stocks/number of declining stocks, as well as the ratio of new highs to new lows on the NYSE

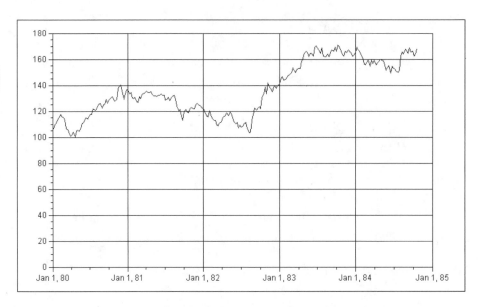

Figure 14.3 The S&P 500 Index for the period of interest.

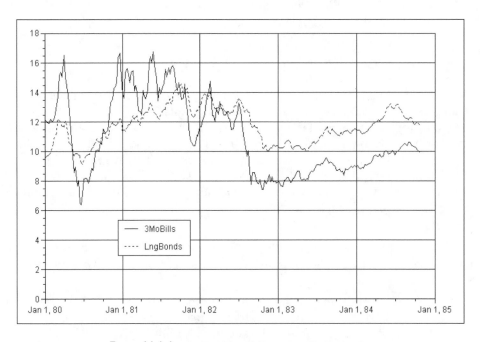

Figure 14.4 Long-term and short-term interest rates.

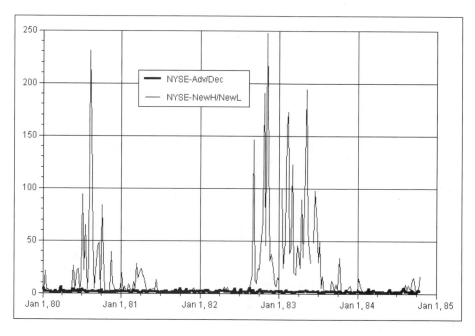

FIGURE 14.5 BREADTH INDICATORS ON THE NYSE: ADVANCING /DECLINING ISSUES
AND NEW HIGHS/NEW LOWS.

A sample of a few lines looks like the following data in Table 14.1. Note that
the order of parameters is the same as listed above.

TABLE 14.1 RAW DATA

Date	3Mo TBills	30YrTBonds	NYSE-Adv/Dec	NYSE-NewH/NewL	SP-Close
1/4/80	12.11	9.64	4.209459	2.764706	106.52
1/11/80	11.94	9.73	1.649573	21.28571	109.92
1/18/80	11.9	9.8	0.881335	4.210526	111.07
1/25/80	12.19	9.93	0.793269	3.606061	113.61
2/1/80	12.04	10.2	1.16293	2.088235	115.12
2/8/80	12.09	10.48	1.338415	2.936508	117.95
2/15/80	12.31	10.96	0.338053	0.134615	115.41
2/22/80	13.16	11.25	0.32381	0.109091	115.04
2/29/80	13.7	12.14	1.676895	0.179245	113.66

3/7/80	15.14	12.1	0.282591	0	106.9
3/14/80	15.38	12.01	0.690286	0.011628	105.43
3/21/80	15.05	11.73	0.486267	0.027933	102.31
3/28/80	16.53	11.67	5.247191	0.011628	100.68
4/3/80	15.04	12.06	0.983562	0.117647	102.15
4/11/80	14.42	11.81	1.565854	0.310345	103.79
4/18/80	13.82	11.23	1.113287	0.146341	100.55
4/25/80	12.73	10.59	0.849807	0.473684	105.16
5/2/80	10.79	10.42	1.147465	1.857143	105.58
5/9/80	9.73	10.15	0.513052	0.473684	104.72
5/16/80	8.6	9.7	1.342444	6.75	107.35
5/23/80	8.95	9.87	3.110825	26	110.62

Highlight Features in the Data

For each of the five inputs, we want use a function to highlight rate of change types of features. We will use the following function (as originally proposed by Jurik) for this purpose.

$$ROC(n) = (input(t) - BA(t - n)) / (input(t) + BA(t - n))$$

where: input(t) is the input's current value and BA($t - n$) is a five unit block average of adjacent values centered around the value n periods ago.

Now we need to decide how many of these features we need. Since we are making a prediction 10 weeks into the future, we will take data as far back as 10 weeks also. This will be ROC(10). We will also use one other rate of change, ROC(3). We have now added 5*2 = 10 inputs to our network, for a total of 15. All of the preprocessing can be done with a spreadsheet.

Here's what we get (Table 14.2) after doing the block averages. Example :
BA3MoBills for 1/18/80 = (3MoBills(1/4/80) + 3MoBills(1/11/80) +
3MoBills(1/18/80) + 3MoBills(1/25/80) + 3MoBills(2/1/80))/5.

TABLE 14.2 DATA AFTER DOING BLOCK AVERAGES

Date	3MoBills	LngBonds	NYSE-Adv/Dec	NYSE-NewH/NewL	SP-Close	BA3MoB	BALngBnd	BAA/D	BAH/L	BAClose
1/4/80	12.11	9.64	4.209459	2.764706	106.52					
1/11/80	11.94	9.73	1.649573	21.28571	109.92					
1/18/80	11.9	9.8	0.881335	4.210526	111.07	12.036	9.86	1.739313	6.791048	111.248
1/25/80	12.19	9.93	0.793269	3.606061	113.61	12.032	10.028	1.165104	6.825408	113.534
2/1/80	12.04	10.2	1.16293	2.088235	115.12	12.106	10.274	0.9028	2.595189	114.632
2/8/80	12.09	10.48	1.338415	2.936508	117.95	12.358	10.564	0.791295	1.774902	115.426
2/15/80	12.31	10.96	0.338053	0.134615	115.41	12.66	11.006	0.968021	1.089539	115.436
2/22/80	13.16	11.25	0.32381	0.109091	115.04	13.28	11.386	0.791953	0.671892	113.792
2/29/80	13.7	12.14	1.676895	0.179245	113.66	13.938	11.692	0.662327	0.086916	111.288
3/7/80	15.14	12.1	0.282591	0	106.9	14.486	11.846	0.69197	0.065579	108.668
3/14/80	15.38	12.01	0.690286	0.011628	105.43	15.16	11.93	1.676646	0.046087	105.796
3/21/80	15.05	11.73	0.486267	0.027933	102.31	15.428	11.914	1.537979	0.033767	103.494
3/28/80	16.53	11.67	5.247191	0.011628	100.68	15.284	11.856	1.794632	0.095836	102.872
4/3/80	15.04	12.06	0.983562	0.117647	102.15	14.972	11.7	1.879232	0.122779	101.896
4/11/80	14.42	11.81	1.565854	0.310345	103.79	14.508	11.472	1.95194	0.211929	102.466
4/18/80	13.82	11.23	1.113287	0.146341	100.55	13.36	11.222	1.131995	0.581032	103.446
4/25/80	12.73	10.59	0.849807	0.473684	105.16	12.298	10.84	1.037893	0.652239	103.96
5/2/80	10.79	10.42	1.147465	1.857143	105.58	11.134	10.418	0.993211	1.94017	104.672
5/9/80	9.73	10.15	0.513052	0.473684	104.72	10.16	10.146	1.392719	7.110902	106.686
5/16/80	8.6	9.7	1.342444	6.75	107.35	7.614	8.028	1.222757	7.016165	85.654
5/23/80	8.95	9.87	3.110825	26	110.62	5.456	5.944	0.993264	6.644737	64.538

Now let's look at the rest of this table, which is made up of the new 10 values
of ROC indicators (Table 14.3).

TABLE 14.3 ADDED RATE OF CHANGE (ROC) INDICATORS

Date	ROC3_3Mo	ROC3_Bond	ROC3_A/D	ROC3_H/L	ROC3_SPC	ROC10_3Mo	ROC10_Bnd	ROC10_AD	ROC10_HL	ROC10_SP
1/4/80										
1/11/80										
1/18/80										
1/25/80										
2/1/80										
2/8/80	0.002238	0.030482	-0.13026	-0.39625	0.029241					
2/15/80	0.011421	0.044406	-0.55021	-0.96132	0.008194					
2/22/80	0.041716	0.045345	-0.47202	-0.91932	0.001776					
2/29/80	0.0515	0.069415	0.358805	-0.81655	-0.00771					
3/7/80	0.089209	0.047347	-0.54808	-1	-0.03839					
3/14/80	0.073273	0.026671	-0.06859	-0.96598	-0.03814					
3/21/80	0.038361	0.001622	-0.15328	-0.51357	-0.04203					
3/28/80	0.065901	-0.00748	0.766981	-0.69879	-0.03816	0.15732	0.084069	0.502093	-0.99658	-0.04987
4/3/80	-0.00397	0.005419	-0.26054	0.437052	-0.01753	0.111111	0.091996	-0.08449	-0.96611	-0.05278
4/11/80	-0.03377	-0.00438	0.008981	0.803743	0.001428	0.087235	0.069553	0.268589	-0.78638	-0.04964
4/18/80	-0.0503	-0.02712	-0.23431	0.208545	-0.01141	0.055848	0.030559	0.169062	-0.84766	-0.06888
4/25/80	-0.08093	-0.0498	-0.37721	0.58831	0.015764	0.002757	-0.01926	-0.06503	-0.39396	-0.04658
5/2/80	-0.14697	-0.04805	-0.25956	0.795146	0.014968	-0.10345	-0.0443	0.183309	0.468658	-0.03743
5/9/80	-0.15721	-0.05016	-0.37625	-0.10178	0.00612	-0.17779	-0.0706	-0.127	0.689919	-0.03041
5/16/80	-0.17695	-0.0555	0.127944	0.823772	0.016043	-0.25496	-0.0996	0.319735	0.980756	-0.0061
5/23/80	-0.10874	-0.02701	0.515983	0.86112	0.027628	-0.25757	-0.0945	0.299569	0.996461	0.02229

N O T E   Note that you don't get completed rows until 3/28/90, since we have a ROC indicator dependent on a Block Average value 10 weeks before it. The first block average value is generated 1/1/80, two weeks after the start of the data set. All of this indicates that you will need to discard the first 12 values in the dataset to get complete rows, also called complete *facts*.

Normalizing the Range

We now have values in the original five data columns that have a very large range. We would like to reduce the range by some method. We use the following function:

new value = (old value - Mean)/ (Maximum Range)

This relates the distance from the mean for a value in a column as a fraction of the Maximum range for that column. You should note the value of the Maximum range and Mean, so that you can un-normalize the data when you get a result.

The Target

We've taken care of all our inputs, which number 15. The final piece of information is the target. The objective as stated at the beginning of this exercise is to predict the percentage change 10 weeks into the future. We need to time shift the S&P 500 close 10 weeks back, and then calculate the value as a percentage change as follows:

Result = 100 X ((S&P 10 weeks ahead) - (S&P this week))/(S&P this week).

This gives us a value that varies between -14.8 to and + 33.7. This is not in the form we need yet. As you recall, the output comes from a **sigmoid** function that is restricted to 0 to +1. We will first add 14.8 to all values and then scale them by a factor of 0.02. This will result in a scaled target that varies from 0 to 1.

scaled target = (result + 14.8) X 0.02

The final data file with the scaled target shown along with the scaled original six columns of data is shown in Table 14.4.

TABLE 14.4 NORMALIZED RANGES FOR ORIGINAL COLUMNS AND SCALED TARGET

Date	S_3MOBill	S_LngBnd	S_A/D	S_H/L	S_SPC	Result	Scaled Target
3/28/80	0.534853	-0.01616	0.765273	-0.07089	-0.51328	12.43544	0.544709
4/3/80	0.391308	0.055271	-0.06356	-0.07046	-0.49236	12.88302	0.55366
4/11/80	0.331578	0.009483	0.049635	-0.06969	-0.46901	9.89498	0.4939
4/18/80	0.273774	-0.09674	-0.03834	-0.07035	-0.51513	15.36549	0.60331
4/25/80	0.168765	-0.21396	-0.08956	-0.06903	-0.44951	11.71548	0.53031
5/2/80	-0.01813	-0.2451	-0.0317	-0.06345	-0.44353	11.61205	0.528241
5/9/80	-0.12025	-0.29455	-0.15503	-0.06903	-0.45577	16.53934	0.626787
5/16/80	-0.22912	-0.37696	0.006205	-0.04372	-0.41833	12.51048	0.54621
5/23/80	-0.1954	-0.34583	0.349971	0.033901	-0.37179	9.573314	0.487466

Storing Data in Different Files

You need to place the first 200 lines in a training.dat file (provided for you in the accompanying diskette) and the subsequent 40 lines of data in the another test.dat file for use in testing. You will read more about this shortly. There is also more data than this provided on this diskette in raw form for you to do further experiments.

Training and Testing

With the training data available, we set up a simulation. The number of inputs are 15, and the number of outputs is 1. A total of three layers are used with a middle layer of size 5. This number should be made as small as possible with acceptable results. The optimum sizes and number of layers can only be found by much trial and error. After each run, you can look at the error from the training set and from the test set.

Using the Simulator to Calculate Error

You obtain the error for the test set by running the simulator in **Training** mode (you need to temporarily copy the test data with expected outputs to the training.dat file) for one cycle with weights loaded from the weights file. Since this is the last and only cycle, weights do not get modified, and you can get a reading of the average error. Refer to Chapter 13 for more information on the simulator's **Test** and **Training** modes. This approach has been taken with five runs of the simulator for 500 cycles each. Table 14.5 summarizes the results along with the parameters used. The error gets better and better with each run up to run # 4. For run #5, the training set error decreases, but the test set error increases, indicating the onset of memorization. Run # 4 is used for the final network results, showing test set RMS error of 13.9% and training set error of 6.9%.

Table 14.5 Results of Training the Backpropagation Simulator for Predicting the Percentage Change in the S&P 500 Index

Run#	Tolerance	Beta	Alpha	NF	max cycles	cycles run	training set error	test set error
1	0.001	0.5	0.001	0.0005	500	500	0.150938	0.25429
2	0.001	0.4	0.001	0.0005	500	500	0.114948	0.185828

3	0.001	0.3	0	0	500	500	0.0936422	0.148541
4	0.001	0.2	0	0	500	500	0.068976	0.139230
5	0.001	0.1	0	0	500	500	0.0621412	0.143430

N O T E If you find the test set error does not decrease much, whereas the training set error continues to make substantial progress, then this means that memorization is starting to set in (run#5 in example). It is important to monitor the test set(s) that you are using while you are training to make sure that good, generalized learning is occurring versus memorizing or overfitting the data. In the case shown, the test set error continued to improve until run#5, where the test set error degraded. You need to revisit the 12-step process to forecasting model design to make any further improvements beyond what was achieved.

To see the exact correlation, you can copy any period you'd like, with the expected value output fields deleted, to the test.dat file. Then you run the simulator in **Test** mode and get the output value from the simulator for each input vector. You can then compare this with the expected value in your training set or test set.

Now that you're done, you need to un-normalize the data back to get the answer in terms of the change in the S&P 500 index. What you've accomplished is a way in which you can get data from a financial newspaper, like *Barron's* or *Investor's Business Daily*, and feed the current week's data into your trained neural network to get a prediction of what the S&P 500 index is likely to do ten weeks from now.

Here are the steps to un-normalize:

1. Take the predicted scaled target value and calculate, the result value as **Result = (Scaled target/0.02) - 14.8**

2. Take the result from step 1 (which is the percentage change 10 weeks from now) and calculate the projected value, **Projected S&P 10 weeks from now = (This week's S&P value)(1+ Result/100)**

Only the Beginning

This is only a very brief illustration (not meant for trading !) of what you can do with neural networks in financial forecasting. You need to further analyze the data, provide more predictive indicators, and optimize/redesign your neural network architecture to get better generalization and lower error. You need to present many, many more test cases representing different market conditions to have a robust predictor that can be traded with. A graph of the expected and predicted output for the test set and the training set is shown in Figure 14.6. Here, the normalized values are used for the output. Note that the error is about 13.9% on average over the test set and 6.9% over the training set. You can see that the test set did well in the beginning, but showed great divergence in the last few weeks.

The preprocessing steps shown in this chapter should serve as one example of the kinds of steps you can use. There are a vast variety of analysis and statistical methods that can be used in preprocessing. For applying fuzzy data, you can use a program like the fuzzifier program that was developed in Chapter 3 to preprocess some of the data.

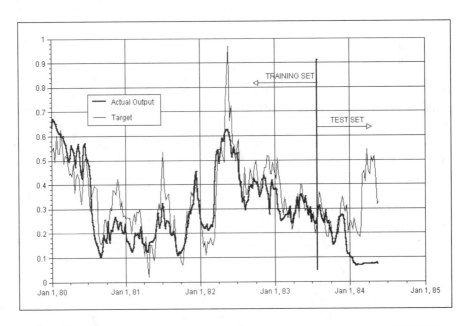

FIGURE 14.6 COMPARISON OF PREDICTED VERSUS ACTUAL FOR THE TRAINING AND TEST DATA SETS.

What's Next?

There are many other experiments you can do from here on. The example chosen was in the field of financial forecasting. But you could certainly try the simulator on other problems like sales forecasting or perhaps even weather forecasting. The key to all of the applications though, is how you present and enhance data, and working through parameter selection by trial and error. Before concluding this chapter, we will cover more topics in pre-processing and present some case studies in financial forecasting. You should consider the suggestions made in the 12-step approach to forecasting model design and research some of the resources listed at the end of this chapter if you have more interest in this area.

TECHNICAL ANALYSIS AND NEURAL NETWORK PREPROCESSING

We cannot overstate the importance of preprocessing in developing a fore-casting model. There is a large body of information related to the study of financial market behavior called Technical Analysis. You can use the math-ematical studies defined by Technical Analysis to preprocess your input data to reveal predictive features. We will present a sampling of Technical Analysis studies that can be used, with formulae and graphs.

Moving Averages

Moving averages are used very widely to capture the underlying trend of a price move. Moving averages are simple filters that average data over a mov-ing window. Popular moving averages include 5-, 10-, and 20 period moving averages. The formula is shown below for a simple moving average, SMA:

$$SMA_t = (P_t + P_{t-1} + ... \ P_{t-n} \ )/ \ n$$
where n = the number of time periods back
$\quad P_{-n}$ = price at n time periods back

An exponential moving average is a weighted moving average that places more weight on the most recent data. The formula for this indicator, EMA is as follows:

$$EMA_t = (1 - a)P_t + a (EMA_{t-1})$$
where a = smoothing constant (typical 0.10)
$\quad P_t$ = price at time t

Momentum and Rate of Change

Momentum is really velocity, or rate of price change with time. The formula for this is

$M_t = (P_t - P_{t-a})$
where a = lookback parameter
for a 5-day momentum value, a = 5

The Rate of Change indicator is actually a ratio. It is the current price divided by the price some interval, *a*, ago divided by a constant. Specifically,

$ROC = P_t / P_{t-a}$ x 1000

Relative Strength Index

The Relative Strength Index (RSI) is the strength of up closes versus down closes over a certain time interval. It is calculated over a time interval T as :

$RSI = 100 - [100 / (1 + RS)]$
where RS = average of x days' up closes/ average of x days' down closes

A typical time interval, *T*, is 14 days. The assumption with the use of RSI is that higher values up closes relative to down closes indicates a strong market, and the opposite indicates weak markets.

Percentage R

This indicator measures where in a recent range of prices today's price falls. The indicator assumes that prices regress to their mean. A low %R indicates that prices are hitting the ceiling of a range, while a high %R indicates that prices are at their low in a range. The formula is:

$\%R = 100 \times (HighX - P)/(HighX - LowX)$
where HighX is the highest price over the price interval of interest
   LowX is the lowest price over the price interval of interest
   P is the current price

Herrick Payoff Index

This indicator makes use of other market data that is available besides price information. It uses the *volume* of the security, which, for a stock, is the number of shares traded for a stock during a particular interval. It also uses the *open interest,* which is the value of the total number of open trades at a particular time. For a commodity future, this is the number of open short and long positions. This study attempts to measure the flow of money in and out of a market. The formula for this is as follows (note that a tick is the smallest permissible move in a given market) :

Let MP = mean price over a particular interval

OI = the larger of yesterday's or today's open interest

then

$$K = [(MP_t - MP_{t-1}) \times \text{dollar value of 1 tick move} \times \text{volume}] \times [1 +/- 2/OI]$$
$$HPI_t = HPI_{t-1} + [0.1 \times (K - HPI_{t-1})] / 100{,}000$$

MACD

The MACD (moving average convergence divergence) indicator is the difference between two moving averages, and it tells you when short-term overbought or oversold conditions exist in the market. The formula is as follows:

Let OSC = EMA1 - EMA2,

where   EMA1 is for one smoothing constant and time period, for example 0.15 and 12   weeks

EMA2 is for another smoothing constant and time period, for example 0.075 and 26   weeks

then

$$MACD_t = MACD_{t-1} + K \times (OSC_t - MACD_{t-1})$$
where K is a smoothing constant, for example, 0.2

The final formula effectively does another exponential smoothing on the difference of the two moving averages, for example, over a 9-week period.

"Stochastics"

This indicator has absolutely nothing to do with stochastic processes. The reason for the name is a mystery, but the indicator is composed of two parts: %K and %D, which is a moving average of %K. The crossover of these lines indicates overbought and oversold areas. The formulas follow:

$$\text{Raw \%K} = 100 \times (P - \text{LowX})/(\text{HighX} - \text{LowX})$$
$$\%K_t = [(\%K_{t-1} \times 2) + \text{Raw \%K}_t]/3$$
$$\%D_t = [(\%D_{t-1} \times 2) + \%K_t]/3$$

On-Balance Volume

The on-balance volume (OBV) indicator was created to try to uncover accumulation and distribution patterns of large player in the stock market. This is a cumulative sum of volume data, specified as follows:

If today's close is greater than yesterday's close

$$OBV_t = OBV_{t-1} + 1$$

If today's close is less than yesterday's close

$$OBV_t = OBV_{t-1} - 1$$

The absolute value of the index is not important; attention is given only to the direction and trend.

Accumulation-Distribution

This indicator does for price what OBV does for volume.

If today's close is greater than yesterday's close:

$$AD_t = AD_{t-1} + (\text{Close}_t - \text{Low}_t)$$

If today's close is less than yesterday's close

$$AD_t = AD_{t-1} + (\text{High}_t - \text{Close}_t)$$

Now let's examine how these indicators look. Figure 14.7 shows a *bar chart,* which is a chart of price data versus time, along with the following indicators:

- Ten-unit moving average
- Ten-unit exponential moving average
- Momentum
- MACD
- Percent R

The time period shown is 5 minute bars for the S&P 500 September 1995 Futures contract. The top of each bar indicates the highest value ("high") for that time interval, the bottom indicates the lowest value("low"), and the horizontal lines on the bar indicate the initial ("open") and final ("close") values for the time interval.

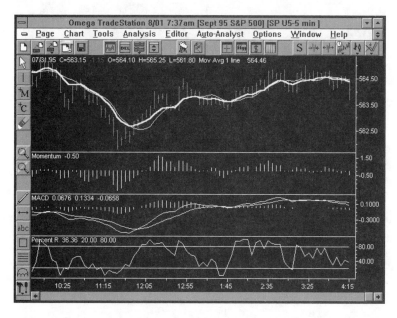

FIGURE 14.7 FIVE MINUTE BAR CHART OF THE S&P 500 SEPT 95 FUTURES CONTRACT
WITH SEVERAL TECHNICAL INDICATORS DISPLAYED.

Figure 14.8 shows another bar chart for Intel Corporation stock for the period from December 1994 to July 1995, with each bar representing a day of activity. The following indicators are displayed also.

- Rate of Change
- Relative Strength
- Stochastics
- Accumulation-Distribution

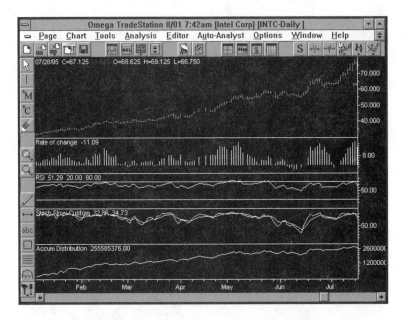

FIGURE 14.8 DAILY BAR CHART OF INTEL CORPORATION
WITH SEVERAL TECHNICAL INDICATORS DISPLAYED.

You have seen a few of the hundreds of technical indicators that have been invented to date. New indicators are being created rapidly as the field of Technical Analysis gains popularity and following. There are also pattern recognition studies, such as formations that resemble flags or pennants as well as more exotic types of studies, like *Elliot wave* counts. You can refer to books on Technical Analysis (e.g., Murphy) for more information about these and other studies.

Neural preprocessing with Technical Analysis tools as well as with traditional engineering analysis tools such as Fourier series, Wavelets, and Fractals can be very useful in finding predictive patterns for forecasting.

WHAT OTHERS HAVE REPORTED

In this final section of the chapter, we outline some case studies documented in periodicals and books, to give you an idea of the successes or failures to date with neural networks in financial forecasting. Keep in mind that the very best (= most profitable) results are usually never reported (so as not to lose a competitive edge) ! Also, remember that the market inefficiencies exploited yesterday may no longer be the same to exploit today.

Can a Three-Year-Old Trade Commodities?

Well, Hillary Clinton can certainly trade commodities, but a three-year-old, too? In his paper, "Commodity Trading with a Three Year Old," J. E. Collard describes a neural network with the supposed intelligence of a three-year-old. The application used a feedforward backpropagation network with a 37-30-1 architecture. The network was trained to buy ("go long") or sell ("go short") in the live cattle commodity futures market. The training set consisted of 789 facts for trading days in 1988, 1989, 1990, and 1991. Each input vector consisted of 18 fundamental indicators and six market technical variables (Open, High, Low, Close, Open Interest, Volume). The network could be trained for the correct output on all but 11 of the 789 facts.

The fully trained network was used on 178 subsequent trading days in 1991. The cumulative profit increased from $0 to $1547.50 over this period by trading one live cattle contract. The largest loss in a trade was $601.74 and the largest gain in a trade was $648.30.

Forecasting Treasury Bill and Treasury Note Yields

Milam Aiken designed a feedforward backpropagation network that predicted Treasury Bill Rates and compared the forecast he obtained with forecasts made by top U.S. economists. The results showed the neural network, given the same data, made better predictions (.18 versus .71 absolute error). Aiken used 250 economic data series to see correlation to T-Bills and used only the series that showed leading correlation: Dept. of Commerce Index of Leading Economic Indicators, the Center for International Business Cycle Research (CIBCR) Short Leading Composite Index, and the CIBCR Long Leading Composite Index. Prior data for these three indicators for the past four years (total 12 inputs) was used to predict the average annual T-Bill rate (one output) for the current year.

Guido Deboeck and Masud Cader designed profitable trading systems for two-year and 10-year treasury securities. They used feedforward neural networks with a learning algorithm called *extended-delta-bar-delta* (*EDBD*), which is a variant of backpropagation. Training samples composed of 100 facts were selected from 1120 trading days spanning from July 1 1989 to June 30, 1992. The test period consisted of more than 150 trading days from July 1, 1992 to December 30, 1992. Performance on the test set was monitored every N thousand training cycles, and the training procedure was stopped when performance degraded on the test set. (This is the same procedure we used when developing a model for the S&P 500.)

A criterion used to judge model performance was the ratio of the average profit divided by the maximum *drawdown,* which is the largest unrealized loss seen during the trading period. A portfolio of separate designed trading systems for two-year and 10-year securities gave the following performance: Over a period of 4.5 years, the portfolio had 133 total trades with 65% profitable trades and the maximum drawdown of 64 *basis points,* or thousands of units for bond yields. The total gain was 677 basis points over that period with a maximum gain in one trade of 52 basis points and maximum loss in one trade of 47 basis points.

The stability and robustness of this system was checked by using over 1000 moving time windows of 3-month, 6-month, and 12-month duration over the 4.5-year interval and noting the standard deviations in profits and maximum drawdown. The maximum drawdown varied from 30 to 48 basis points.

Neural Nets versus Box-Jenkins Time-Series Forecasting

Ramesh Sharda and Rajendra Patil used a standard 12-12-1 feedforward backpropagation network and compared the results with Box-Jenkins methodology for time-series forecasting. Box-Jenkins forecasting is traditional time-series forecasting technique. The authors used 75 different time series for evaluation. The results showed that neural networks achieved better MAPE (mean absolute percentage error) with a mean over all 75 time series MAPEs of 14.67 versus 15.94 for the Box-Jenkins approach.

Neural Nets versus Regression Analysis

Leorey Marquez et al. compared neural network modeling with standard regression analysis. The authors used a feedforward backpropagation network with a structure of 1-6-1. They used three functional forms found in regression analysis:

1. $Y = B0 + B1 \, X + e$
2. $Y = B0 + B1 \log(X) + e$
3. $Y = B0 + B1/X + e$

For each of these forms, 100 pairs of (x,y) data were generated for this "true" model.

Now the neural network was trained on these 100 pairs of data. An additional 100 data points were generated by the network to test the forecasting ability of the network. The results showed that the neural network achieved a MAPE within 0.6% of the true model, which is a very good result. The neural network model approximated the linear model best. An experiment was also done with intentional mis-specification of some data points. The neural network model did well in these cases also, but comparatively worse for the reciprocal model case.

Hierarchical Neural Network

Mendelsohn developed a multilevel neural network as shown in Figure 14.9. Here five neural networks are arranged such that four network outputs feed that final network. The four networks are trained to produce the High, Low, short-term trend, and medium-term trend for a particular financial instrument. The final network takes these four outputs as input and produces a turning point indicator.

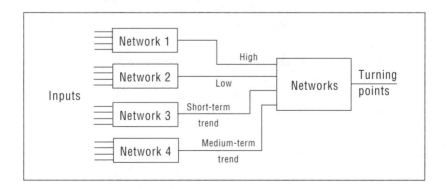

FIGURE 14.9 HIERARCHICAL NEURAL NETWORK SYSTEM TO PREDICT TURNING POINTS.

Each network was trained and tested with 1200 fact days spanning 1988 to 1992 (33% used for testing). Preprocessing was accomplished by using differences of the inputs and with some technical analysis studies:

- Moving averages
- Exponential moving averages
- Stochastic indicators

For the network that produces a predicted High value, the average error ranged between 7.04% and 7.65% for various financial markets over the test period, including Treasury Bonds, Eurodollar, Japanese Yen, and S&P 500 futures contracts.

The Walk-Forward Methodology of Market Prediction

A methodology that is sometimes used in neural network design is *walk-forward* training and testing. This means that you choose an interval of time (e.g., six months) over which you train a neural network and test the network over a subsequent interval. You then move the training window and testing window forward one month, for example, and repeat the exercise. You do this for the time period of interest to see your forecasting results. The advantage of this approach is that you maximize the network's ability to model the recent past in making a prediction. The disadvantage is that the network forget characteristics of the market that happened prior to the training window.

Takashi Kimoto et al. used the walk forward methodology in designing a trading system for Fujitsu and Nikko Securities. They also, like Mendelsohn, use a hierarchical neural network composed of individual feedforward neural networks. Prediction of the TOPIX, which is the Japanese equivalent of the Dow Jones Industrial Average, was performed for 33 months from January 1987 to September 1980. Four networks were used in the first level of the hierarchy trained on price data and economic data. The results were fed to a final network that generated buy and sell signals. The performance of the trading system achieved a result that is 20% better than a buy and hold strategy for the TOPIX.

Dual Confirmation Trading System

Jeremy Konstenius, discusses a trading system for the S&P 400 index with a *holographic neural network*, which is unlike the feedforward backpropagation neural network. The holographic network uses complex numbers for data input and output from neurons, which are mathematically more complex than feedforward network neurons. The author uses two trained networks to forecast the next day's direction based on data for the past 10 days. Each network uses input data that is *detrended,* by subtracting a moving average from the data. Network 1 uses detrended closing values. Network 2 uses detrended High values. If both networks agree, or confirm each other, then a trade is made. There is no trade otherwise.

Network 1 showed an accuracy of 61.9% for the five-month test period (the training period spanned two years prior to the test period), while Network 2 also showed an accuracy of 61.9%. Using the two networks together, Konstenius achieved an accuracy of 65.82%.

A Turning Point Predictor

This neural network approach is discussed by Michitaka Kosaka et al. (1991).

They discuss applying the feedforward backpropagation network to develop buy/sell signals for securities. You would gather time-series data on stock prices, and want to find trends in the data so that changes in the direction of the trend provide you the turning points, which you interpret as signals to buy or sell.

You will need to list these factors that you think have any influence on the price of a security you are studying. You need to also determine how you measure these factors. You then formulate a nonlinear function combining the factors on your list and the past however many prices of your security (your time series data).

The function has the form, as Michitaka Kosaka, et al. (1991) put it,

$p(t + h) = F(x(t), x(t - 1), \ldots, f_1, f_2, \ldots)$
where
$f_1, f_2,$ represent factors on your list,
$x(t)$ is the price of your stock at time t,
$p(t + h)$ is the turning point of security price at time $t + h$, and
$p(t + h) = -1$ for a turn from downward to upward,
$p(t + h) = +1$ for a turn from upward to downward,
$p(t + h) = 0$ for no change and therefore no turn

Here you vary h through the values 1, 2, etc. as you move into the future one day (time period) at a time. Note that the detailed form of the function **F** is not given. This is for you to set up as you see fit.

You can set up a similar function for $x(t + h)$, the stock price at time $t + h$, and have a separate network computing it using the backpropagation paradigm. You will then be generating future prices of the stock and the future buy/sell signals hand in hand, but parallel.

Michitaka Kosaka, et al. (1991) report that they used time-series data over five years to identify the network model, and time-series data over one year to evaluate the model's forecasting performance, with a success rate of 65% for turning points.

The S&P 500 and Sunspot Predictions

Michael Azoff in his book on time-series forecasting with neural networks (see references) creates neural network systems for predicting the S&P 500 index as well as for predicting chaotic time series, such as sunspot occurrences. Azoff uses feedforward backpropagation networks, with a training algorithm called *adaptive steepest descent,* a variation of the standard algorithm. For the sunspot time series, and an architecture of 6-5-1, and a ratio of training vectors to trainable weights of 5.1, he achieves training set error of 12.9% and test set error of 21.4%. This series was composed of yearly sunspot numbers for the years 1706 to 1914. Six years of consecutive annual data were input to the network.

One network Azoff used to forecast the S&P 500 index was a 17-7-1 network. The training vectors to trainable weights ratio was 6.1. The achieved training set error was 3.29%, and on the test set error was 4.67%. Inputs to this network included price data, a *volatility* indicator, which is a function of the range of price movement, and a *random walk* indicator, a technical analysis study.

A Critique of Neural Network Time-Series Forecasting for Trading

Michael de la Maza and Deniz Yuret, managers for the Redfire Capital Management Group, suggest that risk-adjusted return, and not mean-squared error should be the metric to optimize in a neural network application for trading. They also point out that with neural networks, like with statistical methods such as linear regression, data facts that seem unexplainable can't be ignored even if you want them to be. There is no equivalent for a "don't care," condition for the output of a neural network. This type of condition may be an important option for trading environments that have no "discoverable regularity" as the authors put it, and therefore are really not tradable. Some solutions to the two problems posed are given as follows:

- Use an algorithm other than backpropagation, which allows for maximization of risk-adjusted return, such as simulated annealing or genetic algorithms.

- Transform the data input to the network so that minimizing mean-squared error becomes equivalent to maximizing risk-adjusted return.

- Use a hierarchy (see hierarchical neural network earlier in this section) of neural networks, with each network responsible for detecting features or regularities from one component of the data.

411

Resource Guide for Neural Networks and Fuzzy Logic in Finance

412

Here is a sampling of resources compiled from trade literature:

We do not take responsibility for any errors or omissions.

NOTE

Magazines

Technical Analysis of Stocks and Commodities
Technical Analysis, Inc., 3517 S.W. Alaska St., Seattle, WA 98146.

Futures
Futures Magazine, 219 Parkade, Cedar Falls, IA 50613.

AI in Finance
Miller Freeman Inc, 600 Harrison St., San Francisco, CA 94107

NeuroVest Journal
P.O. Box 764, Haymarket, VA 22069

IEEE Transactions on Neural Networks
IEEE Service Center, 445 Hoes Lane, P.O. Box 1331, Piscataway, NJ 08855

Particularly worthwhile is an excellent series of articles by consultant Murray Ruggiero Jr., in *Futures* magazine on neural network design and trading system design in issues spanning July '94 through June '95.

Books

Azoff, Michael, *Neural Network Time Series Forecasting of Financial Markets*, John Wiley and Sons, New York, 1994.

Lederman, Jess, *Virtual Trading*, Probus Publishing, 1995.

Trippi, Robert, *Neural Networks in Finance and Investing*, Probus Publishing, 1993.

Book Vendors

Traders Press, Inc. (800) 927-8222
P.O. Box 6206, Greenville, SC 29606

Traders' Library (800) 272-2855

9051 Red Branch Rd., Suite M, Columbia, MD 21045

Consultants

Mark Jurik

Jurik Research

P.O. Box 2379, Aptos, CA 95001

Hy Rao

Via Software Inc, v: (609) 275-4786, fax: (609) 799-7863

BEI Suite 480, 660 Plainsboro Rd., Plainsboro, NJ 08536

ViaSW@aol.com

Mendelsohn Enterprises Inc.

25941 Apple Blossom Lane

Wesley Chapel, FL 33544

Murray Ruggiero Jr.

Ruggiero Associates,

East Haven, CT

The Schwartz Associates (800) 965-4561

801 West El Camino Real, Suite 150, Mountain View, CA 94040

Historical Financial Data Vendors

CSI (800) CSI-4727

200 W. Palmetto Park Rd., Boca Raton, FL 33432

Free Financial Network, New York (212) 838-6324

Genesis Financial Data Services (800) 808-DATA

411 Woodmen, Colorado Springs, CO 80991

Pinnacle Data Corp. (800) 724-4903

460 Trailwood Ct., Webster, NY 14580

Stock Data Corp. (410) 280-5533

905 Bywater Rd., Annapolis, MD 21401

Technical Tools Inc. (800) 231-8005

334 State St., Suite 201, Los Altos, CA 94022

Tick Data Inc. (800) 822-8425

720 Kipling St., Suite 115, Lakewood, CO 80215

Worden Bros., Inc. (800) 776-4940

4905 Pine Cone Dr., Suite 12, Durham, NC 27707

Preprocessing Tools for Neural Network Development

NeuralApp Preprocessor for Windows

Via Software Inc., v: (609) 275-4786 fax: (609) 799-7863

BEI Suite 480, 660 Plainsboro Rd., Plainsboro, NJ 08536

ViaSW@aol.com

Stock Prophet

Future Wave Software (310) 540-5373

1330 S. Gertruda Ave., Redondo Beach, CA 90277

Wavesamp & Data Decorrelator & Reducer

TSA (800) 965-4561

801 W. El Camino Real, #150, Mountain. View, CA 94040

Genetic Algorithms Tool Vendors

C Darwin

ITANIS International Inc.

1737 Holly Lane, Pittsburgh, PA 15216

EOS

Man Machine Interfaces Inc. (516) 249-4700

555 Broad Hollow Rd., Melville, NY 11747

Evolver

Axcelis, Inc. (206) 632-0885

4668 Eastern Ave. N., Seattle, WA 98103

Fuzzy Logic Tool Vendors

CubiCalc

HyperLogic Corp. (619) 746-2765

1855 East Valley Pkwy., Suite 210, Escondido, CA 92027

TILSHELL

Togai InfraLogic Inc.

5 Vanderbilt, Irvine, CA 92718

Neural Network Development Tool Vendors

Braincel

Promised Land Technologies (203) 562-7335

195 Church St., 8th Floor, New Haven, CT 06510

BrainMaker

California Scientific Software (916) 478-9040

10024 Newtown Rd., Nevada City, CA 95959

ForecastAgent for Windows, ForecastAgent for Windows 95

Via Software Inc., v: (609) 275-4786 fax: (609) 799-7863

BEI Suite 480, 660 Plainsboro Rd., Plainsboro, NJ 08536

ViaSW@aol.com

InvestN 32

RaceCom, Inc. (800) 638-8088

555 West Granada Blvd., Suite E-10, Ormond Beach, FL 32714

NetCaster, DataCaster

Maui Analysis & Synthesis Technologies (808) 875-2516

590 Lipoa Pkwy., Suite 226, Kihei, HI 96753

NeuroForecaster

NIBS Pte. Ltd. (65) 344-2357

62 Fowlie Rd., Republic of Singapore 1542

NeuroShell

Ward Systems Group (301)662-7950

Executive Park West, 5 Hillcrest Dr., Frederick, MD 21702

NeuralWorks Predict

NeuralWare Inc. (412) 787-8222

202 Park West Dr., Pittsburgh, PA 15276

N-Train

Scientific Consultant Services (516) 696-3333

20 Stagecoach Rd., Selden, NY 11784

SUMMARY

416

This chapter presented a neural network application in financial forecasting. As an example of the steps needed to develop a neural network forecasting model, the change in the Standard & Poor's 500 stock index was predicted 10 weeks out based on weekly data for five indicators. Some examples of preprocessing of data for the network were shown as well as issues in training.

At the end of the training period, it was seen that memorization was taking place, since the error in the test data degraded, whereas the error in the training set improved. It is important to monitor the error in the test data (without weight changes) while you are training to ensure that generalization ability is maintained. The final network resulted in average RMS error of 6.9 % over the training set and 13.9% error over the test set.

This chapter's example in forecasting highlights the ease of use and wide applicability of the backpropagation algorithm for large, complex problems and data sets. Several examples of research in financial forecasting were presented with a number of ideas and real-life methodologies presented.

Technical Analysis was briefly discussed with examples of studies that can be useful in preprocessing data for neural networks.

A Resource guide was presented for further information on financial applications of neural networks.

CHAPTER 15

APPLICATION TO NONLINEAR OPTIMIZATION

INTRODUCTION

Nonlinear optimization is an area of operations research, and efficient algorithms for some of the problems in this area are hard to find. In this chapter, we describe the traveling salesperson problem and discuss how this problem is formulated as a nonlinear optimization problem in order to use neural networks (Hopfield and Kohonen) to find an optimum solution. We start with an explanation of the concepts of *linear*, *integer linear* and *nonlinear* optimization.

An optimization problem has an **objective** function and a set of *constraints* on the variables. The problem is to find the values of the variables that lead to an optimum value for the objective function, while satisfying all the constraints. The **objective** function may be a linear function in the variables, or it may be a nonlinear function. For example, it could be a function expressing the total cost of a particular production plan, or a function giving the net profit from a group of products that share a given set of resources. The objective may be to find the minimum value for the objective function, if, for example, it represents cost, or to find the maximum value of a profit function. The resources shared by the products in their manufacturing are usually in limited supply or have some other restrictions on their availability. This consideration leads to the specification of the constraints for the problem.

Each constraint is usually in the form of an equation or an inequality. The left side of such an equation or inequality is an expression in the variables for the problem, and the right-hand side is a constant. The constraints are said to be linear or nonlinear depending on whether the expression on the left-hand side is a linear function or nonlinear function of the variables. A *linear programming problem* is an optimization problem with a *linear* objective function as well as a set of *linear* constraints. An *integer linear* programming problem is a linear programming problem where the variables are required to have integer values. A *nonlinear optimization problem* has one or more of the constraints nonlinear and/or the objective function is nonlinear.

Here are some examples of statements that specify objective functions and constraints:

- Linear objective function: Maximize $Z = 3X_1 + 4X_2 + 5.7X_3$
- Linear equality constraint: $\quad 13X_1 - 4.5X_2 + 7X_3 = 22$
- Linear inequality constraint: $3.6X_1 + 8.4X_2 - 1.7X_3 \leq 10.9$
- Nonlinear objective function: Minimize $Z = 5X^2 + 7XY + Y^2$
- Nonlinear equality constraint: $4X + 3XY + 7Y + 2Y^2 = 37.6$
- Nonlinear inequality constraint: $4.8X + 5.3XY + 6.2Y^2 \geq 34.56$

An example of a linear programming problem is the **blending problem**. An example of a blending problem is that of making different flavors of ice cream blending different ingredients, such as sugar, a variety of nuts, and so on, to produce different amounts of ice cream of many flavors. The objective in the problem is to find the amounts of individual flavors of ice cream to produce with given supplies of all the ingredients, so the total profit is maximized.

A nonlinear optimization problem example is the **quadratic programming problem**. The constraints are all linear but the objective function is a quadratic form. A quadratic form is an expression of two variables with 2 for the sum of the exponents of the two variables in each term.

An example of a quadratic programming problem, is a simple investment strategy problem that can be stated as follows. You want to invest a certain amount in a growth stock and in a speculative stock, achieving at least 25% return. You want to limit your investment in the speculative stock to no more than 40% of the total investment. You figure that the expected return on the growth stock is 18%, while that on the speculative stock is 38%. Suppose G and S represent the proportion of your investment in the growth stock and the speculative stock, respectively. So far you have specified the following constraints. These are linear constraints:

$G + S = 1$	This says the proportions add up to 1.
$S \leq 0.4$	This says the proportion invested in speculative stock is no more than 40%.
$1.18G + 1.38S \geq 1.25$	This says the expected return from these investments should be at least 25%.

Now the objective function needs to be specified. You have specified already the expected return you want to achieve. Suppose that you are a conservative investor and want to minimize the variance of the return. The variance works out as a quadratic form. Suppose it is determined to be:

$$2G^2 + 3S^2 - GS$$

This quadratic form, which is a function of G and S, is your objective function that you want to minimize subject to the (linear) constraints previously stated.

NEURAL NETWORKS FOR OPTIMIZATION PROBLEMS

It is possible to construct a neural network to find the values of the variables that correspond to an optimum value of the **objective** function of a problem. For example, the neural networks that use the *Widrow-Hoff learning rule* find the minimum value of the **error** function using the *least mean squared error*. Neural networks such as the feedforward backpropagation network use the *steepest descent* method for this purpose and find a local minimum of the error, if not the global minimum. On the other hand, the Boltzmann machine or the Cauchy machine uses statistical methods and probabilities and achieves success in finding the global minimum of an **error** function. So we have an idea of how to go about using a neural network to find an optimum value of a function. The question remains as to how the constraints of an optimization problem should be treated in a neural network operation. A good example in answer to this question is the *traveling salesperson problem*. Let's discuss this example next.

TRAVELING SALESPERSON PROBLEM

The traveling salesperson problem is well-known in optimization. Its mathematical formulation is simple, and one can state a simple solution strategy also. Such a strategy is often impractical, and as yet, there is no efficient algorithm for this problem that consistently works in all instances. The traveling salesperson problem is one among the so- called *NP-complete problems*, about which you will read more in what follows. That means that any algorithm for this problem is going to be impractical with certain examples. The neural network approach tends to give solutions with less computing time than other available algorithms for use on a digital computer. The problem is defined as follows. A traveling salesperson has a number of cities to visit. The sequence in which the salesperson visits different cities is called a *tour*. A tour should be such that every city on the list is visited once and only once, except that he returns to the city from which he starts. The goal is to find a tour that minimizes the total distance the salesperson travels, among all the tours that satisfy this criterion.

A simple strategy for this problem is to enumerate all feasible tours—a tour is feasible if it satisfies the criterion that every city is visited but once—to calculate the total distance for each tour, and to pick the tour with the smallest total distance. This simple strategy becomes impractical if the number of cities is large. For example, if there are 10 cities for the traveling salesperson to visit (not counting home), there are 10! = 3,628,800 possible tours, where 10! denotes the factorial of 10—the product of all the integers from 1 to 10—and is the number of distinct permutations of the 10 cities. This number grows to over 6.2 billion with only 13 cities in the tour, and to over a trillion with 15 cities.

The TSP in a Nutshell

For n cities to visit, let X_{ij} be the variable that has value 1 if the salesperson goes from city i to city j and value 0 if the salesperson does not go from city i to city j. Let d_{ij} be the distance from city i to city j. The traveling salesperson problem (TSP) is stated as follows:

Minimize the linear objective function: $\sum\limits_{i=1}^{n} \sum\limits_{j=1}^{n} X_{ij}\, d_{ij}$

subject to:

$$\sum\limits_{\substack{i=1 \\ i \neq j}}^{n} X_{ij} = 1 \text{ for each } j=1, \ldots, n \text{ (linear constraint)}$$

$$\sum\limits_{\substack{i=1 \\ j \neq i}}^{n} X_{ij} = 1 \text{ for each } i=1, \ldots, n \text{ (linear constraint)}$$

$$\sum X_{ij} = 0 \text{ for 1 for all i and j (integer constraint)}$$

This is a 0-1 integer linear programming problem.

Solution via Neural Network

This section shows how the linear and integer constraints of the TSP are absorbed into an objective function that is nonlinear for solution via Neural network.

The first consideration in the formulation of an optimization problem is the identification of the underlying variables and the type of values they can have. In a traveling salesperson problem, each city has to be visited once and only once, except the city started from. Suppose you take it for granted that the last leg of the tour is the travel between the last city visited and the city from which the tour starts, so that this part of the tour need not be explicitly included in the formulation. Then with n cities to be visited, the only information needed for any city is the position of that city in the order of visiting cities in the tour. This suggests that an ordered n-tuple is associated with each city with some element equal to 1, and the rest of the $n - 1$ elements equal to 0. In a neural network representation, this requires n neurons associated with one city. Only one of these n neurons corresponding to the position of the city, in the order of cities in the tour, fires or has output 1. Since there are n cities to be visited, you need n^2 neurons in the network. If these neurons are all arranged in a square array, you need a single 1 in each row and in each column of this array to indicate that each city is visited but only once.

Let x_{ij} be the variable to denote the fact that city i is the jth city visited in a tour. Then x_{ij} is the output of the jth neuron in the array of neurons corresponding to the ith city. You have n^2 such variables, and their values are binary, 0 or 1. In addition, only n of these variables should have value 1 in the solution. Furthermore, exactly one of the x's with the same first subscript (value of i) should have value 1. It is because a given city can occupy only one position in the order of the tour. Similarly, exactly one of the x's with the same second subscript (value of j) should have value 1. It is because a given position in the tour can be only occupied by one city. These are the constraints in the problem. How do you then describe the tour? We take as the starting city for the tour to be city 1 in the array of cities. A tour can be given by the sequence $1, a, b, c, ..., q$, indicating that the cities visited in the tour in order starting at 1 are, $a, b, c, ..., q$ and back to 1. Note that the sequence of subscripts $a, b, ..., q$ is a permutation of 2, 3, ... $n - 1$, $x_{a1}=1$, $x_{b2}=1$, etc.

Having frozen city 1 as the first city of the tour, and noting that distances are symmetric, the distinct number of tours that satisfy the constraints is not $n!$, when there are n cities in the tour as given earlier. It is much less, namely, $n!/2n$. Thus when n is 10, the number of distinct feasible tours is 10!/20, which is 181,440. If n is 15, it is still over 43 billion, and it exceeds a trillion with 17 cities in the tour. Yet for practical purposes there is not much comfort knowing that for the case of 13 cities, 13! is over 6.2 billion and 13!/26 is only 239.5 million—it is still a tough combinatorial problem.

The cost of the tour is the total distance traveled, and it is to be minimized. The total distance traveled in the tour is the sum of the distances from each city to the next. The **objective** function has one term that corresponds to the total distance traveled in the tour. The other terms, one for each constraint, in the **objective** function are expressions, each attaining a minimum value if and only if the corresponding constraint is satisfied. The **objective** function then takes the following form. Hopfield and Tank formulated the problem as one of minimizing energy. Thus it is, customary to refer to the value of the objective function of this problem, while using Hopfield-like neural network for its solution, as the energy level, E, of the network. The goal is to minimize this energy level.

In formulating the equation for E, one uses constant parameters A_1, A_2, A_3, and A_4 as coefficients in different terms of the expression on the right-hand side of the equation. The equation that defines E is given as follows. Note that the notation in this equation includes d_{ij} for the distance from city i to city j.

$$E = A_1 \, \Sigma_i \, \Sigma_k \, \Sigma_{j\neq k} \, x_{ik} \, x_{ij} + A_2 \, \Sigma_i \, \Sigma_k \, \Sigma_{j\neq k} \, x_{ki} \, x_{ji} + A_3 [(\, \Sigma_i \, \Sigma_k \, x_{ik}) - n]^2 + A_4 \, \Sigma_k \, \Sigma_{j\neq k} \, \Sigma_i \, d_{kj} \, x_{ki}(x_{j\cdot i+1} + x_{j\cdot i-1})$$

Our first observation at this point is that E is a **nonlinear** function of the x's, as you have quadratic terms in it. So this formulation of the traveling salesperson problem renders it a **nonlinear optimization problem**.

All the summations indicated by the occurrences of the summation symbol Σ, range from 1 to n for the values of their respective indices. This means that the same summand such as $x_{12}x_{33}$ also as $x_{33}x_{12}$, appears twice with only the factors interchanged in their order of occurrence in the summand. For this reason, many authors use an additional factor of 1/2 for each term in the expression for E. However, when you minimize a quantity z with a given set of values for the variables in the expression for z, the same values for these variables minimize any whole or fractional multiple of z, as well.

The third summation in the first term is over the index j, from 1 to n, but excluding whatever value k has. This prevents you from using something like $x_{12}x_{12}$. Thus, the first term is an abbreviation for the sum of $n^2(n - 1)$ terms with no two factors in a summand equal. This term is included to correspond to the constraint that no more than one neuron in the same row can output a 1. Thus, you get 0 for this term with a valid solution. This is also true for the second term in the right-hand side of the equation for E.

Note that for any value of the index i, x_{ii} has value 0, since you are not making a move like, from city i to the same city i in any of the tours you consider as a solution to this problem. The third term in the expression for E has a minimum value of 0, which is attained if and only if exactly n of the n^2 x's have value 1 and the rest 0.

The last term expresses the goal of finding a tour with the least total distance traveled, indicating the shortest tour among all possible tours for the traveling salesperson. Another important issue about the values of the subscripts on the right-hand side of the equation for E is, what happens to $i + 1$, for example, when i is already equal to n, and to $i-1$, when i is equal to 1. The $i + 1$ and $i - 1$ seem like impossible values, being out of their allowed range from 1 to n. The trick is to replace these values with their moduli with respect to n. This means, that the value $n + 1$ is replaced with 1, and the value 0 is replaced with n in the situations just described.

Modular values are obtained as follows. If we want, say 13 modulo 5, we subtract 5 as many times as possible from 13, until the remainder is a value between 0 and 4, 4 being $5 - 1$. Since we can subtract 5 twice from 13 to get a remainder of 3, which is between 0 and 4, 3 is the value of 13 modulo 5. Thus $(n + 3)$ modulo n is 3, as previously noted. Another way of looking at these results is that 3 is 13 modulo 5 because, if you subtract 3 from 13, you get a number divisible by 5, or which has 5 as a factor. Subtracting 3 from $n + 3$ gives you n, which has n as a factor. So 3 is $(n + 3)$ modulo n. In the case of -1, by subtracting $(n - 1)$ from it, we get $-n$, which can be divided by n getting -1. So $(n - 1)$ is the value of (-1) modulo n.

Example of a Traveling Salesperson Problem for Hand Calculation

Suppose there are four cities in the tour. Call these cities, C_1, C_2, C_3, and C_4. Let the matrix of distances be the following matrix D.

$$
D = \begin{matrix}
0 & 10 & 14 & 7 \\
10 & 0 & 6 & 12 \\
14 & 6 & 0 & 9 \\
7 & 12 & 9 & 0
\end{matrix}
$$

From our earlier discussion on the number of valid and distinct tours, we infer that there are just three such tours. Since it is such a small number, we can afford to enumerate the three tours, find the energy values associated with them, and pick the tour that has the smallest energy level among the three. The three tours are:

- **Tour 1.** $1 - 2 - 3 - 4 - 1$ In this tour city 2 is visited first, followed by city 3, from where the salesperson goes to city 4, and then returns to city 1. For city 2, the corresponding ordered array is (1, 0, 0, 0), because city 2 is the first in this permutation of cities. Then $x_{21} = 1$, $x_{22} = 0$, $x_{23} = 0$, $x_{24} = 0$. Also (0, 1, 0, 0), (0, 0, 1, 0), and (0, 0, 0, 1) correspond to cities 3, 4, and 1, respectively. The total distance of the tour is $d_{12} + d_{23} + d_{34} + d_{41} = 10 + 6 + 9 + 7 = 32$.
- **Tour 2.** $1 - 3 - 4 - 2 - 1$
- **Tour 3.** $1 - 4 - 2 - 3 - 1$

There seems to be some discrepancy here. If there is one, we need an explanation. The discrepancy is that we can find many more tours that should be valid because no city is visited more than once. You may say they are distinct from the three previously listed. Some of these additional tours are:

- **Tour 4.** $1 - 2 - 4 - 3 - 1$
- **Tour 5.** $3 - 2 - 4 - 1 - 3$
- **Tour 6.** $2 - 1 - 4 - 3 - 2$

There is no doubt that these three tours are distinct from the first set of three tours. And in each of these three tours, every city is visited exactly once, as required in the problem. So they are valid tours as well. Why did our formula give us 3 for the value of the number of possible valid tours, while we are able to find 6?

The answer lies in the fact that if two valid tours are symmetric and have the same energy level, because they have the same value for the total distance traveled in the tour, one of them is in a sense redundant, or one of them can be considered degenerate, using the terminology common to this context. As long as they are valid and give the same total distance, the two tours are not individually interesting, and any one of the two is enough to have. By simple inspection, you find the total distances for the six listed tours. They are 32 for tour 1, 32 also for tour 6, 45 for each of tours 2 and 4, and 39 for each of tours 3 and 5. Notice also that tour 6 is not very different from tour 1. Instead of starting at city 1 as in tour 1, if you start at city 2 and follow tour 1 from there in reverse order of visiting cities, you get tour 6. Therefore, the distance covered is the same for both these tours. You can find similar relationships for other pairs of tours that have the same total distance for the tour. Either by reversing the order of cities in a tour, or by making a circular permutation of the order of the cities in a tour, you get another tour with the same total distance. This way you can find tours. Thus, only three distinct total distances are possible, and among them 32 is the lowest. The tour $1 - 2 - 3 - 4 - 1$ is the shortest and is an optimum solution for this traveling salesperson problem. There is an alternative optimum solution, namely tour 6, with 32 for the total length of the tour. The problem is to find an optimal tour, and not to find all optimal tours if more than one exist. That is the reason only three distinct tours are suggested by the formula for the number of distinct valid tours in this case.

The formula we used to get the number of valid and distinct tours to be 3 is based on the elimination of such symmetry. To clarify all this discussion, you should determine the energy levels of all your six tours identified earlier, hoping to find pairs of tours with identical energy levels.

Note that the first term on the right-hand side of the equation results in 0 for a valid tour, as this term is to ensure there is no more than a single 1 in each row. That is, in any summand in the first term, at least one of the factors, x_{ik} or x_{ij}, where $k \neq j$ has to be 0 for a valid tour. So all those summands are 0, making the first term itself 0. Similarly, the second term is 0 for a valid tour, because in any summand at least one of the factors, x_{ki} or x_{ji}, where $k \neq j$ has to be 0 for a valid tour. In all, exactly 4 of the 16 x's are each 1, making the total of the x's 4. This causes the third term to be 0 for a valid tour. These observations make it clear that it does not matter for valid tours, what values are assigned to the parameters A_1, A_2, and A_3. Assigning large values for these parameters would cause the energy levels, for tours that are not valid, to be much larger than the energy levels for valid tours. Thereby, these tours become unattractive for the solution of the traveling salesperson problem. Let us use the value 1/2 for the parameter A_4.

Let us demonstrate the calculation of the value for the last term in the equation for E, in the case of tour 1. Recall that the needed equation is

$$E = A_1 \sum_i \sum_k \sum_{j \neq k} x_{ik} x_{ij} + A_2 \sum_i \sum_k \sum_{j \neq k} x_{ki} x_{ji} + A_3 [(\sum_i \sum_k x_{ik}) - n]^2 + A_4 \sum_k \sum_{j \neq k} \sum_i d_{kj} x_{ki}(x_{j \cdot i+1} + x_{j \cdot i-1})$$

The last term expands as given in the following calculation:

$$A_4\{ d_{12}[x_{12}(x_{23} + x_{21}) + x_{13}(x_{24} + x_{22}) + x_{14}(x_{21} + x_{23})] +$$
$$d_{13}[x_{12}(x_{33} + x_{31}) + x_{13}(x_{34} + x_{32}) + x_{14}(x_{31} + x_{33})] +$$
$$d_{14}[x_{12}(x_{43} + x_{41}) + x_{13}(x_{44} + x_{42}) + x_{14}(x_{41} + x_{43})] +$$
$$d_{21}[x_{21}(x_{12} + x_{14}) + x_{23}(x_{14} + x_{12}) + x_{24}(x_{11} + x_{13})] +$$
$$d_{23}[x_{21}(x_{32} + x_{34}) + x_{23}(x_{34} + x_{32}) + x_{24}(x_{31} + x_{33})] +$$

$$d_{24}[x_{21}(x_{42} + x_{44}) + x_{23}(x_{44} + x_{42}) + x_{24}(x_{41} + x_{43})] +$$
$$d_{31}[x_{31}(x_{12} + x_{14}) + x_{32}(x_{13} + x_{11}) + x_{34}(x_{11} + x_{13})] +$$
$$d_{32}[x_{31}(x_{22} + x_{24}) + x_{32}(x_{23} + x_{21}) + x_{34}(x_{21} + x_{23})] +$$
$$d_{34}[x_{31}(x_{42} + x_{44}) + x_{32}(x_{43} + x_{41}) + x_{34}(x_{41} + x_{43})] +$$
$$d_{41}[x_{41}(x_{12} + x_{14}) + x_{42}(x_{13} + x_{11}) + x_{43}(x_{14} + x_{12})] +$$
$$d_{42}[x_{41}(x_{22} + x_{24}) + x_{42}(x_{23} + x_{21}) + x_{43}(x_{24} + x_{22})] +$$
$$d_{43}[x_{41}(x_{32} + x_{34}) + x_{42}(x_{33} + x_{31}) + x_{43}(x_{34} + x_{32})] \}$$

When the respective values are substituted for the tour $1 - 2 - 3 - 4 - 1$, the previous calculation becomes:

1/2{10[0(0 + 1) + 0(0 + 0) + 1(1 + 0)] + 14[0(0 + 0) + 0(0 + 1) + 1(0 + 0)] +

7[0(1 + 0) + 0(0 + 0) + 1(0 + 1)] + 10[1(0 + 1) + 0(1 + 0) + 0(0 + 0)] +

6[1(1 + 0) + 0(0 + 1) + 0(0 + 0)] + 12[1(0 + 0) + 0(0 + 0) + 0(0 + 1)] +

14[0(0 + 1) +1(0 + 0) + 0(0 + 0)] + 6[0(0 + 0) + 1(0 + 1) + 0(1 + 0)] +

9[0(0 + 0) + 1(1 + 0) + 0(0 + 1)] + 7[0(0 + 1) + 0(0 + 0) + 1(1 + 0)] +

12[0(0 + 0) + 0(0 + 1) + 1(0 + 0)] + 9[0(1 + 0) + 0(0 + 0) + 1(0 + 1)]}

= 1/2(10 + 0 + 7 + 10 + 6 + 0 + 0 + 6 + 9 + 7 + 0 + 9)

= 1/2(64)

= 32

Table 15.1 contains the values we get for the fourth term on the right-hand side of the equation, and for E, with the six listed tours.

TABLE 15.1 ENERGY LEVELS FOR SIX OF THE VALID TOURS

Tour #	Non-Zero x's	Value for the Last Term	Energy Level	Comment
1	$x_{14}, x_{21}, x_{32}, x_{43}$	32	32	1 - 2 - 3 - 4 - 1 tour
2	$x_{14}, x_{23}, x_{31}, x_{42}$	45	45	1 - 3 - 4 - 2 - 1 tour
3	$x_{14}, x_{22}, x_{33}, x_{41}$	39	39	1 - 4 - 2 - 3 - 1 tour
4	$x_{14}, x_{21}, x_{33}, x_{42}$	45	45	1 - 2 - 4 - 3 - 1 tour
5	$x_{13}, x_{21}, x_{34}, x_{42}$	39	39	3 - 2 - 4 - 1 - 3 tour
6	$x_{11}, x_{24}, x_{33}, x_{42}$	32	32	2 - 1 - 4 - 3 - 2 tour

NEURAL NETWORK FOR TRAVELING SALESPERSON PROBLEM

Hopfield and Tank used a neural network to solve a traveling salesperson problem. The solution you get may not be optimal in certain instances. But by and large you may get a solution close to an optimal solution. One cannot find for the traveling salesperson problem a consistently efficient method of solution, as it has been proved to belong to a set of problems called NP-complete problems. NP-complete problems have the distinction that there is no known algorithm that is efficient and practical, and there is little likelihood that such an algorithm will be developed in the future. This is a caveat to keep in mind when using a neural network to solve a traveling salesperson problem.

Network Choice and Layout

We will describe the use of a Hopfield network to attempt to solve the traveling salesperson problem. There are n^2 neurons in the network arranged in a two-dimensional array of n neurons per row and n per column. The network is fully connected. The connections in the network in each row and in each column are lateral connections. The layout of the neurons in the network with their connections is shown in Figure 15.1 for the case of three cities, for illustration. To avoid cluttering, the connections between diagonally opposite neurons are not shown.

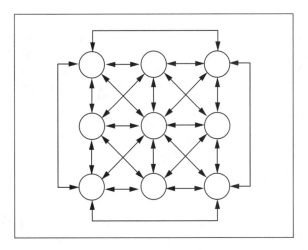

FIGURE 15.1 LAYOUT OF A HOPFIELD NETWORK FOR THE TRAVELING SALESPERSON PROBLEM.

The most important task on hand then is finding an appropriate connection weight matrix. It is constructed taking into account that nonvalid tours should be prevented and valid tours should be preferred. A consideration in this regard is, for example, no two neurons in the same row (column) should fire in the same cycle of operation of the network. As a consequence, the lateral connections should be for inhibition and not for excitation.

In this context, the **Kronecker delta** function is used to facilitate simple notation. The **Kronecker delta** function has two arguments, which are given usually as subscripts of the symbol δ. By definition δ_{ik} has value 1 if $i = k$, and 0 if $i \neq k$. That is, the two subscripts should agree for the Kronecker delta to have value 1. Otherwise, its value is 0.

We refer to the neurons with two subscripts, one for the city it refers to, and the other for the order of that city in the tour. Therefore, an element of the weight matrix for a connection between two neurons needs to have four subscripts, with a comma after two of the subscripts. An example is $w_{ik,lj}$ referring to the weight on the connection between the (ik) neuron and the (lj) neuron. The value of this weight is set as follows:

$$w_{ik,lj} = -A_1\delta_{il}(1-\delta_{kj})-A_2\delta_{kj}(1-\delta_{il})-A_3-A_4\,d_{il}(\delta_{j,\,k+1} + \delta_{j,k-1})$$

Here the negative signs indicate inhibition through the lateral connections in a row or column. The $-A_3$ is a term for global inhibition.

Inputs

The inputs to the network are chosen arbitrarily. If as a consequence of the choice of the inputs, the activations work out to give outputs that add up to the number of cities, an initial solution for the problem, a legal tour, will result. A problem may also arise that the network will get stuck at a local minimum. To avoid such an occurrence, random noise can be added. Usually you take as the input at each neuron a constant times the number of cities and adjust this adding a random number, which can differ for different neurons.

Activations, Outputs, and Their Updating

We denote the activation of the neuron in the ith row and jth column by a_{ij}, and the output is denoted by x_{ij}. A time constant τ, and a gain λ are used as well. A constant m is another parameter used. Also, Δt denotes the increment in time, from one cycle to the next. Keep in mind that the index for the summation Σ ranges from 1 to n, the number of cities. Excluded values of the index are shown by the use of the symbol $\neq$.

The change in the activation is then given by Δa_{ij}, where:

$\Delta a_{ij} = \Delta t \ (Term_1 + Term_2 + Term_3 + Term_4 + Term_5)$
$Term_1 = - a_{ij}/\tau$
$Term_2 = - A_1 \ \Sigma_{k \neq j} x_{ik}$
$Term_3 = - A_2 \ \Sigma_{k \neq i} x_{kj}$
$Term_4 = - A_3(\Sigma_i \Sigma_k x_{ik} - m)$
$Term_5 = - A_4 \ \Sigma_{k \neq i} d_{ik}(x_{k,j+1} + x_{k,j-1})$

To update the activation of the neuron in the ith row and jth column, you take:

$a_{ijnew} = a_{ijold} + \Delta a_{ij}$

The output of a neuron in the ith row and jth column is calculated from:

$x_{ij} = (1 + \tanh(\lambda a_{ij}))/2$

$\dfrac{\tanh X + 1}{2} = \dfrac{1}{1 + e^{-ix}}$, which is the original sigmoid function

NOTE

The function used here is the hyperbolic tangent function. The gain parameter mentioned earlier λ is. The output of each neuron is calculated after updating the activation. Ideally, you want to get the outputs as 0's and 1's, preferably a single one for each row and each column, to represent a tour that satisfies the conditions of the problem. But the hyperbolic tangent function gives a real number, and you have to settle for a close enough value to 1 or 0. You may get, for example, 0.96 instead of 1, or 0.07 instead of 0. The solution is to be obtained from such values by rounding up or down so that 1 or 0 will be used, as the case may be

Performance of the Hopfield Network

Let us now look at Hopfield and Tank's approach at solving the TSP.

Hopfield and Tank Example

The Hopfield network's use in solving the traveling salesperson problem is a pioneering effort in the use of the neural network approach for this problem. Hopfield and Tank's example is for a problem with 10 cities. The parameters used were, $A_1 = 500$, $A_2 = 500$, $A_3 = 200$, $A_4 = 500$, $\tau = 1$, $\lambda = 50$, and $m = 15$. A good solution corresponding to a local minimum for E is the expected, if not the best, solution (global minimum). An annealing process could be considered to move out of any local minimum. As was mentioned before, the traveling salesperson problem is one of those problems for which a single approach cannot be found that will be successful in all cases. There isn't very much guidance as to how to choose the parameters in general for the use of the Hopfield network to solve the traveling salesperson problem.

C++ Implementation of the Hopfield Network for the Traveling Salesperson Problem

We present a C++ program for the Hopfield network operation for the traveling salesperson problem. The header file is in the Listing 15.1, and the source file is in the Listing 15.2. A **tsneuron** class is declared for a neuron and a **network** class for the network. The **network** class is declared a friend in the **tsneuron** class. The program follows the procedure described for setting inputs, connection weights, and updating.

Program Details

The following is a listing of the characteristics of the C++ program along with definitions and/or functions.

- The number of cities, the number of iterations, and the distances between the cities are solicited from the user.

- The distances are taken as **integer** values. If you want to use real numbers as distances, the type for distance matrix needs to be changed to **float**, and corresponding changes are needed for **calcdist** ( ) function, etc.

- **tourcity** and **tourorder** arrays keep track of the cities that have to be covered and the order in which it is to be done.

- A neuron corresponds to each combination of a city and its order in the tour. The ith city visited in the order j , is the neuron corresponding to the element $j + i*n$, in the array for neurons. Here n is the number of cities. The i and the j vary from 0 to $n - 1$. There are n^2 neurons.

- **mtrx** is the matrix giving the weights on the connections between the neurons. It is a square matrix of order n^2 .

- An input vector is generated at random in the function **main ( )**, and is later referred to as **ip**

- **asgninpt ( )** function presents the input vector **ip** to the network and determines the initial activations of the neurons.

- **getacts ( )** function updates the activations of the neurons after each iteration.

- **getouts ( )** function computes the outputs after each iteration. **la** is used as abbreviation for lambda in its argument.

- **iterate ( )** function carries out the number of iterations desired.

- **findtour ( )** function determines the tour orders of cities to be visited using the outputs of the neurons. When used at the end of the iterations, it gives the solution obtained for the traveling salesperson problem.

- **calcdist ( )** function calculates the distance of the tour in the solution.

LISTING 15.1 **HEADER FILE FOR THE C++ PROGRAM FOR THE HOPFIELD NETWORK FOR THE TRAVELING SALESPERSON PROBLEM**

```
//trvslsmn.h   V. Rao,   H. Rao
#include <iostream.h>
#include <stdlib.h>
#include <math.h>
#include <stdio.h>


#define MXSIZ 11



class tsneuron
        {
        protected:
        int cit,ord;
        float output;
        float activation;
        friend class network;

        public:
        tsneuron() { };
```

```
            void getnrn(int,int);
            };

class network
        {
        public:
        int   citnbr;
        float pra,prb,prc,prd,totout,distnce;

        tsneuron (tnrn)[MXSIZ][MXSIZ];
        int dist[MXSIZ][MXSIZ];
        int tourcity[MXSIZ];
        int tourorder[MXSIZ];
        float outs[MXSIZ][MXSIZ];
        float acts[MXSIZ][MXSIZ];
        float mtrx[MXSIZ][MXSIZ];
        float citouts[MXSIZ];
        float ordouts[MXSIZ];

        network() { };
        void getnwk(int,float,float,float,float);
        void getdist(int);
        void findtour();
        void asgninpt(float *);
        void calcdist();
        void iterate(int,int,float,float,float);
        void getacts(int,float,float);
        void getouts(float);

//print functions

        void prdist();
        void prmtrx(int);
        void prtour();
        void practs();
        void prouts();
        };
```

Source File for Hopfield Network for Traveling Salesperson Problem

The following listing gives the source code for the C++ program for the Hopfield network for traveling salesperson problem. The user is prompted to input the number of cities and the maximum number of iterations for the operation of the network.

433

The parameters *a*, *b*, *c*, *d* declared in the function main correspond to A_1, A_2, A_3, and A_4, respectively. These and the parameters *tau*, *lambda*, and *nprm* are all given values in the declaration statements in the function main. If you change these parameter values or change the number of cities in the traveling salesperson problem, the program will compile but may not work as you'd like.

LISTING 15.2 SOURCE FILE FOR THE C++ PROGRAM FOR THE HOPFIELD NETWORK FOR THE TRAVELING SALESPERSON PROBLEM

```
//trvslsmn.cpp V. Rao,  H. Rao
#include "trvslsmn.h"
#include <stdlib.h>
#include <time.h>



//generate random noise



int randomnum(int maxval)
{
// random number generator
// will return an integer up to maxval

return rand() % maxval;
}



//Kronecker delta function

int krondelt(int i,int j)
```

```
        {
        int k;
        k= ((i == j) ? (1):(0));
        return k;
        };

void tsneuron::getnrn(int i,int j)
        {
        cit = i;
        ord = j;
        output = 0.0;
        activation = 0.0;
        };


//distances between cities

void network::getdist(int k)
        {
        citnbr = k;
        int i,j;
        cout<<"\n";

        for(i=0;i<citnbr;++i)
                {
                dist[i][i]=0;
                for(j=i+1;j<citnbr;++j)
                        {
                        cout<<"\ntype distance (integer) from city "<<
                        i<<" to city "<<j<<"\n";
                        cin>>dist[i][j];
                        }
                cout<<"\n";
                }

        for(i=0;i<citnbr;++i)
                {
                for(j=0;j<i;++j)
                        {
                        dist[i][j] = dist[j][i];
```

```
                    }
               }
        prdist();
        cout<<"\n";
        }
```

```
//print distance matrix

void network::prdist()
        {
        int i,j;
        cout<<"\n Distance Matrix\n";

        for(i=0;i<citnbr;++i)
               {
               for(j=0;j<citnbr;++j)
                      {
                      cout<<dist[i][j]<<"   ";
                      }
               cout<<"\n";
               }
        }


//set up network

void network::getnwk(int citynum,float a,float b,float c,float d)
        {
        int i,j,k,l,t1,t2,t3,t4,t5,t6;
        int p,q;
        citnbr = citynum;
        pra = a;
        prb = b;
        prc = c;
        prd = d;
        getdist(citnbr);

        for(i=0;i<citnbr;++i)
```

```
                {
                for(j=0;j<citnbr;++j)
                        {
                        tnrn[i][j].getnrn(i,j);
                        }
                }


        //find weight matrix

        for(i=0;i<citnbr;++i)
                {
                for(j=0;j<citnbr;++j)
                        {
                        p = ((j == citnbr-1) ? (0) : (j+1));
                        q = ((j == 0) ? (citnbr-1) : (j-1));
                        t1 = j + i*citnbr;
                        for(k=0;k<citnbr;++k)
                                {
                                for(l=0;l<citnbr;++l)
                                        {
                                        t2 = l + k*citnbr;
                                        t3 = krondelt(i,k);
                                        t4 = krondelt(j,l);
                                        t5 = krondelt(l,p);
                                        t6 = krondelt(l,q);
                                        mtrx[t1][t2] =
                                        -a*t3*(1-t4) -b*t4*(1-t3)
                                        -c -d*dist[i][k]*(t5+t6);
                                        }
                                }
                        }
                prmtrx(citnbr);
                }

//print weight matrix

void network::prmtrx(int k)
        {
```

```
int i,j,nbrsq;
nbrsq = k*k;
cout<<"\nWeight Matrix\n";
for(i=0;i<nbrsq;++i)
        {
        for(j=0;j<nbrsq;++j)
                {
                if(j%k == 0)
                        {
                        cout<<"\n";
                        }
                cout<<mtrx[i][j]<<"  ";
                }
        cout<<"\n";
        }
    }


//present input to network

void network::asgninpt(float *ip)
    {
    int i,j,k,l,t1,t2;

    for(i=0;i<citnbr;++i)
        {
        for(j =0;j<citnbr;++j)
                {
                acts[i][j] = 0.0;
                }
        }

    //find initial activations
    for(i=0;i<citnbr;++i)
        {
        for(j =0;j<citnbr;++j)
                {
                t1 = j + i*citnbr;
                for(k=0;k<citnbr;++k)
                        {
```

```
                                   for(l=0;l<citnbr;++l)
                                       {
                                       t2 = l + k*citnbr;
                                       acts[i][j] +=
                                       mtrx[t1][t2]*ip[t1];
                                       }
                               }
                       }
               }

       //print activations
       cout<<"\ninitial activations\n";
       practs();
       }


//find activations

void network::getacts(int nprm,float dlt,float tau)
       {
       int i,j,k,p,q;
       float r1, r2, r3, r4,r5;
       r3 = totout - nprm ;

       for(i=0;i<citnbr;++i)
               {
               r4 = 0.0;
               p = ((i == citnbr-1) ? (0) : (i+1));
               q = ((i == 0) ? (citnbr-1) : (i-1));
               for(j=0;j<citnbr;++j)
                       {
                       r1 = citouts[i] - outs[i][j];
                       r2 = ordouts[i] - outs[i][j];
                       for(k=0;k<citnbr;++k)
                               {
                               r4 += dist[i][k] *
                               (outs[k][p] + outs[k][q]);
                               }
                       r5 = dlt*(-acts[i][j]/tau -
```

```
                                pra*r1 -prb*r2 -prc*r3 -prd*r4);
                                acts[i][j] += r5;
                                }
                        }
                }
```

```cpp
//find outputs and totals for rows and columns

void network::getouts(float la)
        {
        float b1,b2,b3,b4;
        int i,j;
        totout = 0.0;

        for(i=0;i<citnbr;++i)
                {
                citouts[i] = 0.0;
                for(j=0;j<citnbr;++j)
                        {
                        b1 = la*acts[i][j];
                        b4 = b1/500.0;
                        b2 = exp(b4);
                        b3 = exp(-b4);
                        outs[i][j] = (1.0+(b2-b3)/(b2+b3))/2.0;
                        citouts[i] += outs[i][j];};
                        totout += citouts[i];
                        }
                for(j=0;j<citnbr;++j)
                        {
                        ordouts[j] = 0.0;
                        for(i=0;i<citnbr;++i)
                                {
                                ordouts[j] += outs[i][j];
                                }
                        }
                }

//find tour
```

```
void network::findtour()
    {
    int i,j,k,tag[MXSIZ][MXSIZ];
    float tmp;
    for (i=0;i<citnbr;++i)
        {
        for(j=0;j<citnbr;++j)
            {
            tag[i][j] = 0;
            }
        }
        for (i=0;i<citnbr;++i)
            {
            tmp = -10.0;
            for(j=0;j<citnbr;++j)
                {
                for(k=0;k<citnbr;++k)
                    {
                    if((outs[i][k] >=tmp)&&
                    (tag[i][k] ==0))
                        tmp = outs[i][k];
                    }
                if((outs[i][j] ==tmp)&&
                (tag[i][j]==0))
                    {
                    tourcity[i] =j;
                    tourorder[j] = i;
                    cout<<"\ntourcity "<<i
                    <<" tour order "<<j<<"\n";
                    for(k=0;k<citnbr;++k)
                        {
                        tag[i][k] = 1;
                        tag[k][j] = 1;
                        }
                    }
                }
            }
        }
```

```
//print outputs

void network::prouts()
        {
        int i,j;
        cout<<"\nthe outputs\n";
        for(i=0;i<citnbr;++i)
                {
                for(j=0;j<citnbr;++j)
                        {
                        cout<<outs[i][j]<<"   ";
                        }
                cout<<"\n";
                }
        }


//calculate total distance for tour

void network::calcdist()
        {
        int i, k, l;
        distnce = 0.0;

        for(i=0;i<citnbr;++i)
                {
                k = tourorder[i];
                l = ((i == citnbr-1 ) ? (tourorder[0]):(tourorder[i+1]));
                distnce += dist[k][l];
                }
        cout<<"\n distance of tour is : "<<distnce<<"\n";
        }


// print tour

void network::prtour()
        {
        int i;
        cout<<"\n the tour :\n";
```

```
for(i=0;i<citnbr;++i)
        {
        cout<<tourorder[i]<<"  ";
        cout<<"\n";
        }
}
```

```
//print activations

void network::practs()
        {
        int i,j;
        cout<<"\n the activations:\n";
        for(i=0;i<citnbr;++i)
                {
                for(j=0;j<citnbr;++j)
                        {
                        cout<<acts[i][j]<<"  ";
                        }
                cout<<"\n";
                }
        }
```

```
//iterate specified number of times

void network::iterate(int nit,int nprm,float dlt,float tau,float la)
        {
        int k;

        for(k=1;k<=nit;++k)
                {
                getacts(nprm,dlt,tau);
                getouts(la);
                }
        cout<<"\n" <<nit<<" iterations completed\n";
        practs();
        cout<<"\n";
        prouts();
        cout<<"\n";
```

```
        }




void main()
        {


//numit = #iterations; n = #cities; u=intial input; nprm - parameter n'
//dt = delta t;
// —————————-
// parameters to be tuned are here:
        int u=1;
        int nprm=10;
        float a=40.0;
        float b=40.0;
        float c=30.0;
        float d=60.0;
        float dt=0.01;
        float tau=1.0;
        float lambda=3.0;
//—————————-
        int i,n2;
        int numit=100;
        int n=4;
        float input_vector[MXSIZ*MXSIZ];

        srand ((unsigned)time(NULL));

        cout<<"\nPlease type number of cities, number of iterations\n";

        cin>>n>>numit;
        cout<<"\n";
        if (n>MXSIZ)
                {
                cout << "choose a smaller n value\n";
                exit(1);
                }
        n2 = n*n;
        for(i=0;i<n2;++i)
```

```
        {
        if(i%n == 0)cout<<"\n";
        input_vector[i] =(u + (float)(randomnum(100)/100.0))/20.0;
        cout<<input_vector[i]<<" ";
        }
```

//create network and operate

```
        network *tntwk = new network;
        if (tntwk==0)
                {
                cout << "not enough memory\n";
                exit(1);
                }
        tntwk->getnwk(n,a,b,c,d);
        tntwk->asgninpt(input_vector);
        tntwk->getouts(lambda);
        tntwk->prouts();
        tntwk->iterate(numit,nprm,dt,tau,lambda);
        tntwk->findtour();
        tntwk->prtour();
        tntwk->calcdist();
        cout<<"\n";
}
```

Output from Your C++ Program for the Traveling Salesperson Problem

A three-city tour problem is trivial, since there is just one value for the total distance no matter how you permute the cities for the order of the tour. In this case the natural order is itself an optimal solution. The program is run for two cases, for illustration. The first run is for a problem with four cities. The second one is for a five-city problem. By the way, the cities are numbered from 0 to n - 1. The same parameter values are used in the two runs. The number of cities, and consequently, the matrix of distances were different. In the first run, the number of iterations asked for is 30, and in the second run it is 40.

The solution you get for the four-city problem is not the one in natural order. The total distance of the tour is 32. The tour in the solution is 1 - 0 - 3 - 2

- 1. This tour is equivalent to the tour in natural order, as you can see by starting at 0 and reading cyclically right to left in the previous sequence of cities.

The solution you get for the five-city problem is the tour 1 - 2 - 0 - 4 - 3 - 1. This reads, starting from 0 as either 0 - 4 - 3 - 1 - 2 - 0, or as 0 - 2 - 1 - 3 - 4 - 0. It is different from the tour in natural order. It has a total distance of 73, as compared to the distance of 84 with the natural order. It is not optimal though, as a tour of shortest distance is 0 - 4 - 2 - 1 - 3 - 0 with total distance 50.

Can the program give you the shorter tour? Yes, the solution can be different if you run the program again because each time you run the program, the input vector is different, as it is chosen at random. The parameter values given in the program are by guess.

Note that the seed for the random number generator is given in the statement

```
srand ((unsigned)time(NULL));
```

The program gives you the order in the tour for each city. For example, if it says **tourcity 1 tour order 2**, that means the second (tour) city is the city visited third (in tour order). Your tour orders are also with values from 0 to $n - 1$, like the cities.

The user input is in italic, and computer output is normal, as you have seen before.

Output for a Four-City Problem

```
Please type number of cities, number of iterations
4 30

0.097 0.0585 0.053 0.078                //input vector—there are 16 neu-
rons in the network
0.0725 0.0535 0.0585 0.0985
0.0505 0.061 0.0735 0.057
0.0555 0.075 0.0885 0.0925

type distance (integer) from city 0 to city 1
10
type distance (integer) from city 0 to city 2
14
type distance (integer) from city 0 to city 3
7
```

```
type distance (integer) from city 1 to city 2
6
type distance (integer) from city 1 to city 3
12

type distance (integer) from city 2 to city 3
9
```

```
 Distance Matrix
0   10   14   7
10   0    6   12
14   6    0    9
7   12   9    0
```

```
 Weight Matrix                    //16x16 matrix of weights. There are 16 neu-
rons in the network.

-30    -70    -70    -70
-70    -630   -30    -630
-70    -870   -30    -70
-30    -70    -70    -630


-70    -30    -70    -70
-630   -70    -630   -30
-870   -70    -870   -70
-70    -30    -70    -30


-70    -70    -30    -70
-30    -630   -70    -630
-30    -870   -70    -70
-70    -70    -30    -630


-70    -70    -70    -30
-630   -30    -630   -70
-870   -30    -870   -70
-630   -30    -630   -30


-70    -630   -30    -630
-30    -70    -70    -70
```

```
-70   -390   -30   -630
-70   -630   -30   -70

-630   -70   -630   -30
-70   -30   -70   -70
-390   -70   -390   -30
-630   -70   -630   -70

-30   -630   -70   -630
-70   -70   -30   -70
-30   -390   -70   -630
-30   -630   -70   -70

-630   -30   -630   -70
-70   -70   -70   -30
-390   -30   -390   -70
-870   -30   -870   -70

-70   -870   -30   -870
-70   -390   -30   -390
-30   -70   -70   -870
-70   -870   -30   -390

-870   -70   -870   -30
-390   -70   -390   -30
-70   -30   -70   -30
-870   -70   -870   -30

-30   -870   -70   -870
-30   -390   -70   -390
-70   -70   -30   -870
-30   -870   -70   -390

-870   -30   -870   -70
-390   -30   -390   -70
-70   -70   -70   -70
-450   -30   -450   -70

-70   -450   -30   -450
```

```
-70   -750   -30   -750
-70   -570   -30   -450
-70   -450   -30   -750


-450   -70   -450   -30
-750   -70   -750   -30
-570   -70   -570   -30
-450   -70   -450   -30


-30   -450   -70   -450
-30   -750   -70   -750
-30   -570   -70   -450
-30   -450   -70   -750


-450   -30   -450   -70
-750   -30   -750   -70
-570   -30   -570   -70
-70   -70   -70   -30


 initial activations

 the activations:
-333.680054   -215.280014   -182.320023   -371.280029
-255.199997   -207.580002   -205.920013   -425.519989
-258.560028   -290.360016   -376.320007   -228.000031
-278.609985   -363   -444.27005   -377.400024


the outputs
0.017913   0.070217   0.100848   0.011483
0.044685   0.076494   0.077913   0.006022
0.042995   0.029762   0.010816   0.060882
0.034115   0.012667   0.004815   0.010678


 30 iterations completed

 the activations:
-222.586884   -176.979172   -195.530823   -380.166107
-164.0271   -171.654053   -214.053177   -421.249023
-158.297867   -218.833755   -319.384827   -245.097473
```

-194.550751   -317.505554   -437.527283   -447.651581

the outputs
0.064704   0.10681   0.087355   0.010333
0.122569   0.113061   0.071184   0.006337
0.130157   0.067483   0.021194   0.050156
0.088297   0.021667   0.005218   0.004624

tourcity 0 tour order 1

tourcity 1 tour order 0

tourcity 2 tour order 3

tourcity 3 tour order 2

 the tour :
1
0
3
2

 distance of tour is : 32

Output for a Five-City Problem

Please type number of cities, number of iterations
5 40

0.0645 0.069 0.0595 0.0615 0.0825   //input vector—there are 25 neurons
in the network
0.074 0.0865 0.056 0.095 0.06
0.0625 0.0685 0.099 0.0645 0.0615
0.056 0.099 0.065 0.093 0.051
0.0675 0.094 0.0595 0.0635 0.0515

```
type distance (integer) from city 0 to city 1
10
type distance (integer) from city 0 to city 2
14
type distance (integer) from city 0 to city 3
7
type distance (integer) from city 0 to city 4
6

type distance (integer) from city 1 to city 2
12
type distance (integer) from city 1 to city 3
9
type distance (integer) from city 1 to city 4
18

type distance (integer) from city 2 to city 3
24
type distance (integer) from city 2 to city 4
16

type distance (integer) from city 3 to city 4
32


 Distance Matrix
0   10  14  7   6
10  0   12  9   18
14  12  0   24  16
7   9   24  0   32
6   18  16  32  0



Weight Matrix    //25x25 matrix of weights. There are 25 neurons in the
network.



-30   -70  -70   -70   -70
```

```
-70   -630   -30    -30   -630
-70    -70   -30    -70    -70
-70   -630   -70   -630    -30
-30   -870   -70    -70    -30


-70    -30   -70    -70    -70
-630   -70  -630    -30    -30
-870   -70   -70    -30    -70
-70    -30  -630    -70   -630
-30    -30   -70    -70    -70


-70    -70   -30    -70    -70
-30   -630   -70   -630    -30
-30    -70   -70    -70    -30
-70    -30   -30   -630    -70
-630   -30   -70    -70    -70


-70    -70   -70    -30    -70
-30    -30  -630    -70   -630
-30    -70   -70    -70    -70
-30   -630   -30    -30   -630
-70   -870   -70   -630    -30


-70    -70   -70    -70    -30
-630   -30   -30   -630    -70
-870   -70  -630    -30    -30
-630   -30   -70    -70    -70
-70    -70  -630    -70   -630


-70   -630   -30    -30   -630
-30    -70   -70    -70    -70
-70   -630   -70   -630    -30
-30    -70   -30    -70    -70
-70   -750   -30   -630    -70


-630   -70  -630    -30    -30
-70    -30   -70    -70    -70
-750   -30  -630    -70   -630
-30    -70   -70    -30    -70
```

```
-70    -30    -30    -30   -630

-30   -630    -70   -630    -30
-70    -70    -30    -70    -70
-30    -30    -30   -630    -70
-630   -70    -70    -70    -30
-70    -30   -630    -30    -30

-30    -30   -630    -70   -630
-70    -70    -70    -30    -70
-30   -630    -30    -30   -630
-70    -70    -70    -70    -70
-30   -750    -70   -870    -30

-630   -30    -30   -630    -70
-70    -70    -70    -70    -30
-750   -70   -870    -30    -30
-870   -70   -750    -30    -30
-750   -30   -870    -70   -870

-70   -870    -30    -30   -870
-70   -750    -30    -30   -750
-30   -870    -70   -870    -30
-30   -750    -70   -750    -30
-30    -70    -30   -870    -70

-870   -70   -870    -30    -30
-750   -70   -750    -30    -30
-70    -30   -870    -70   -870
-30    -30   -750    -70   -750
-30    -70    -30    -30   -870

-30   -870    -70   -870    -30
-30   -750    -70   -750    -30
-70    -30    -30   -870    -70
-870   -30    -30   -750    -70
-750   -70   -870    -30    -30

-30    -30   -870    -70   -870
```

```
-30    -30    -750   -70    -750
-70    -870   -30    -30    -870
-70    -750   -30    -30    -750
-70    -70    -70    -450   -30
```

```
-870   -30    -30    -870   -70
-750   -30    -30    -750   -70
-70    -70    -450   -30    -30
-450   -70    -570   -30    -30
-570   -70    -450   -70    -450
```

```
-70    -450   -30    -30    -450
-70    -570   -30    -30    -570
-70    -450   -70    -450   -30
-30    -570   -70    -570   -30
-30    -1470  -30    -450   -70
```

```
-450   -70    -450   -30    -30
-570   -70    -570   -30    -30
-1470  -30    -450   -70    -450
-30    -30    -570   -70    -570
-30    -30    -30    -30    -450
```

```
-30    -450   -70    -450   -30
-30    -570   -70    -570   -30
-30    -30    -30    -450   -70
-450   -30    -30    -570   -70
-570   -30    -450   -30    -30
```

```
-30    -30    -450   -70    -450
-30    -30    -570   -70    -570
-30    -450   -30    -30    -450
-70    -570   -30    -30    -570
-70    -1470  -70    -390   -30
```

```
-450   -30    -30    -450   -70
-570   -30    -30    -570   -70
-1470  -70    -390   -30    -30
-390   -70    -1110  -30    -30
```

```
-1110   -70   -390   -70   -390

 -70   -390   -30   -30   -390
 -70  -1110   -30   -30  -1110
 -70   -390   -70   -390   -30
 -30  -1110   -70  -1110   -30
 -30   -990   -30   -390   -70

-390   -70   -390   -30   -30
-1110   -70  -1110   -30   -30
-990   -30   -390   -70   -390
 -30   -30  -1110   -70  -1110
 -30   -30   -30   -30   -390

 -30   -390   -70   -390   -30
 -30  -1110   -70  -1110   -30
 -30   -30   -30   -390   -70
-390   -30   -30  -1110   -70
-1110   -30   -390   -30   -30

 -30   -30   -390   -70   -390
 -30   -30  -1110   -70  -1110
 -30   -390   -30   -30   -390
 -70  -1110   -30   -30  -1110
 -70   -990   -30   -30   -990

-390   -30   -30   -390   -70
-1110   -30   -30  -1110   -70
-990   -30   -30   -990   -70
-1950   -30   -30  -1950   -70
 -70   -70   -70   -70   -30

initial activations

 the activations:
-290.894989   -311.190002   -218.365005   -309.344971   -467.774994
-366.299957   -421.254944   -232.399963   -489.249969   -467.399994
-504.375   -552.794983   -798.929871   -496.005005   -424.964935
-374.639984   -654.389832   -336.049988   -612.870056   -405.450012
```

-544.724976 -751.060059   -418.285034   -545.465027   -500.065063

the outputs
0.029577   0.023333   0.067838   0.023843   0.003636
0.012181   0.006337   0.057932   0.002812   0.003652
0.002346   0.001314   6.859939e-05   0.002594   0.006062
0.011034   0.000389   0.017419   0.000639   0.00765
0.001447   0.000122   0.006565   0.001434   0.002471

40 iterations completed

 the activations:
-117.115494   -140.58519   -85.636215   -158.240143   -275.021301
-229.135956   -341.123871   -288.208496   -536.142212   -596.154297
-297.832794   -379.722595   -593.842102   -440.377625   -442.091064
-209.226883   -447.291016   -283.609589   -519.441101   -430.469696
-338.93219   -543.509766   -386.950531   -538.633606   -574.604492

the outputs
0.196963   0.156168   0.263543   0.130235   0.035562
0.060107   0.016407   0.030516   0.001604   0.000781
0.027279   0.010388   0.000803   0.005044   0.004942
0.07511   0.004644   0.032192   0.001959   0.005677
0.016837   0.001468   0.009533   0.001557   0.001012

tourcity 0 tour order 2

tourcity 1 tour order 0

tourcity 2 tour order 1

tourcity 3 tour order 4

tourcity 4 tour order 3

 the tour :
1

2
0
4
3

distance of tour is : 73

Other Approaches to Solve the Traveling Salesperson Problem

The following describes a few other methods for solving the traveling salesperson problem.

Anzai's Presentation

Yuichiro Anzai describes the Hopfield network for the traveling salesperson problem in a slightly different way. For one thing, a **global inhibition term** is not used. A **threshold** value is associated with each neuron, added to the **activation**, and taken as the average of A_1 and A_2, using our earlier notation. The **energy** function is formulated slightly differently, as follows:

$$E = A_1 \Sigma_i (\Sigma_k x_{ik} - 1)^2 + A_2 \Sigma_i (\Sigma_k x_{ki} - 1)^2 +$$
$$A_4 \Sigma_k \Sigma_{j \neq k} \Sigma_{i \neq k} d_{kj} x_{ki} (x_{j \cdot i+1} + x_{j \cdot i-1})$$

The first term is 0 if the sum of the outputs is 1 in each column. The same is true for the second term with regard to rows.

The output is calculated using a parameter λ, here called the *reference activation level*, as:

$$x_{ij} = (1 + \tan \tanh(a_{ij}/\lambda)) / 2$$

The parameters used are $A_1 = 1/2$, $A_2 = 1/2$, $A_4 = 1/2$, $\Delta t = 1$, $\tau = 1$, and $\lambda = 1$. An attempt is made to solve the problem for a tour of 10 cities. The solution obtained is not crisp, in the sense that exactly one 1 occurs in each row and each column, there are values of varying magnitude with one dominating value in each row and column. The prominent value is considered to be part of the solution.

Kohonen's Approach for the Traveling Salesperson Problem

Kohonen's self-organizing maps can be used for the traveling salesperson problem. We summarize the discussion of this approach described in Eric

Davalo's and Patrick Naim's work. Each city considered for the tour is referenced by its x and y coordinates. To each city there corresponds a neuron. The neurons are placed in a single array, unlike the two-dimensional array used in the Hopfield approach. The first and the last neurons in the array are considered to be neighbors.

There is a weight vector for each neuron, and it also has two components. The weight vector of a neuron is the image of the neuron in the map, which is sought to self-organize. There are as many input vectors as there are cities, and the coordinate pair of a city constitutes an input vector. A neuron with a weight vector closest to the input vector is selected. The weights of neurons in a neighborhood of the selected neuron are modified, others are not. A gradually reducing scale factor is also used for the modification of weights.

One neuron is created first, and its weight vector has 0 for its components. Other neurons are created one at a time, at each iteration of learning. Neurons may also be destroyed. The creation of the neuron and destruction of the neuron suggest adding a city provisionally to the final list in the tour and dropping a city also provisionally from that list. Thus the possibility of assigning any neuron to two inputs or two cities is prevented. The same is true about assigning two neurons to the same input.

As the input vectors are presented to the network, if an unselected neuron falls in the neighborhood of the closest twice, it is created. If a created neuron is not selected in three consecutive iterations for modification of weights, along with others being modified, it is destroyed.

That a tour of shortest distance results from this network operation is apparent from the fact that the closest neurons are selected. It is reported that experimental results are very promising. The computation time is small, and solutions somewhat close to the optimal are obtained, if not the optimal solution itself. As was before about the traveling salesperson problem, this is an NP-complete problem and near efficient (leading to suboptimal solutions, but faster) approaches to it should be accepted.

Algorithm for Kohonen's Approach

A gain parameter λ and a scale factor q are used while modifying the weights. A value between 0.02 and 0.2 was tried in previous examples for q. A distance of a neuron from the selected neuron is defined to be an integer between 0 and $n - 1$, where n is the number of cities for the tour. This means that these distances are not necessarily the actual distances between the cities. They could be made representative of the actual distances in some way. One such attempt is described in the following discussion on

C++ implementation. This distance is denoted by d_j for neuron j. A **squashing** function similar to the **Gaussian density** function is also used.

The details of the algorithm are in a paper referenced in Davalo. The steps of the algorithm to the extent given by Davalo are:

- Find the weight vector for which the distance from the input vector is the smallest
- Modify the weights using

$$w_{jnew} = w_{jold} + (I_{new} - w_{jold})g(\lambda, d_j),$$
$$\text{where } g(\lambda, dj) = \exp(-dj^2/\lambda)/\sqrt{2}$$

- Reset λ as $\lambda(1 - q)$

C++ Implementation of Kohonen's Approach

Our C++ implementation of this algorithm (described above) is with small modifications. We create but do not destroy neurons explicitly. That is, we do not count the number of consecutive iterations in which a neuron is not selected for modification of weights. This is a consequence of our not defining a neighborhood of a neuron. Our example is for a problem with five neurons, for illustration, and because of the small number of neurons involved, the entire set is considered a neighborhood of each neuron.

When all but one neuron are created, the remaining neuron is created without any more work with the algorithm, and it is assigned to the input, which isn't corresponded yet to a neuron. After creating $n - 1$ neurons, only one unassigned input should remain.

In our C++ implementation, the distance matrix for the distances between neurons, in our example, is given as follows, following the stipulation in the algorithm that these values should be integers between 0 and $n - 1$.

$$d = \begin{matrix} 0 & 1 & 2 & 3 & 4 \\ 1 & 0 & 1 & 2 & 3 \\ 2 & 1 & 0 & 1 & 2 \\ 3 & 2 & 1 & 0 & 1 \\ 4 & 3 & 2 & 1 & 0 \end{matrix}$$

We also ran the program by replacing the previous matrix with the following matrix and obtained the same solution. The actual distances between the cities are about four times the corresponding values in this matrix, more or less. We have not included the output from this second run of the program.

$$d = \begin{matrix} 0 & 1 & 3 & 3 & 2 \\ 1 & 0 & 3 & 2 & 1 \\ 3 & 3 & 0 & 4 & 2 \\ 3 & 2 & 4 & 0 & 1 \\ 2 & 1 & 2 & 1 & 0 \end{matrix}$$

In our implementation, we picked a function similar to the **Gaussian density** function as the **squashing** function. The **squashing** function used is:

$$f(d,\lambda) = \exp(\,-d^2/\lambda)\,/\,\sqrt{(2\pi)}$$

Header File for C++ Program for Kohonen's Approach

Listing 15.3 contains the header file for this program, and listing 15.4 contains the corresponding source file:

LISTING 15.3 HEADER FILE FOR C++ PROGRAM FOR KOHONEN'S APPROACH

```
//tsp_kohn.h V.Rao, H.Rao
#include<iostream.h>
#include<math.h>

#define MXSIZ 10
#define pi 3.141592654

class city_neuron
      {
        protected:
              double x,y;
              int mark,order,count;
              double weight[2];
              friend class tspnetwork;
        public:
              city_neuron(){};
              void get_neuron(double,double);
      };
```

```
class tspnetwork
      {
      protected:
              int chosen_city,order[MXSIZ];
              double gain,input[MXSIZ][2];
              int citycount,index,d[MXSIZ][MXSIZ];
              double gain_factor,diffsq[MXSIZ];
              city_neuron (cnrn)[MXSIZ];

      public:
              tspnetwork(int,double,double,double,double*,double*);
              void get_input(double*,double*);
              void get_d();
              void find_tour();
              void associate_city();
              void modify_weights(int,int);
              double wtchange(int,int,double,double);
              void print_d();
              void print_input();
              void print_weights();
              void print_tour();
      };
```

Source File Listing

The following is the source file listing for the Kohonen approach to the traveling salesperson problem.

LISTING 15.4 SOURCE FILE FOR C++ PROGRAM FOR KOHONEN'S APPROACH

```
//tsp_kohn.cpp   V.Rao, H.Rao
#include "tsp_kohn.h"

void city_neuron::get_neuron(double a,double b)
      {
      x = a;
      y = b;
      mark = 0;
      count = 0;
      weight[0] = 0.0;
```

```
        weight[1] = 0.0;
        };

tspnetwork::tspnetwork(int k,double f,double q,double h,double *ip0,dou-
ble *ip1)
        {
        int i;
        gain = h;
        gain_factor = f;
        citycount = k;

        // distances between neurons as integers between 0 and n-1

        get_d();
        print_d();
        cout<<"\n";

        // input vectors

        get_input(ip0,ip1);
        print_input();

        // neurons in the network

        for(i=0;i<citycount;++i)
                {
                order[i] = citycount+1;
                diffsq[i] = q;
                cnrn[i].get_neuron(ip0[i],ip1[i]);
                cnrn[i].order = citycount +1;
                }
        }

void tspnetwork::associate_city()
        {
        int i,k,j,u;
        double r,s;

        for(u=0;u<citycount;u++)
```

```
        {
//start a new iteration with the input vectors
for(j=0;j<citycount;j++)
        {
        for(i=0;i<citycount;++i)
                {
                if(cnrn[i].mark==0)
                        {
                        k = i;
                        i =citycount;
                        }
                }

        //find the closest neuron

        for(i=0;i<citycount;++i)
                {
                r = input[j][0] - cnrn[i].weight[0];
                s = input[j][1] - cnrn[i].weight[1];
                diffsq[i] = r*r +s*s;
                if(diffsq[i]<diffsq[k]) k=i;
                }

        chosen_city = k;
        cnrn[k].count++;
        if((cnrn[k].mark<1)&&(cnrn[k].count==2))
                {
                //associate a neuron with a position
                cnrn[k].mark = 1;
                cnrn[k].order = u;
                order[u] = chosen_city;
                index = j;
                gain *= gain_factor;

                //modify weights
                modify_weights(k,index);
                print_weights();
                j = citycount;
                }
```

```
                                    }
                            }
                    }


void tspnetwork::find_tour()
            {
            int i;
            for(i=0;i<citycount;++i)
                    {
                    associate_city();
                    }

            //associate the last neuron with remaining position in
        // tour
                    for(i=0;i<citycount;++i)
                            {
                            if( cnrn[i].mark ==0)
                                    {
                                    cnrn[i].order = citycount-1;
                                    order[citycount-1] = i;
                                    cnrn[i].mark = 1;
                                    }
                            }

                            //print out the tour.
                            //First the neurons in the tour order
                            // Next cities in the tour
                            //order with their x,y coordinates

                            print_tour();
                    }

void tspnetwork::get_input(double *p,double *q)
            {
            int i;

            for(i=0;i<citycount;++i)
                    {
```

```
            input[i][0] = p[i];
            input[i][1] = q[i];
            }
    }


//function to compute distances (between 0 and n-1) between
//neurons

void tspnetwork::get_d()
        {
        int i,j;

        for(i=0;i<citycount;++i)
                {
                for(j=0;j<citycount;++j)
                        {
                        d[i][j] = (j-i);
                        if(d[i][j]<0) d[i][j] = d[j][i];
                        }
                }
        }


//function to find the change in weight component

double tspnetwork::wtchange(int m,int l,double g,double h)
        {
        double r;
        r = exp(-d[m][l]*d[m][l]/gain);
        r *= (g-h)/sqrt(2*pi);
        return r;
        }


//function to determine new weights

void tspnetwork::modify_weights(int jj,int j)
        {
        int i;
        double t;
        double w[2];
```

```
        for(i=0;i<citycount;++i)
            {
            w[0] = cnrn[i].weight[0];
            w[1] = cnrn[i].weight[1];
            //determine new first component of weight
            t = wtchange(jj,i,input[j][0],w[0]);
            w[0] = cnrn[i].weight[0] +t;
            cnrn[i].weight[0] = w[0];

            //determine new second component of weight
            t = wtchange(jj,i,input[j][1],w[1]);
            w[1] = cnrn[i].weight[1] +t;
            cnrn[i].weight[1] = w[1];
            }
        }


//different print routines

void tspnetwork::print_d()
        {
        int i,j;
        cout<<"\n";

        for(i=0;i<citycount;i++)
            {
            cout<<" d: ";
            for(j=0;j<citycount;j++)
                {

                cout<<d[i][j]<<"   ";
                }
            cout<<"\n";
            }
        }


void tspnetwork::print_input()
        {
        int i,j;
```

```
        for(i=0;i<citycount;i++)
            {
            cout<<"input : ";
            for(j=0;j<2;j++)
                {
                cout<<input [i][j]<<"    ";
                }
            cout<<"\n";
            }
    }


void tspnetwork::print_weights()
    {
    int i,j;
    cout<<"\n";

        for(i=0;i<citycount;i++)
            {
            cout<<" weight: ";
            for(j=0;j<2;j++)
                {
                cout<<cnrn[i].weight[j]<<"    ";
                }
            cout<<"\n";
            }
    }


void tspnetwork::print_tour()
    {
    int i,j;
    cout<<"\n tour : ";

        for(i=0;i<citycount;++i)
            {
            cout<<order[i]<<" -> ";
            }
        cout<<order[0]<<"\n\n";
```

```
for(i=0;i<citycount;++i)
        {
        j = order[i];
        cout<<"("<<cnrn[j].x<<", "<<cnrn[j].y<<") -> ";
        }
    j= order[0];
    cout<<"("<<cnrn[j].x<<", "<<cnrn[j].y<<")\n\n";
    }

void main()
        {
        int nc= 5;//nc = number of cities
        double q= 0.05,h= 1.0,p= 1000.0;

        double input2[][5]=
{7.0,4.0,14.0,0.0,5.0,3.0,6.0,13.0,12.0,10.0};
        tspnetwork tspn2(nc,q,p,h,input2[0],input2[1]);
        tspn2.find_tour();


        }
```

Output from a Sample Program Run

The program, as mentioned, is created for the Kohonen approach to the traveling salesperson problem for five cities. There is no user input from the keyboard. All parameter values are given to the program with appropriate statements in the function main. A scale factor of 0.05 is given to apply to the gain parameter, which is given as 1. Initially, the distance of each neuron weight vector from an input vector is set at 1000, to facilitate finding the closest for the first time. The cities with coordinates (7,3), (4,6), (14,13), (0,12), (5,10) are specified for input vectors.

The tour found is not the one in natural order, namely $0 \to 1 \to 2 \to 3 \to 4 \to 0$, with a distance of 43.16. The tour found has the order $0 \to 3 \to 1 \to 4 \to 2 \to 0$, which covers a distance of 44.43, which is slightly higher, as shown in Figure 15.2. The best tour, $0 \to 2 \to 4 \to 3 \to 1 \to 0$ has a total distance of 38.54.

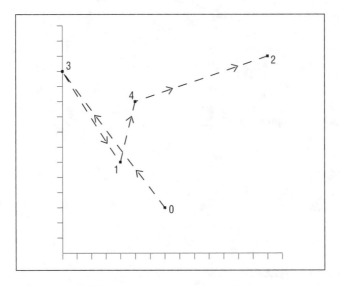

FIGURE 15.2 City placement and tour found for TSP.

Table 15.2 gives for the five-city example, the 12 (5!/10) distinct tour distances and corresponding representative tours. These are not generated by the program, but by enumeration and calculation by hand. This table is provided here for you to see the different solutions for this five-city example of the traveling salesperson problem.

TABLE 15.2 Distances and Representative Tours for Five-City Example

Distance	Tour	Comment
49.05	0-3-2-1-4-0	worst case
47.59	0-3-1-2-4-0	
45.33	0-2-1-4-3-0	
44.86	0-2-3-1-4-0	
44.43	0-3-1-4-2-0	tour given by the program
44.30	0-2-1-3-4-0	
43.29	0-1-4-2-3-0	
43.16	0-1-2-3-4-0	
42.73	0-1-2-4-3-0	
42.26	0-1-3-2-4-0	
40.00	0-1-4-3-2-0	
38.54	0-2-4-3-1-0	optimal tour

There are 12 different distances you can get for tours with these cities by hand calculation, and four of these are higher and seven are lower than the one you find from this program. The worst case tour ($0 \rightarrow 3 \rightarrow 2 \rightarrow 1 \rightarrow 4 \rightarrow 0$) gives a distance of 49.05, and the best, as you saw above, 38.54. The solution from the program is at about the middle of the best and worst, in terms of total distance traveled.

The output of the program being all computer generated is given below as follows:

```
d: 0   1   2   3   4
d: 1   0   1   2   3
d: 2   1   0   1   2
d: 3   2   1   0   1
d: 4   3   2   1   0

input : 7    3
input : 4    6
input : 14   13
input : 0    12
input : 5    10

weight: 1.595769     2.393654
weight: 3.289125e-09    4.933688e-09
weight: 2.880126e-35    4.320189e-35
weight: 1.071429e-78    1.607143e-78
weight: 1.693308e-139   2.539961e-139


weight: 1.595769     2.393654
weight: 5.585192     5.18625
weight: 2.880126e-35    4.320189e-35
weight: 1.071429e-78    1.607143e-78
weight: 1.693308e-139   2.539961e-139


weight: 1.595769     2.393654
weight: 5.585192     5.18625
weight: 5.585192     5.18625
weight: 1.071429e-78    1.607143e-78
weight: 1.693308e-139   2.539961e-139


weight: 1.595769     2.393654
```

```
weight: 5.585192      5.18625
weight: 5.585192      5.18625
weight: 5.585192      5.18625
weight: 1.693308e-139      2.539961e-139

weight: 1.595769      2.393654
weight: 5.585192      5.18625
weight: 5.585192      5.18625
weight: 5.585192      5.18625
weight: 5.585192      5.18625


tour : 0 -> 3-> 1 -> 4 -> 2-> 0


(7, 3) -> (0, 12) -> (4, 6) -> (5, 10) -> (14, 13) -> (7, 3)
```

Optimizing a Stock Portfolio

Development of a neural network approach to a stock selection process in securities trading is similar to the application of neural networks to nonlinear optimization problems. The seminal work of Markowitz in making a mathematical formulation of an **objective** function in the context of *portfolio selection* forms a basis for such a development. There is risk to be minimized or put a cap on, and there are profits to be maximized. Investment capital is a limited resource naturally.

The **objective** function is formulated in such a way that the optimal portfolio minimizes the **objective** function. There would be a term in the objective function involving the product of each pair of stock prices. The covariance of that pair of prices is also used in the **objective** function. A product renders the objective function a quadratic. There would of course be some linear terms as well, and they represent the individual stock prices with the stock's average return as coefficient in each such term. You already get the idea that this optimization problem falls into the category of quadratic programming problems, which result in real number values for the variables in the optimal solution. Some other terms would also be included in the objective function to make sure that the constraints of the problem are satisfied.

A practical consideration is that a real number value for the amount of a stock may be unrealistic, as fractional numbers of stocks may not be purchased. It makes more sense to ask that the variables be taking 0 or 1 only. The implication then is that either you buy a stock, in which case you include it in the portfolio, or you do not buy at all. This is what is usually called a *zero-one programming problem*. You also identify it as a combinatorial problem.

You already saw a combinatorial optimization problem in the traveling salesperson problem. The constraints were incorporated into special terms in the **objective** function, so that the only function to be computed is the **objective** function.

Deeming the **objective** function as giving the energy of a network in a given state, the simulated annealing paradigm and the Hopfield network can be used to solve the problem. You then have a neural network in which each neuron represents a stock, and the size of the layer is determined by the number of stocks in the pool from which you want to build your stock portfolio. The paradigm suggested here strives to minimize the energy of the machine. The **objective** function needs therefore to be stated for minimization to get the best portfolio possible.

Tabu Neural Network

Tabu search, popularized by Fred Glover with his contributions, is a paradigm that has been used successfully in many optimization problems. It is a method that can steer a search procedure from a limited domain to an extended domain, so as to seek a solution that is better than a local minimum or a local maximum.

Tabu search (TS), suggests that an adaptive memory and a *responsive exploration* need to be part of an algorithm. Responsive exploration exploits the information derivable from a selected strategy. Such information may be more substantial, even if the selected strategy is in some sense a bad strategy, than what you can get even in a good strategy that is based on randomness. It is because there is an opportunity provided by such information to intelligently modify the strategy. You can get some clues as to how you can modify the strategy.

When you have a paradigm that incorporates adaptive memory, you see the relevance of associating a neural network:. a TANN is a Tabu neural network. Tabu search and Kohonen's self-organizing map have a common approach in that they work with "neighborhoods." As a new neighborhood is determined, TS prohibits some of the earlier solutions, as it classifies them as *tabu*. Such solutions contain attributes that are identified as *tabu active*.

Tabu search, has STM and LTM components as well. The short-term memory is sometimes called *recency-based* memory. While this may prove adequate to find good solutions, the inclusion of long-term memory makes the search method much more potent. It also does not necessitate longer runs of the search process.

Some of the examples of applications using Tabu search are:

- Training neural nets with the reactive Tabu search
- Tabu Learning: a neural network search method for solving nonconvex optimization problems
- Massively parallel Tabu search for the quadratic assignment problem
- Connection machine implementation of a Tabu search algorithm for the traveling salesman problem
- A Tabu search procedure for multicommodity location/allocation with balancing requirements

SUMMARY

The traveling salesperson problem is presented in this chapter as an example of nonlinear optimization with neural networks. Details of formulation are given of the energy function and its evaluation. The approaches to the solution of the traveling salesperson problem using a Hopfield network and using a Kohonen self-organizing map are presented. C++ programs are included for both approaches.

The output with the C++ program for the Hopfield network refers to examples of four- and five-city tours. The output with the C++ program for the Kohonen approach is given for a tour of five cities, for illustration. The solution obtained is good, if not optimal. The problem with the Hopfield approach lies in the selection of appropriate values for the parameters. Hopfield's choices are given for his 10-city tour problem. The same values for the parameters may not work for the case of a different number of cities. The version of this approach given by Anzai is also discussed briefly.

Use of neural networks for nonlinear optimization as applied to portfolio selection is also presented in this chapter. You are introduced to Tabu search and its use in optimization with neural computing.

CHAPTER 16

APPLICATIONS OF FUZZY LOGIC

INTRODUCTION

Up until now, we have discussed how fuzzy logic could be used in conjunction with neural networks: We looked at a fuzzifier in Chapter 3 that takes crisp input data and creates fuzzy outputs, which then could be used as inputs to a neural network. In chapter 9, we used fuzzy logic to create a special type of associative memory called a FAM (fuzzy associative memory). In this chapter, we focus on applications of fuzzy logic by itself. This chapter starts with an overview of the different types of application areas for fuzzy logic. We then present two application domains of fuzzy logic: fuzzy control systems, and fuzzy databases and quantification. In these sections, we also introduce some more concepts in fuzzy logic theory.

A Fuzzy Universe of Applications

Fuzzy logic is being applied to a wide variety of problems. The most pervasive field of influence is in control systems, with the rapid acceptance of fuzzy logic controllers (FLCs) for machine and process control. There are a number of other areas where fuzzy logic is being applied. Here is a brief list adapted from Yan, et al., with examples in each area:

- **Biological and Medical Sciences** Fuzzy logic based diagnosis systems, cancer research, fuzzy logic based manipulation of prosthetic devices, fuzzy logic based analysis of movement disorders, etc.

- **Management and Decision Support** Fuzzy logic based factory site selection, fuzzy logic aided military decision making (sounds scary, but remember that the fuzzy in fuzzy logic applies to the imprecision in the data and not in the logic), fuzzy logic based decision making for marketing strategies, etc.

- **Economics and Finance** Fuzzy modeling of complex marketing systems, fuzzy logic based trading systems, fuzzy logic based cost-benefit analysis, fuzzy logic based investment evaluation, etc.

- **Environmental Science** Fuzzy logic based weather prediction, fuzzy logic based water quality control, etc.

- **Engineering and Computer Science** Fuzzy database systems, fuzzy logic based prediction of earthquakes, fuzzy logic based automation of nuclear plant control, fuzzy logic based computer network design, fuzzy logic based evaluation of architectural design, fuzzy logic control systems, etc.

- **Operations Research** Fuzzy logic based scheduling and modeling, fuzzy logic based allocation of resources, etc.

- **Pattern Recognition and Classification** Fuzzy logic based speech recognition, fuzzy logic based handwriting recognition, fuzzy logic based facial characteristic analysis, fuzzy logic based military command analysis, fuzzy image search, etc.

- **Psychology** Fuzzy logic based analysis of human behavior, criminal investigation and prevention based on fuzzy logic reasoning, etc.

- **Reliability and Quality Control** Fuzzy logic based failure diagnosis, production line monitoring and inspection, etc.

We will now move to one of the two application domains that we will discuss in depth, Fuzzy Databases. Later in the chapter, we examine the second application domain, Fuzzy Control Systems.

SECTION I: A LOOK AT FUZZY DATABASES AND QUANTIFICATION

In this section, we want to look at some ways in which fuzzy logic may be applied to databases and operations with databases. Standard databases have *crisp* data sets, and you create unambiguous relations over the data sets. You also make queries that are specific and that do not have any ambiguity. Introducing ambiguity in one or more of these aspects of standard databases leads to ideas of how fuzzy logic can be applied to databases. Such application of fuzzy logic could mean that you get databases that are easier to query and easier to interface to. A fuzzy search, where search criteria are not precisely bounded, may be more appropriate than a crisp search. You can recall any number of occasions when you tend to make ambiguous queries, since you are not certain of what you need. You also tend to make ambiguous queries when a "ball park" value is sufficient for your purposes.

In this section, you will also learn some more concepts in fuzzy logic. You will see these concepts introduced where they arise in the discussion of ideas relating to fuzzy databases. You may at times see somewhat of a digression in the middle of the fuzzy database discussion to fuzzy logic topics. You may skip to the area where the fuzzy database discussion is resumed and refer back to the skipped areas whenever you feel the need to get clarification of a concept.

We will start with an example of a standard database, relations and queries. We then point out some of the ways in which fuzziness can be introduced.

Databases and Queries

Imagine that you are interested in the travel business. You may be trying to design special tours in different countries with your own team of tour guides, etc. , and you want to identify suitable persons for these positions. Initially, let us say, you are interested in their own experiences in traveling, and the knowledge they possess, in terms of geography, customs, language, and special occasions, etc. The information you want to keep in your database may be something like, who the person is, the person's citizenship, to where the person traveled, when such travel occurred, the length of stay at that destination, the person's languages, the languages the person understands, the number of trips the person made to each place of travel, etc. Let us use some abbreviations:

cov—country visited

lov—length of visit (days)

nov—number of visits including previous visits

ctz—citizenship

yov—year of visit

lps—language (other than mother tongue) with proficiency to speak

lpu—language with only proficiency to understand

hs—history was studied (1—yes, 0—no)

Typical entries may appear as noted in Table 16.1.

TABLE 16.1 EXAMPLE DATABASE

Name	age	ctz	cov	lov	nov	yov	lps	lpu	hs
John Smith	35	U.S.	India	4	1	1994		Hindi	1
John Smith	35	U.S.	Italy	7	2	1991	Italian		1
John Smith	35	U.S.	Japan	3	1	1993			0

When a query is made to list persons that visited India or Japan after 1992 for 3 or more days, John Smith's two entries will be included. The conditions stated for this query are straightforward, with lov ≥ 3 and yov > 1992 and (cov = India or cov = Japan).

Relations in Databases

A relation from this database may be the set of quintuples, (name, age, cov, lov, yov). Another may be the set of triples, (name, ctz, lps). The quintuple (John Smith, 35, India, 4, 1994) belongs to the former relation, and the triple (John Smith, U.S., Italian) belongs to the latter. You can define other relations, as well.

Fuzzy Scenarios

Now the query part may be made fuzzy by asking to list young persons who recently visited Japan or India for a few days. John Smith's entries may or may not be included this time since it is not clear if John Smith is considered young, or whether 1993 is considered recent, or if 3 days would qualify as a few days for the query. This modification of the query illustrates one of three scenarios in which fuzziness can be introduced into databases and their use.

This is the case where the database and relations are standard, but the queries may be fuzzy. The other cases are: one where the database is fuzzy, but the queries are standard with no ambiguity; and one where you have both a fuzzy database and some fuzzy queries.

Fuzzy Sets Revisited

We will illustrate the concept of fuzziness in the case where the database and the queries have fuzziness in them. Our discussion is guided by the reference Terano, Asai, and Sugeno. First, let us review and recast the concept of a fuzzy set in a slightly different notation.

If **a, b, c,** and **d** are in the set **A** with 0.9, 0.4, 0.5, 0, respectively, as degrees of membership, and in **B** with 0.9, 0.6, 0.3, 0.8, respectively, we give these fuzzy sets **A** and **B** as **A** = { 0.9/a, 0.4/b, 0.5/c} and **B** = {0.9/a, 0.6/b, 0.3/c, 0.8/d}. Now **A**∪**B** = {0.9/a, 0.6/b, 0.5/c, 0.8/d} since you take the larger of the degrees of membership in **A** and **B** for each element. Also, **A**∩**B** = {0.9/a, 0.4/b, 0.3/c} since you now take the smaller of the degrees of membership in **A** and **B** for each element. Since **d** has **0** as degree of membership in **A** (it is therefore not listed in **A**), it is not listed in **A**∩**B**.

Let us impart fuzzy values (FV) to each of the attributes, age, lov, nov, yov, and hs by defining the sets in Table 16.2.

TABLE 16.2 FUZZY VALUES FOR EXAMPLE SETS

Fuzzy Value	Set
FV(age)	{ very young, young, somewhat old, old }
FV(nov)	{ never, rarely, quite a few, often, very often }
FV(lov)	{ barely few days, few days, quite a few days, many days }
FV(yov)	{distant past, recent past, recent }
FV(hs)	{ barely, adequately, quite a bit, extensively }

The attributes of name, citizenship, country of visit are clearly not candidates for having fuzzy values. The attributes of lps, and lpu, which stand for language in which speaking proficiency and language in which understand-

477

ing ability exist, can be coupled into another attribute called *flp* (foreign language proficiency) with fuzzy values. We could have introduced in the original list an attribute called *lpr* (language with proficiency to read) along with *lps* and *lpu*. As you can see, these three can be taken together into the fuzzy-valued attribute of foreign language proficiency. We give below the fuzzy values of flp.

FV(flp) = {not proficient, barely proficient, adequate, proficient, very proficient }

Note that each fuzzy value of each attribute gives rise to a fuzzy set, which depends on the elements you consider for the set and their degrees of membership.

Now let us determine the fuzzy sets that have John Smith as an element, besides possibly others. We need the values of degrees of membership for John Smith (actually his attribute values) in various fuzzy sets. Let us pick them as follows:

Age: $m_{\text{very young}}(35) = 0$ (degree of membership of 35 in very young is 0. We will employ this notation from now on).

$m_{\text{young}}(35) = 0.75$

$m_{\text{somewhat old}}(35) = 0.3$

$m_{\text{old}}(35) = 0$

Assume that similar values are assigned to the degrees of membership of values of John Smith's attributes in other fuzzy sets. Just as John Smith's age does not belong in the fuzzy sets young and old, some of his other attribute values do not belong in some of the other fuzzy sets. The following is a list of fuzzy sets in which John Smith appears:

```
age_young = {0.75/35, ...}
age_somewhat old = {0.3/35, ... }
```

A similar statement attempted for the number of visits may prompt you to list **nov_rarely** = {0.7/1, 0.2/2}, and **nov_quite** a few = {0.3/2, .6/3, ...}. But you readily realize that the number of visits by itself does not mean much unless it is referenced with the country of visit. A person may visit one country very often, but another only rarely. This suggests the notion of a fuzzy relation, which is also a fuzzy set.

What follows is an explanation of relations and discussion of fuzzy relations. If you want to skip this part for now, you may go to the "Fuzzy Queries" section a few pages later in this chapter.

N O T E

479

Fuzzy Relations

A standard relation from set A to set B is a subset of the Cartesian product of A and B, written as $A \times B$. The elements of $A \times B$ are ordered pairs (a, b) where a is an element of A and b is an element of B. For example, the ordered pair (Joe, Paul) is an element of the Cartesian product of the set of fathers, which includes Joe and the set of sons which includes Paul. Or, you can consider it as an element of the Cartesian product of the set of men with itself. In this case, the ordered pair (Joe, Paul) is in the subset which contains (a, b), if a is the father of b. This subset is a relation on the set of men. You can call this relation "father."

A fuzzy relation is similar to a standard relation, except that the resulting sets are fuzzy sets. An example of such a relation is 'much_more_educated'. This fuzzy set may look something like,

```
much_more_educated = { ..., 0.2/(Jeff, Steve), 0.7/(Jeff, Mike), ... }
```

Matrix Representation of a Fuzzy Relation

A fuzzy relation can be given as a matrix also when the underlying sets, call them domains, are finite. For example, let the set of men be $S = \{$ Jeff, Steve, Mike $\}$, and let us use the same relation, much_more_educated. For each element of the Cartesian product $S \times S$, we need the degree of membership in this relation. We already have two such values, $m_{much\_more\_educated}$(Jeff, Steve) = 0.2, and $m_{much\_more\_educated}$(Jeff, Mike) = 0.7. What degree of membership in the set should we assign for the pair (Jeff, Jeff)? It seems reasonable to assign a 0. We will assign a 0 whenever the two members of the ordered pair are the same. Our relation much_more_educated is given by a matrix that may look like the following:

```
                       0/(Jeff, Jeff)   0.2/(Jeff, Steve)  0.7/(Jeff, Mike)
much_more_educated =   0.4/(Steve, Jeff)  0/(Steve, Steve) 0.3/(Steve, Mike)
                       0.1/(Mike, Jeff) 0.6/(Mike, Steve)   0/(Mike, Mike)
```

Note that the first row corresponds to ordered pairs with Jeff as the first member, second column corresponds to those with Steve as the second member, and so on. The main diagonal has ordered pairs where the first and second members are the same.

N O T E

Properties of Fuzzy Relations

A relation on a set, that is a subset of a Cartesian product of some set with itself, may have some interesting properties. It may be *reflexive*. For this you need to have 1 for the degree of membership of each main diagonal entry. Our example here is evidently not reflexive.

A relation may be *symmetric*. For this you need the degrees of membership of each pair of entries symmetrically situated to the main diagonal to be the same value. For example (Jeff, Mike) and (Mike, Jeff) should have the same degree of membership. Here they do not, so our example of a relation is not symmetric.

A relation may be *antisymmetric*. This requires that if a is different from b and the degree of membership of the ordered pair (a, b) is not 0, then its mirror image, the ordered pair (b, a), should have 0 for degree of membership. In our example, both (Steve, Mike) and (Mike, Steve) have positive values for degree of membership; therefore, the relation much_more_educated over the set {Jeff, Steve, Mike} is not antisymmetric also.

A relation may be *transitive*. For transitivity of a relation, you need the following condition, illustrated with our set {Jeff, Steve, Mike}. For brevity, let us use r in place of much_more_educated, the name of the relation:

$$\min (m_r(\text{Jeff, Steve}) , m_r(\text{Steve, Mike})) \leq m_r(\text{Jeff, Mike})$$
$$\min (m_r(\text{Jeff, Mike}) , m_r(\text{Mike, Steve})) \leq m_r(\text{Jeff, Steve})$$
$$\min (m_r(\text{Steve, Jeff}) , m_r(\text{Jeff, Mike})) \leq m_r(\text{Steve, Mike})$$
$$\min (m_r(\text{Steve, Mike}) , m_r(\text{Mike, Jeff})) \leq m_r(\text{Steve, Jeff})$$
$$\min (m_r(\text{Mike, Jeff}) , m_r(\text{Jeff, Steve})) \leq m_r(\text{Mike, Steve})$$
$$\min (m_r(\text{Mike, Steve}) , m_r(\text{Steve, Jeff})) \leq m_r(\text{Mike, Jeff})$$

In the above listings, the ordered pairs on the left-hand side of an occurrence of $\leq$ are such that the second member of the first ordered pair matches the first member of the second ordered pair, and also the right-hand side ordered pair is made up of the two nonmatching elements, in the same order.

In our example,

min (m$_r$(Jeff, Steve) , m$_r$(Steve, Mike)) = min (0.2, 0.3) = 0.2
m$_r$(Jeff, Mike) = 0.7 > 0.2

For this instance, the required condition is met. But in the following:

min (m$_r$(Jeff, Mike), m$_r$(Mike, Steve)) = min (0.7, 0.6) = 0.6
m$_r$(Jeff, Steve) = 0.2 < 0.6

The required condition is violated, so the relation much_more_educated is not transitive.

 NOTE If a condition defining a property of a relation is not met even in one instance, the relation does not possess that property. Therefore, the relation in our example is not reflexive, not symmetric, not even antisymmetric, and not transitive.

If you think about it, it should be clear that when a relation on a set of more than one element is symmetric, it cannot be antisymmetric also, and vice versa. But a relation can be both not symmetric and not antisymmetric at the same time, as in our example.

An example of reflexive, symmetric, and transitive relation is given by the following matrix:

```
1       0.4     0.8
0.4     1       0.4
0.8     0.4     1
```

Similarity Relations

A reflexive, symmetric, and transitive fuzzy relation is said to be a *fuzzy equivalence relation*. Such a relation is also called a *similarity relation*. When you have a similarity relation *s*, you can define the *similarity class* of an element *x* of the domain as the fuzzy set in which the degree of membership of *y* in the domain is m$_s$(*x*, *y*). The similarity class of *x* with the relation *s* can be denoted by [*x*]$_s$.

Resemblance Relations

Do you think similarity and resemblance are one and the same? If x is similar to y, does it mean that x resembles y? Or does the answer depend on what sense is used to talk of similarity or of resemblance? In everyday jargon, Bill may be similar to George in the sense of holding high office, but does Bill resemble George in financial terms? Does this prompt us to look at a 'resemblance relation' and distinguish it from the 'similarity relation'? Of course.

Recall that a fuzzy relation that is reflexive, symmetric, and also transitive is called similarity relation. It helps you to create similarity classes. If the relation lacks any one of the three properties, it is not a similarity relation. But if it is only not transitive, meaning it is both reflexive and symmetric, it is still not a similarity relation, but it is a resemblance relation. An example of a resemblance relation, call it t, is given by the following matrix.

Let the domain have elements a, b, and c:

$$t = \begin{array}{ccc} 1 & 0.4 & 0.8 \\ 0.4 & 1 & 0.5 \\ 0.8 & 0.5 & 1 \end{array}$$

This fuzzy relation is clearly reflexive, and symmetric, but it is not transitive. For example:

$$\min (m_t(a, c) , m_t(c, b)) = \min (0.8, 0.5) = 0.5 ,$$

but the following:

$$m_t(a, b) = 0.4 < 0.5 ,$$

is a violation of the condition for transitivity. Therefore, t is not a similarity relation, but it certainly is a resemblance relation.

Fuzzy Partial Order

One last definition is that of a *fuzzy partial order*. A fuzzy relation that is reflexive, antisymmetric, and transitive is a fuzzy partial order. It differs from a similarity relation by requiring antisymmetry instead of symmetry. In the context of crisp sets, an equivalence relation that helps to generate *equivalence* classes is also a reflexive, symmetric, and transitive relation. But those *equivalence* classes are disjoint, unlike similarity classes with fuzzy relations. With crisp sets, you can define a partial order, and it serves as a basis for making comparison of elements in the domain with one another.

Fuzzy Queries

At this point, our digression from the discussion of fuzzy data bases is finished. Let us now recall, for immediate reference, the entries in the definitions we listed earlier in Table 16.3:

TABLE 16.3 FUZZY VALUES FOR EXAMPLE SETS

Fuzzy Value	Set
FV(age)	{ very young, young, somewhat old, old }
FV(nov)	{ never, rarely, quite a few, often, very often }
FV(lov)	{ barely few days, few days, quite a few days, many days }
FV(yov)	{distant past, recent past, recent }
FV(hs)	{ barely, adequately, quite a bit, extensively }
FV(flp)	{not proficient, barely proficient, adequate, proficient, very proficient }

If you refer to Tables 16.1 and 16.3, you will see that a fuzzy query defines a fuzzy set. For example, suppose you ask for a list of young people with enough knowledge of Italy, and who know Italian almost like a native. You want the intersection of the fuzzy sets, **age_young**, **hs_adequately**, and **flp_very proficient**. John Smith is in the fuzzy set **age_young** with a 0.75 degree of membership, as noted before. Suppose he is in **hs_adequately** with a 0.2 degree of membership, and in **flp_very proficient** with a 0.9 degree of membership. We then put him in the fuzzy set for this query. The degree of membership for John Smith is the smallest value among 0.75, 0.2, and 0.9, since we are taking the intersection of fuzzy sets. So 0.2/John Smith will be part of the enumeration of the query fuzzy set.

Note that you can use unknown in place of the value of an attribute for an item in the database. If you do not know John Smith's age, you can enter unknown in that field of John Smith's record. You may say his age is **unknown**, even though you have a rough idea that he is about 35. You would then be able to assign some reasonable degrees of membership of John Smith in fuzzy sets like **age_young** or **age_somewhat old**.

Extending Database Models

One way of extending a model into a fuzzy model, as far as databases are concerned, is to make use of similarity relations and to extend the operations with them as Buckles and Perry do, such as **PROJECT**. First there are the *domains*, for the database. In our example relating to travel and tour guides above, the domains are:

D1 = { John Smith, ... }, the set of persons included in the   database,

D2 = {India, Italy, Japan, ... }, the set of countries of visit,

D3 = {Hindi, Italian, Japanese, ... }, the set of foreign languages,

D4 = {U.S., ... }, the set of countries of citizenship,

D5 = set of ages,

D6 = set of years,

D7 = set of number of visits,

D8 = set of length of visits.

Note that the enumerated sets are shown with '...' in the sets, to indicate that there may be more entries listed, but we are not giving a complete list. In practice, you will make a complete enumeration unlike in our example. The domains D5, D6, D7, and D8 can also be given with all their elements listed, since in practice, these sets are finite. Conceptually though, they are infinite sets; for example, D6 is the set of positive integers.

Next, you have the similarity relations that you define. And then there are the operations.

Example

Consider a standard database given below. It can be called also the relation R1, from set D1 = {Georgette, Darrell, Ernie , Grace } to set D2 = {Spanish, Italian, French, Japanese, Chinese, Russian} as shown in Table 16.4:

TABLE 16.4 EXAMPLE FOR RELATION R1, WITH DOMAINS D1 AND D2

D1	D2
Georgette	Spanish
Georgette	Italian
Darrell	French

Darrell	Spanish
Darrell	Japanese
Ernie	Spanish
Ernie	Russian
Grace	Chinese
Grace	French

Let's say a fuzzy database has D1 and D2 as domains. Suppose you have a similarity relation S1 on domain D1 given by the matrix:

$$
\begin{matrix}
1 & 0.4 & 0.8 & 0.7 \\
0.4 & 1 & 0.5 & 0.3 \\
0.8 & 0.5 & 1 & 0.4 \\
0.7 & 0.3 & 0.4 & 1
\end{matrix}
$$

Recall that the entries in this matrix represent the degrees of membership of ordered pairs in the relation S1. For example, the first row third column element, which is 0.8, in the matrix refers to the degree of membership of the pair (Georgette, Ernie), since the rows and columns refer to Georgette, Darrell, Ernie and Grace in that order.

The result of the operation:

PROJECT (R1 over D1) with LEVEL(D1) = 0.6

is the relation R2 given in Table 16.5.

TABLE 16.5 RELATION R2, WHICH IS THE RESULT OF **PROJECT** OPERATION

D1

{Georgette, Ernie}
{Georgette, Grace}
Darrell

This projection operation with a condition on the level works as follows. First, the column for D1 is to be picked, of the two columns that R1 shows. Repetitions are removed. The condition says that if two elements of D1 have a similarity of 0.6 or higher, they should be treated as the same. Even though similarity levels of pairs (Georgette, Ernie) and (Georgette, Grace) are both greater than 0.6, the pair (Ernie, Grace) has similarity of 0.4 only so that we do not treat the three as the same.

This type of table is not constructed in the standard database model, so this is part of an extension of the standard model.

Suppose you recast the information in the relation R1 and call it R3, shown in Table 16.6.

TABLE 16.6 EXAMPLE FOR RELATION R3, WITH DOMAINS D1 AND D2

D1	D2
Georgette	{Spanish, Italian}
Darrell	{French, Spanish, Japanese}
Grace	{Chinese, French}
Ernie	{Russian, Spanish}

This kind of a table also is not found in standard databases (where there are groups with more than one element used in the relation), and is an example of an *extended model.*

Possibility Distributions

As an alternative to using similarity relations for introducing fuzziness into a database model, you can, following Umano, et al., use a *possibility distribution-relational model.* The possibility distributions represent the fuzzy values of attributes in the data. An example of a possibility distribution is the fuzzy set you saw before, **nov_rarely** = {0.7/1, 0.2/2}, where **nov_rarely** stands for number of visits considered to be "rarely."

Example

An example of a database on the lines of this model is shown in Table 16.7:

TABLE 16.7 EXAMPLE OF POSSIBILITY DISTRIBUTION RELATIONAL MODEL

Name	Number of Visits Outside the U.S.	Citizenship	Name of Companion on Latest Visit
Peter	3	U.S.	Barbara
Roberto	$\{10, 11\}_p$	Spain	Anne
Andre	2	unknown	Carol
Raj	14	$\{India, U.S.\}_p$	Uma
Alan	unknown	U.S.	undefined
James	many	U.K.	null

A standard database cannot look like this. Entries like **many** and $\{10, 11\}_p$ clearly suggest fuzziness. The entry $\{10, 11\}_p$, is a *possibility distribution*, suggesting that the number of visits outside the United States made by Roberto is either 10 or 11. Similarly, Raj's citizenship is India or United States, but not dual citizenship in both. Andre's citizenship and Alan's number of visits outside the United States are not known, and they can have any values. The possibilities cannot be narrowed down as in the case of Raj's citizenship and Roberto's frequency of visits outside the United States The entry **undefined** is used for Alan because he always traveled alone, he never took a companion.

James' number of visits is fuzzy. He traveled many times. A fuzzy set for many will provide the possibility distribution. It can be defined, for example, as:

many = {0.2/6, 0.5/7, 0.8/8, 1/9, 1/10, ...}

The name of the companion on James' latest visit outside the United States is entered as **null** because we do not know on the one hand whether he never took a companion, in which case we could have used **undefined** as in Alan's case, and on the other whom he took as a companion if he did take one, in which case we could have used **unknown**. Simply put, we use **null** when we do not know enough to use either **unknown** or **undefined**.

Queries

Let us turn our attention now to how queries are answered with this type of a database model. Suppose you want a list of U.S. citizens in your database. Peter and Alan clearly satisfy this condition on citizenship. Andre and Raj can only be said to possibly satisfy this condition. But Roberto and James clearly do not satisfy the given condition. This query itself is crisp and not fuzzy (either you belong to the list of U.S. citizens or you don't). The answer therefore should be a crisp set, meaning that unless the degree of membership is 1, you will not list an element. So you get the set containing Peter and Alan only.

A second query could be for people who made more than a few visits outside the United States Here the query is fuzzy. James with many visits outside United States seems to clearly satisfy the given condition. It is perhaps reasonable to assume that each element in the fuzzy set for **many** appears in the fuzzy set **more than a few** with a degree of membership 1. Andre's 2 may be a number that merits 0 degree of membership in **more than a few**. The other numbers in the database are such that they possibly satisfy the given condition to different degrees. You can see that the answer to this fuzzy query is a fuzzy set. Now, we switch gears a little, to talk more on fuzzy theory. This will help with material to follow.

FUZZY EVENTS, MEANS AND VARIANCES

Let us introduce you to fuzzy events, fuzzy means, and fuzzy variances. These concepts are basic to make a study of *fuzzy quantification* theories. You will see how a variable's probability distribution and its fuzzy set are used together. We will use an example to show how fuzzy means and fuzzy variances are calculated.

Example: XYZ Company Takeover Price

Suppose you are a shareholder of company XYZ and that you read in the papers that its takeover is a prospect. Currently the company shares are selling at $40 a share. You read about a hostile takeover by a group prepared to pay $100 a share for XYZ. Another company whose business is related to XYZ's business and has been on friendly terms with XYZ is offering $85 a share. The employees of XYZ are concerned about their job security and have organized themselves in preparing to buy the company collectively for $60 a share. The buyout price of the company shares is a variable with these three possible values, viz., 100, 85, and 60. The board of directors of XYZ have to make the ultimate decision regarding whom they would sell the company. The probabilities are 0.3, 0.5, and 0.2 respectively, that the board decides to sell at $100 to the first group, to sell at $85 to the second, and to let the employees buy out at $60.

Thus, you get the probability distribution of the takeover price to be as follows:

price	100	85	60
probability	0.3	0.5	0.2

From standard probability theory, this distribution gives a mean(or expected price) of:

$$100 \times 0.3 + 85 \times 0.5 + 60 \times 0.2 = 84.5$$

and a variance of :

$$(100-84.5)^2 \times 0.3 + (85-84.5)^2 \times 0.5 + (60-84.5)^2 \times 0.2 = 124.825$$

Suppose now that a security analyst specializing in takeover situations feels that the board hates a hostile takeover but to some extent they cannot resist the price being offered. The analyst also thinks that the board likes to keep some control over the company, which is possible if they sell the company to their friendly associate company. The analyst recognizes that the Board is sensitive to the fears and apprehensions of their loyal employees with whom they built a healthy relationship over the years, and are going to consider the offer from the employees.

The analyst's feelings are reflected in the following fuzzy set:

$$\{0.7/100, 1/85, 0.5/60\}$$

You recall that this notation says, the degree of membership of 100 in this set is 0.7, that of 85 is 1, and the degree of membership is 0.5 for the value 60.

The fuzzy set obtained in this manner is also called a fuzzy event. A different analyst may define a different fuzzy event with the takeover price values. You, as a shareholder, may have your own intuition that suggests a different fuzzy event. Let us stay with the previous fuzzy set that we got from a security analyst, and give it the name A.

Probability of a Fuzzy Event

At this point, we can calculate the probability for the fuzzy event A by using the takeover prices as the basis to correspond the probabilities in the probability distribution and the degrees of membership in A. In other words, the degrees of membership, 0.7, 1, and 0.5 are treated as having the probabilities 0.3, 0.5, and 0.2, respectively. But we want the probability of the fuzzy event A, which we calculate as the expected **degree of membership** under the probability distribution we are using.

Our calculation gives the following:

$$0.7 \times 0.3 + 1 \times 0.5 + 0.5 \times 0.2 = 0.21 + 0.5 + 0.1 = 0.81$$

Fuzzy Mean and Fuzzy Variance

Our next step is to calculate the fuzzy mean of the takeover price. Let us call it **A_fuzzy_mean**, to make reference to the fuzzy event used.

The calculation is as follows:

$$\text{A\_fuzzy\_mean} = (1/0.81) \times (100 \times 0.7 \times 0.3 + 85 \times 1 \times 0.5 + 60 \times 0.5 \times 0.2) = 85.8$$

To get the fuzzy variance of the takeover price, you need to use values like $(100-85.8)^2$, which is the square of the deviation of the takeover price 100 from the fuzzy mean. A simpler way to calculate, which is mathematically equivalent, is to first take the fuzzy expected value of the square of the takeover price and then to subtract the square of the fuzzy mean. To make it easier, let us use p as the variable that represents the takeover price.

The calculations are as below:

$$\text{A\_fuzzy\_expected } p^2 = 1/(0.81) \times (100^2 \times 0.7 \times 0.3 + 85^2 \times 1 \times 0.5 + 60^2 \times 0.5 \times 0.2) = 7496.91$$
$$\text{A\_fuzzy\_variance} = 7496.91 - 85.8^2 = 7496.91 - 7361.64 = 135.27$$

Fuzzy logic is thus introduced into the realm of probability concepts such as events, and statistical concepts such as mean and variance. Further, you can talk of fuzzy conditional expectations and fuzzy posterior probabilities, etc. enabling you to use fuzziness in Bayesian concepts, regression analysis, and so on. You will then be delving into the field of fuzzy quantification theories. In what follows, we continue our discussion with fuzzy conditional expectations.

Conditional Probability of a Fuzzy Event

Suppose you, as a shareholder of the XYZ company in the previous example come up with the fuzzy set, we will call the fuzzy event:

B = {0.8/100, 0.4/85, 0.7/60}

The probability for your fuzzy event is as follows:

0.8 x 0.3 + 0.4 x 0.5 + 0.7 x 0.2 = 0.58

B_fuzzy_mean and **B_fuzzy_variance** of the takeover price of XYZ stock work out as 85.17 and 244.35, respectively. But you want to see how these values change if at all when you take **A**, the analyst's fuzzy event, as a given. You are then asking to determine the conditional probability of your fuzzy event, and your conditional fuzzy mean and fuzzy variance as well.

The conditional probability of your fuzzy event is calculated as follows:

(1/0.81) x (0.8 x 0.7 x 0.3 + 0.4 x 1 x 0.5 + 0.7 x 0.5 x 0.2) = 0.54

This value is smaller than the probability you got before when you did not take the analyst's fuzzy event as given. The a priori probability of fuzzy event **B** is 0.58, while the a posteriori probability of fuzzy event **B** given the fuzzy event **A** is 0.54.

Conditional Fuzzy Mean and Fuzzy Variance

The conditional **B_fuzzy_mean** of the takeover price with fuzzy event **A** as given works out as :

(1/0.54) x (100 x 0.8 x 0.7 x 0.3 + 85 x 0.4 x 1 x 0.5 + 60 x 0.7 x 0.5 x 0.2) = 70.37

and the conditional **B_fuzzy_variance** of the takeover price with fuzzy event **A**, as given, amounts to 1301.76, which is over five times as large as when you did not take the analyst's fuzzy event as given.

Linear Regression a la Possibilities

When you see the definitions of fuzzy means and fuzzy variances, you may think that regression analysis can also be dealt with in the realm of fuzzy logic. In this section we discuss what approach is being taken in this regard.

First, recall what regression analysis usually means. You have a set of x-values and a corresponding set of y values, constituting a number of sample observations on variables X and Y. In determining a linear regression of Y on X, you are taking Y as the dependent variable, and X as the independent variable, and the linear regression of Y on X is a linear equation expressing Y in terms of X. This equation gives you the line 'closest' to the sample points (the scatter diagram) in some sense. You determine the coefficients in the equation as those values that minimize the sum of squares of deviations of the actual y values from the y values from the line. Once the coefficients are determined, you can use the equation to estimate the value of Y for any given value of X. People use regression equations for forecasting.

Sometimes you want to consider more than one independent variable, because you feel that there are more than one variable which collectively can explain the variations in the value of the dependent variable. This is your *multiple regression model*. Choosing your independent variables is where you show your modeling expertise when you want to explain what happens to Y, as X varies.

In any case, you realize that it is an optimization problem as well, since the minimization of the sum of squares of deviations is involved. Calculus is used to do this for Linear Regression. Use of calculus methods requires certain continuity properties. When such properties are not present, then some other method has to be used for the optimization problem.

The problem can be formulated as a **linear programming** problem, and techniques for solving **linear programming** problems can be used. You take this route for solving a linear regression problem with fuzzy logic.

In a previous section, you learned about possibility distributions. The linear regression problem with fuzzy logic is referred to as a linear possibility regression problem. The model, following the description of it by Tarano, Asai, and Sugeno, depends upon a reference function **L**, and fuzzy numbers in the form of ordered pairs **(a, b)**. We will present fuzzy numbers in the next section and then return to continue our discussion of the linear possibility regression model.

Fuzzy Numbers

A fuzzy number is an ordered pair of numbers **(a, b)** with respect to a reference function **L**, which gives you the membership function. Here **b** has to be a positive number.

Here are the properties of **L**, the reference function:

1. It is a function of one variable and is symmetric about **0**. That is, $L(x) = L(-x)$.

2. It has a value of **1** at **x = 0**. In other words, $L(0) = 1$.

3. It is generally decreasing, when **x** is **0** or a positive number, meaning that its value drops as the value of its argument is increased. For example, $L(2) < L(1)$. Thus $L(x)$ has its values less than 1 for positive values of x. It does not make sense to have negative numbers as values of **L**, and so you ignore the values of **x** that cause such values for **L**.

4. The maximum value for the function is **1**, at **x = 0**. It has a sort of a bell-shaped curve.

If **A** is a fuzzy number with the ordered pair (a, b) with $b > 0$, and if the reference function is **L**, you do the following:

You write the fuzzy number as **A = (a, b)$_L$**. You get the membership of any element **x**, by taking the quantity **(x - a) / b** (which reminds you of how you get a **z-score**), and evaluate the reference function **L** at this argument. That is:

$$m_A(x) = L((x-a) / b)$$

Examples of a reference function **L** are:

Example 1:	$L(x) = \max (0, 1 - x^2)$

You obtain the following shown in Table 16.8.

TABLE 16.8 REFERENCE FUNCTION L

x	L(x)
-2	0
-1	0
-0.5	0.75
0	1
0.5	0.75
1	0
2	0

This function is not *strictly* decreasing though. It remains a constant at 0 for $x > 1$.

Example 2:	$L(x) = 1/(1 + x^2)$

You get the values shown in Table 16.9.

TABLE 16.9 REFERENCE FUNCTION L

x	L(x)
-7	0.02
-2	0.2
-1	0.5
-0.5	0.8
0	1
0.5	0.8
1	0.5
2	0.2
7	0.02

Let us now determine the membership of 3 in the fuzzy number A = (4, 10)$_L$, where **L** is the function in the second example above, viz., $1/(1 + x^2)$.

First, you get $(x- 4) / 10 = (3 - 4) / 10 = - 0.1$. Use this as the argument of the reference function **L**. $1/ (1 + (- 0.1)^2)$ gives 0.99. This is expressed as follows:

$$m_A(3) = L((3-4) / 10) = 1/ (1 + (- 0.1)^2) = 0.99$$

You can verify the values, $m_A(0) = 0.862$, and $m_A(10) = 0.735$.

Triangular Fuzzy Number

With the right choice of a reference function, you can get a symmetrical fuzzy number **A**, such that when you plot the membership function in **A**, you get a triangle containing the pairs $(x, m_A(x))$, with $m_A(x) > 0$. An example is $A = (5, 8)_L$, where $L = max(1 - |x|, 0)$.

The numbers **x** that have positive values for $m_A(x)$ are in the interval (-3, 13). Also, $m_A(-3) = 0$, and $m_A(13) = 0$. However, $m_A(x)$ has its maximum value at $x = 5$. Now, if **x** is less than -3 or greater than 13, the value of **L** is zero, and you do not consider such a number for membership in **A**. So all the elements for which membership in **A** is nonzero are in the triangle.

This *triangular fuzzy number* is shown in Figure 16.1. The height of the triangle is 1, and the width is 16, twice the number 8, midpoint of the base is at 5. The pair of numbers 5 and 8 are the ones defining the symmetrical fuzzy number **A**. The vertical axis gives the membership, so the range for this is from 0 to 1.

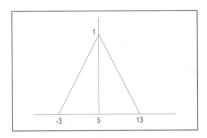

FIGURE 16.1 TRIANGULAR MEMBERSHIP FUNCTION.

Linear Possibility Regression Model

Assume that you have **(n + 1)-tuples** of values of $x_1, ... x_n$, and **y**. That is, for each **i**, **i** ranging from **1** to **k**, you have $(x_1, ... , x_n, y)$, which are **k** sample observations on $X_1, ... , X_n$, and **Y**. The linear possibility regression model is formulated differently depending upon whether the data collected is crisp or fuzzy.

Let us give such a model below, for the case with crisp data. Then the fuzziness lies in the coefficients in the model. You use symmetrical fuzzy numbers, $A_j = (a_j, b_j)_L$. The linear possibility regression model is formulated as:

$$Y_j = A_0 + A_1 X_{j1} + ... + A_n X_{jn}$$

The value of Y from the model is a fuzzy number, since it is a function of the fuzzy coefficients. The fuzzy coefficients A_j are chosen as those that minimize the width (the base of the triangle) of the fuzzy number Y_j. But A_j is also determined by how big the membership of observed y_j is to be, in the fuzzy number Y_j. This last observation provides a constraint for the linear programming problem which needs to be solved to find the linear possibility regression. You select a value d, and ask that $m_Y(y) \geq d$.

We close this section by observing that linear possibility regression gives triangular fuzzy numbers for Y, the dependent variable. It is like doing interval estimation, or getting a regression band. Readers who are seriously interested in this topic should refer to Terano, et al. (see references).

SECTION II: FUZZY CONTROL

This section discusses the fuzzy logic controller (FLC), its application and design. Fuzzy control is used in a variety of machines and processes today, with widespread application especially in Japan. A few of the applications in use today are in the list in Table 16.10, adapted from Yan, et al.

TABLE 16.10 APPLICATIONS OF FUZZY LOGIC CONTROLLERS (FLCS) AND FUNCTIONS PERFORMED

Application	FLC function(s)
Video camcorder	Determine best focusing and lighting when there is movement in the picture
Washing machine	Adjust washing cycle by judging the dirt, size of the load, and type of fabric
Television	Adjust brightness, color, and contrast of picture to please viewers
Motor control	Improve the accuracy and range of motion control under unexpected conditions
Subway train	Increase the stable drive and enhance the stop accuracy by evaluating the passenger traffic conditions. Provide a smooth start and smooth stop.
Vacuum cleaner	Adjust the vacuum cleaner motor power by judging the amount of dust and dirt and the floor characteristics
Hot water heater	Adjust the heating element power according to the temperature and the quantity of water being used
Helicopter control	Determine the best operation actions by judging human instructions and the flying conditions including wind speed and direction

DESIGNING A FUZZY LOGIC CONTROLLER

A fuzzy logic controller diagram was shown in Chapter 3. Let us redraw it now and discuss a design example. Refer to Figure 16.2. For the purpose of discussion, let us assume that this FLC controls a hot water heater. The hot water heater has a knob, HeatKnob(0-10) on it to control the heating element power, the higher the value, the hotter it gets, with a value of 0 indicating the heating element is turned off. There are two sensors in the hot water heater, one to tell you the temperature of the water (TempSense), which varies from 0 to 125° C, and the other to tell you the level of the water in the tank (LevelSense), which varies from 0 = empty to 10 = full. Assume that there is an automatic flow control that determines how much cold water (at temperature 10° C) flows into the tank from the main water supply; whenever the level of the water gets below 40, the flow control turns on, and turns off when the level of the water gets above 95.

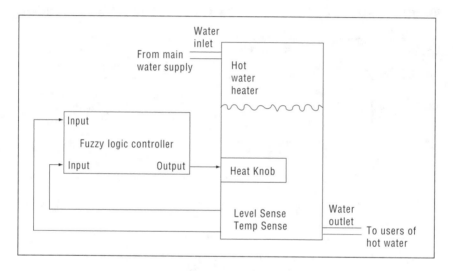

FIGURE 16.2 FUZZY CONTROL OF A WATER HEATER.

The design objective can be stated as:

Keep the water temperature as close to 80° C as possible, in spite of changes in the water flowing out of the tank, and cold water flowing into the tank.

Step One: Defining Inputs and Outputs for the FLC

The range of values that inputs and outputs may take is called the *universe of discourse.* We need to define the universe of discourse for all of the inputs and outputs of the FLC, which are all crisp values. Table 16.11 shows the ranges:

TABLE 16.11 UNIVERSE OF DISCOURSE FOR INPUTS AND OUTPUTS FOR FLC

Name	Input/Output	Minimum value	Maximum value
LevelSense	I	0	10
HeatKnob	O	0	10
TempSense	I	0	125

Step Two: Fuzzify the Inputs

The inputs to the FLC are the LevelSense and the TempSense. We can use triangular membership functions to fuzzify the inputs, just as we did in Chapter 3, when we constructed the fuzzifier program. There are some general guidelines you can keep in mind when you determine the range of the fuzzy variables as related to the crisp inputs (adapted from Yan, et al.):

1. Symmetrically distribute the fuzzified values across the universe of discourse.

2. Use an odd number of fuzzy sets for each variable so that some set is assured to be in the middle. The use of 5 to 7 sets is fairly typical.

3. Overlap adjacent sets (by 15% to 25% typically) .

Both the input variables LevelSense and TempSense are restricted to positive values. We use the following fuzzy sets to describe them:

XSmall, Small, Medium, Large, XLarge

In Table 16.12 and Figure 16.3, we show the assignment of ranges and triangular fuzzy membership functions for LevelSense. Similarly, we assign ranges and triangular fuzzy membership functions for TempSense in Table 16.13 and Figure 16.4. The optimization of these assignments is often done through trial and error for achieving optimum performance of the FLC.

TABLE 16.12 FUZZY VARIABLE RANGES FOR LEVELSENSE

Crisp Input Range	Fuzzy Variable
0–2	XSmall
1.5–4	Small
3–7	Medium
6–8.5	Large
7.5–10	XLarge

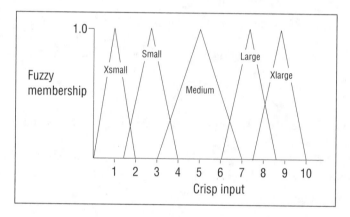

FIGURE 16.3 FUZZY MEMBERSHIP FUNCTIONS FOR LEVELSENSE.

TABLE 16.13 FUZZY VARIABLE RANGES FOR TEMPSENSE

Crisp Input Range	Fuzzy Variable
0–20	XSmall
10–35	Small
30–75	Medium
60–95	Large
85–125	XLarge

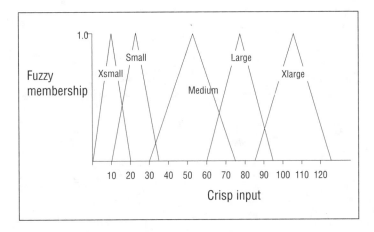

FIGURE 16.4 FUZZY MEMBERSHIP FUNCTIONS FOR TEMPSENSE.

Step Three: Set Up Fuzzy Membership Functions for the Output(s)

In our example, we have just one output, which is the HeatKnob. We need to assign fuzzy memberships to this variable just as we did for the inputs. This is shown in Table 16.14 and Figure 16.5. We use different variable names to make the example clearer later on.

TABLE 16.14 FUZZY VARIABLE RANGES FOR HEATKNOB

Crisp Input Range	Fuzzy Variable
0–2	VeryLittle
1.5–4	ALittle
3–7	AGoodAmount
6–8.5	ALot
7.5–10	AWholeLot

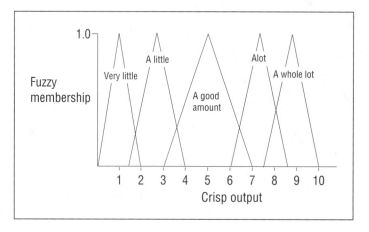

FIGURE 16.5 FUZZY MEMBERSHIP FUNCTIONS FOR THE OUTPUT HEATKNOB.

Step Four: Create a Fuzzy Rule Base

Now that you have the inputs and the output defined in terms of fuzzy variables, you need only specify what actions to take under what conditions; that is, you need to construct a set of rules that describe the operation of the FLC. These rules usually take the form of IF–THEN rules and can be obtained from a human expert (heuristics), or can be supplied from a neural network that infers the rules from behavior of the system. We mentioned this idea in Chapter 3.

Let us construct a rule base for our design example. For the two inputs, we define the matrix shown in Table 16.15. Our heuristic guidelines in determining this matrix are the following statements and their converses:

1. When the temperature is low, the HeatKnob should be set higher than when the temperature is high.

2. When the volume of water is low, the HeatKnob does not need to be as high as when the volume of water is high.

N O T E

In FLCs, we do not need to specify all the boxes in the matrix. That is perfectly fine. No entry signifies that no action is taken, for example, in the column for SenseTemp=XL, no action is required since the temperature is already at or above the target temperature.

TABLE 16.15 FUZZY RULE BASE FOR THE EXAMPLE DESIGN, OUTPUT HEATKNOB

SenseTemp-> Sense Level \/	XS	S	M	L	XL
XS	AGoodAmount	ALittle	VeryLittle		
S	ALot	AGoodAmount	VeryLittle	VeryLittle	
M	AWholeLot	ALot	AGoodAmount	VeryLittle	
L	AWholeLot	ALot	ALot	ALittle	
XL	AWholeLot	ALot	ALot	AGoodAmount	

Let us examine a couple of typical entries in the table: For SenseLevel = Medium (M) and SenseTemp = XSmall (XS), the output is HeatKnob = AWholeLot. Now for the same temperature, as the water level rises, the setting on HeatKnob also should rise to compensate for the added volume of water. You can see that for SenseLevel = Large(L), and SenseTemp = XSmall(XS), the output is HeatKnob = AWholeLot. You can verify that the rest of the table is created by similar reasoning.

Creating IF–THEN Rules

We can now translate the table entries into IF - THEN rules. We take these directly from Table 16.15:

1. IF SenseTemp IS XSmall AND SenseLevel IS XSmall THEN SET HeatKnob TO AGoodAmount

2. IF SenseTemp IS XSmall AND SenseLevel IS Small THEN SET HeatKnob TO ALot

3. IF SenseTemp IS XSmall AND SenseLevel IS Medium THEN SET HeatKnob TO AWholeLot

4. IF SenseTemp IS XSmall AND SenseLevel IS Large THEN SET HeatKnob TO AWholeLot

5. IF SenseTemp IS XSmall AND SenseLevel IS XLarge THEN SET HeatKnob TO AWholeLot

6. IF SenseTemp IS Small AND SenseLevel IS XSmall THEN SET HeatKnob TO ALittle

7. IF SenseTemp IS Small AND SenseLevel IS Small THEN SET HeatKnob TO AGoodAmount

8. IF SenseTemp IS Small AND SenseLevel IS Medium THEN SET HeatKnob TO ALot

9. IF SenseTemp IS Small AND SenseLevel IS Large THEN SET HeatKnob TO ALot

10. IF SenseTemp IS Small AND SenseLevel IS XLarge THEN SET HeatKnob TO ALot

11. IF SenseTemp IS Medium AND SenseLevel IS XSmall THEN SET HeatKnob TO VeryLittle

12. IF SenseTemp IS Medium AND SenseLevel IS Small THEN SET HeatKnob TO VeryLittle

13. IF SenseTemp IS Medium AND SenseLevel IS Medium THEN SET HeatKnob TO AGoodAmount

14. IF SenseTemp IS Medium AND SenseLevel IS Large THEN SET HeatKnob TO ALot

15. IF SenseTemp IS Medium AND SenseLevel IS XLarge THEN SET HeatKnob TO ALot

16. IF SenseTemp IS Large AND SenseLevel IS Small THEN SET HeatKnob TO VeryLittle

17. IF SenseTemp IS Large AND SenseLevel IS Medium THEN SET HeatKnob TO VeryLittle

18. IF SenseTemp IS Large AND SenseLevel IS Large THEN SET HeatKnob TO ALittle

19. IF SenseTemp IS Large AND SenseLevel IS XLarge THEN SET HeatKnob TO AGoodAmount

Remember that the output and inputs to the fuzzy rule base are fuzzy variables. For any given crisp input value, there may be fuzzy membership in several fuzzy input variables (determined by the fuzzification step). And each of these fuzzy input variable activations will cause different fuzzy output cells to *fire*, or be activated. This brings us to the final step, defuzzification of the output into a crisp value.

Step Five: Defuzzify the Outputs

In order to control the HeatKnob, we need to obtain a crisp dial setting. So far, we have several of the IF–THEN rules of the fuzzy rule base firing at once, because the inputs have been fuzzified. How do we arrive at a single crisp output number ? There are actually several different strategies for this; we will consider two of the most common, the *center of area* (*COA*) or *cen-*

troid method, and the *fuzzy Or* method. The easiest way to understand the process is with an example.

Assume that at a particular point in time, LevelSense = 7.0 and TempSense = 65. These are the crisp inputs directly from the sensors. With fuzzification (refer to Chapter 3 for a review), assume that you get the following fuzzy memberships:

---

crisp input — LevelSense = 7.0

fuzzy outputs with membership values -

Medium: 0.4

Large: 0.6

all others : 0.0

crisp input — TempSense=65

fuzzy outputs with membership values -

Medium: 0.75

Large: 0.25

all others: 0.0

This results in four rules firing:

1. TempSense = Medium (0.75) AND LevelSense = Medium (0.4)
2. TempSense = Large (0.25) AND LevelSense = Medium (0.4)
3. TempSense = Medium (0.75) AND LevelSense = Large (0.6)
4. TempSense = Large (0.25) AND LevelSense = Large (0.6)

First you must determine, for each of the AND clauses in the IF–THEN rules, what the output should be. This is done by the *conjunction* or minimum operator. So for each of these rules you have the following firing strengths:

1. $(0.75) \wedge (0.4) = 0.4$
2. $(0.25) \wedge (0.4) = 0.25$
3. $(0.75) \wedge (0.6) = 0.6$
4. $(0.25) \wedge (0.6) = 0.25$

By using the fuzzy rule base and the strengths assigned previously, we find the rules recommend the following output values (with strengths) for HeatKnob:

1. AGoodAmount (0.4)
2. VeryLittle (0.25)
3. ALot (0.6)
4. ALittle (0.25)

Now we must combine the recommendations to arrive at a single crisp value. First, we will use the fuzzy Or method of defuzzification. Here we use a *disjunction* or maximum operator to combine the values. We obtain the following:

$$(0.4) \lor (0.25) \lor (0.6) \lor (0.25) = 0.6$$

The crisp output value for HeatKnob would then be this membership value multiplied by the range of the output variable, or $(0.6) (10-0) = 6.0$.

Another way of combining the outputs is with the centroid method. With the centroid method, there are two variations, the *overlap composition* method and the *additive composition* method. To review, we have the following output values and strengths.

1. AGoodAmount (0.4)
2. VeryLittle (0.25)
3. ALot (0.6)
4. ALittle (0.25)

We use the strength value and fill in the particular triangular membership function to that strength level. For example, for the first rule, we fill the **triangular membership** function, **AGoodAmount** to the 0.4 level. We then cut off the top of the triangle (above 0.4). Next we do the same for the other rules. Finally we align the truncated figures as shown in Figure 16.6 and combine them according to the overlap composition method or the additive composition method (Kosko). You can see the difference in these two composition methods in Figure 16.6. The overlap method simply superimposes all of the truncated triangles onto the same area. You can lose information with this method. The additive method adds the geometrical figures on top of each other.

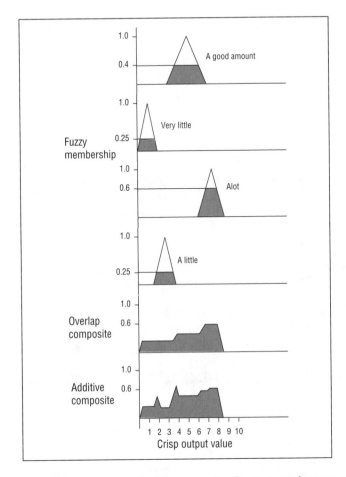

FIGURE 16.6 DEFUZZIFICATION WITH THE CENTROID APPROACH: OVERLAP AND ADDITIVE COMPOSITION.

In order to get a crisp value, you take the centroid or *center of gravity,* of the resulting geometrical figure. Let us do this for the overlap method figure. The centroid is a straight edge that can be placed through the figure to have it perfectly balanced; there is equal area of the figure on either side of the straight edge, as shown in Figure 16.7. Splitting up the geometry into pieces and summing all area contributions on either side of the centroid, we get a value of 5.2 for this example. This is already in terms of the crisp output value range:

HeatKnob = 5.2

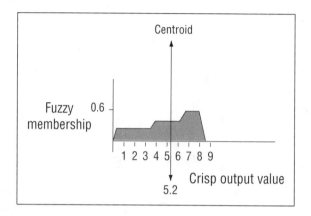

FIGURE 16.7 FINDING THE CENTROID.

This completes the design for the simple example we chose. We conclude with a list of the advantages and disadvantages of FLCs.

ADVANTAGES AND DISADVANTAGES OF FUZZY LOGIC CONTROLLERS

The following list adapted from McNeill and Thro shows the advantages and disadvantages to FLCs for control systems as compared to more traditional control systems.

Advantages:

- Relates input to output in linguistic terms, easily understood
- Allows for rapid prototyping because the system designer doesn't need to know everything about the system before starting
- Cheaper because they are easier to design
- Increased robustness
- Simplify knowledge acquisition and representation
- A few rules encompass great complexity
- Can achieve less overshoot and oscillation
- Can achieve steady state in a shorter time interval

Disadvantages:

- Hard to develop a model from a fuzzy system
- Require more fine tuning and simulation before operational

- Have a stigma associated with the word *fuzzy* (at least in the Western world); engineers and most other people are used to crispness and shy away from fuzzy control and fuzzy decision making

SUMMARY

Fuzzy logic applications are many and varied. You got an overview of the different applications areas that exist for fuzzy logic, from the control of washing machines to fuzzy logic based cost benefit analysis. Further you got details on two application domains: fuzzy databases and fuzzy control.

This chapter dealt with extending database models to accommodate fuzziness in data attribute values and in queries as well. You saw fuzzy relations; in particular, similarity and resemblance relations, and similarity classes were reviewed. You found how possibility distributions help define fuzzy databases. You also learned what fuzzy events are and how to calculate fuzzy means, fuzzy variances, and fuzzy conditional expectations. Concepts related to linear possibility regression model were presented.

The chapter presented the design of a simple fuzzy logic control (FLC) system to regulate the temperature of water in a hot water heater. The components of the FLC were discussed along with design procedures. The advantages of FLC design include rapid prototyping ability and the capability to solve very nonlinear control problems without knowing details of the nonlinearities.

CHAPTER 17

FURTHER APPLICATIONS

INTRODUCTION

In this chapter, we present the outlines of some applications of neural networks and fuzzy logic. Most of the applications fall into a few main categories according to the paradigms they are based on. We offer a sampling of topics of research as found in the current literature, but there are literally thousands of applications of neural networks and fuzzy logic in science, technology and business with more and more applications added as time goes on.

Some applications of neural networks are for adaptive control. Many such applications benefit from adding fuzziness also. Steering a car or backing up a truck with a fuzzy controller is an example. A large number of applications are based on the backpropagation training model. Another category of applications deals with classification. Some applications based on expert systems are augmented with a neural network approach. Decision support systems are sometimes designed this way. Another category is made up of *optimizers*, whose purpose is to find the maximum or the minimum of a function.

You will find other neural network applications related to finance presented toward the end of Chapter 14.

N O T E

COMPUTER VIRUS DETECTOR

IBM Corporation has applied neural networks to the problem of detecting and correcting computer viruses. IBM's AntiVirus program detects and eradicates new viruses automatically. It works on boot-sector types of viruses and keys off of the stereotypical behaviors that viruses usually exhibit. The feedforward backpropagation neural network was used in this application.

New viruses discovered are used in the training set for later versions of the program to make them "smarter." The system was modeled after knowledge about the human immune system: IBM uses a decoy program to "attract" a potential virus, rather than have the virus attack the user's files. These decoy programs are then immediately tested for infection. If the behavior of the decoy program seems like the program was infected, then the virus is detected on that program and removed wherever it's found.

MOBILE ROBOT NAVIGATION

C. Lin and C. Lee apply a multivalued Boltzmann machine, modeled by them, using an artificial magnetic field approach. They define attractive and repulsive magnetic fields, corresponding to goal position and obstacle, respectively. The weights on the connections in the Boltzmann machine are none other than the magnetic fields.

They divide a two-dimensional traverse map into small grid cells. Given the goal cell and obstacle cells, the problem is to navigate the two-dimensional mobile robot from an unobstructed cell to the goal quickly, without colliding with any obstacle. An attracting artificial magnetic field is built for the goal location. They also build a repulsive artificial magnetic field around the boundary of each obstacle. Each neuron, a grid cell, will point to one of its eight neighbors, showing the direction for the movement of the robot. In other words, the Boltzmann machine is adapted to become a compass for the mobile robot.

A CLASSIFIER

James Ressler and Marijke Augusteijn study the use of neural networks to the problem of weapon to target assignment. The neural network is used as a filter to remove unfeasible assignments, where feasibility is determined in terms of the weapon's ability to hit a given target if fired at a specific instant. The large number of weapons and threats along with the limitation on the amount of time lend significance to the need for reducing the number of assignments to consider.

The network's role here is classifier, as it needs to separate the infeasible assignments from the feasible ones. Learning has to be quick, and so Ressler and Augusteijn prefer to use an architecture called the *cascade-correlation* learning architecture, over backpropagation learning. Their network is dynamic in that the number of hidden layer neurons is determined during the training phase. This is part of a class of algorithms that change the architecture of the network during training.

A TWO-STAGE NETWORK FOR RADAR PATTERN CLASSIFICATION

Mohammad Ahmadian and Russell Pimmel find it convenient to use a multistage neural network configuration, a two-stage network in particular, for classifying patterns. The patterns they study are geometrical features of simulated radar targets.

Feature extraction is done in the first stage, while classification is done in the second. Moreover, the first stage is made up of several networks, each for extracting a different estimable feature. Backpropagation is used for learning in the first stage. They use a single network in the second stage. The effect of noise is also studied.

CRISP AND FUZZY NEURAL NETWORKS FOR HANDWRITTEN CHARACTER RECOGNITION

Paul Gader, Magdi Mohamed, and Jung-Hsien Chiang combine a fuzzy neural network and a crisp neural network for the recognition of handwritten alphabetic characters. They use backpropagation for the crisp neural network and a clustering algorithm called K-nearest neighbor for the fuzzy network. Their consideration of a fuzzy network in this study is prompted by their belief that if some ambiguity is possible in deciphering a character, such ambiguity should be accurately represented in the output. For example, a handwritten "u" could look like a "v" or "u." If present, the authors feel that this ambiguity should be translated to the classifier output.

Feature extraction was accomplished as follows: character images of size 24x16 pixels were used. The first stage of processing extracted eight feature images from the input image, two for each direction (north, northeast, northwest, and east). Each feature image uses an integer at each location that represents the length of the longest bar that fits at that point in that direction. These are referred to as *bar features.* Next 8x8 overlapping zones are used on the feature images to derive feature vectors. These are made by taking the summed values of the values in a zone and dividing this by the maximum possible value in the zone. Each feature image results in a 15,120 element feature vectors.

Data was obtained from the U.S. Postal Office, consisting of 250 characters. Results showed 97.5% and 95.6% classification rates on training and test sets, respectively, for the neural network. The fuzzy network resulted in 94.7% and 93.8% classification rates, where the desired output for many characters was set to ambiguous.

Noise Removal with a Discrete Hopfield Network

Arun Jagota applies what is called a HcN, a special case of a discrete Hopfield network, to the problem of recognizing a degraded printed word. HcN is used to process the output of an Optical Character Recognizer, by attempting to remove noise. A dictionary of words is stored in the HcN and searched.

Object Identification by Shape

C. Ganesh, D. Morse, E. Wetherell, and J. Steele used a neural network approach to an object identification system, based on the shape of an object and independent of its size. A two-dimensional grid of ultrasonic data represents the height profile of an object. The data grid is compressed into a smaller set that retains the essential features. Backpropagation is used. Recognition on the order of approximately 70% is achieved.

Detecting Skin Cancer

F. Ercal, A. Chawla, W. Stoecker, and R. Moss study a neural network approach to the diagnosis of malignant melanoma. They strive to discriminate tumor images as malignant or benign. There are as many as three categories of benign tumors to be distinguished from malignant melanoma. Color images of skin tumors are used in the study. Digital images of tumors are classified. Backpropagation is used. Two approaches are taken to reduce training time. The first approach involves using fewer hidden layers, and the second involves randomization of the order of presentation of the training set.

EEG DIAGNOSIS

Fred Wu, Jeremy Slater, R. Eugene Ramsay, and Lawrence Honig use a feed-forward backpropagation neural network as a classifier in EEG diagnosis. They compare the performance of the neural network classifier to that of a nearest neighbor classifier. The neural network classifier shows a classifier accuracy of 75% for Multiple Sclerosis patients versus 65% for the nearest neighbor algorithm.

TIME SERIES PREDICTION WITH RECURRENT AND NONRECURRENT NETWORKS

Sathyanarayan Rao and Sriram Sethuraman take a recurrent neural network and a feedforward network and train then in parallel. A recurrent neural network has feedback connections from the output neurons back to input neurons to model the storage of temporal information. A modified back-propagation algorithm is used for training the recurrent network, called the *real-time recurrent learning algorithm.* They have the recurrent neural network store past information, and the feedforward network do the learning of nonlinear dependencies on the current samples. They use this scheme because the recurrent network takes more than one time period to evaluate its output, whereas the feedforward network does not. This hybrid scheme overcomes the latency problem for the recurrent network, providing immediate nonlinear evaluation from input to output.

SECURITY ALARMS

Deborah Frank and J. Bryan Pletta study the application of neural networks for alarm classification based on their operation under varying weather conditions. Performance degradation of a security system when the environment changes is a cause for losing confidence in the system itself. This problem is more acute with portable security systems.

They investigated the problem using several networks, ranging from backpropagation to learning vector quantization. Data was collected using many scenarios, with and without the coming of an intruder, which can be a vehicle or a human.

They found a 98% probability of detection and 9% nuisance alarm rate over all weather conditions.

CIRCUIT BOARD FAULTS

Anthony Mason, Nicholas Sweeney, and Ronald Baer studied the neural network approach in two laboratory experiments and one field experiment, in diagnosing faults in circuit boards.

Test point readings were expressed as one vector. A fault vector was also defined with elements representing possible faults. The two vectors became a training pair. Backpropagation was used.

WARRANTY CLAIMS

Gary Wasserman and Agus Sudjianto model the prediction of warranty claims with neural networks. The nonlinearity in the data prompted this approach.

The motivation for the study comes from the need to assess warranty costs for a company that offers extended warranties for its products. This is another application that uses backpropagation. The architecture used was 2-10-1.

WRITING STYLE RECOGNITION

J. Nellis and T. Stonham developed a neural network character recognition system that adapts dynamically to a writing style.

They use a hybrid neural network for hand-printed character recognition, that integrates image processing and neural network architectures. The neural network uses random access memory (RAM) to model the functionality of an individual neuron. The authors use a transform called the *five-way* image processing transform on the input image, which is of size 32x32 pixels. The transform converts the high spatial frequency data in a character into four low frequency representations. What they achieve by this are position invariance, and a ratio of black to white pixels approaching 1, rotation invariance, and capability to detect and correct breaks within characters.

The transformed data are input to the neural network that is used as a classifier and is called a *discriminator*.

A particular writing style that has less variability and therefore fewer subclasses is needed to classify the style. Network size will also reduce confusion, and conflicts lessen.

517

Commercial Optical Character Recognition

Optical character recognition (OCR) is one of the most successful commercial applications of neural networks. Caere Corporation brought out its neural network product in 1992, after studying more than 100,000 examples of fax documents. Caere's AnyFax technology combines neural networks with expert systems to extract character information from Fax or scanned images. Calera, another OCR vendor, started using neural networks in 1984 and also benefited from using a very large (more than a million variations of alphanumeric characters) training set.

ART-EMAP and Object Recognition

A neural network architecture called ART-EMAP (Gail Carpenter and William Ross) integrates Adaptive Resonance Theory (ART) with spatial and temporal evidence integration for predictive mapping (EMAP). The result is a system capable of complex 3-D object recognition. A vision system that samples two-dimensional perspectives of a three-dimensional object is created that results in 98% correct recognition with an average of 9.2 views presented on noiseless test data, and 92% recognition with an average of 11.2 views presented on noisy test data. The ART-EMAP system is an extension of *ARTMAP,* which is a neural network architecture that performs supervised learning of recognition categories and multidimensional maps in response to input vectors. A fuzzy logic extension of ARTMAP is called *Fuzzy ARTMAP,* which incorporates two fuzzy modules in the ART system.

Summary

A sampling of current research and commercial applications with neural networks and fuzzy logic technology is presented. Neural networks are applied toward a wide variety of problems, from aiding medical diagnosis to detecting circuit faults in printed circuit board manufacturing. Some of the problem areas where neural networks and fuzzy logic have been successfully applied are:

- Filtering
- Image processing
- Intelligent control
- Machine vision
- Motion analysis
- Optimization
- Pattern recognition
- Prediction
- Time series analysis
- Speech synthesis
- Machine learning and robotics
- Decision support systems
- Classification
- Data compression
- Functional approximation

The use of fuzzy logic and neural networks in software and hardware systems can only increase!

COMPILING YOUR PROGRAMS

All of the programs included in the book have been compiled and tested on Turbo C++, Borland C++, and Microsoft C++/Visual C++ with either the small or medium memory model. You should not have any problems in using other compilers, since standard I/O routines are used. Your target should be a DOS executable. With the backpropagation simulator of Chapters 7, 13, and 14 you may run into a memory shortage situation. You should unload any TSR (Terminate and Stay Resident) programs and/or choose smaller architectures for your networks. By going to more hidden layers with fewer neurons per layer, you may be able to reduce the overall memory requirements.

The programs in this book make heavy use of floating point calculations, and you should compile your programs to take advantage of a math coprocessor, if you have one installed in your computer.

The organization of files on the accompanying diskette are according to chapter number. You will find relevant versions of files in the corresponding chapter directory.

Most of the files are self-contained, or include other needed files in them. You will not require a makefile to build the programs. Load the main file for example for backpropagation, the backprop.cpp file, into the development environment editor for your compiler and build a .exe file. That's it!

MATHEMATICAL BACKGROUND

DOT PRODUCT OR SCALAR PRODUCT OF TWO VECTORS

Given vectors $\mathbf{U}$ and $\mathbf{V}$, where $\mathbf{U} = (u_1, ..., u_n)$ and $\mathbf{V} = (v_1, ..., v_n)$, their dot product or scalar product is $\mathbf{U} \bullet \mathbf{V} = u_1v_1 + ... + u_nv_n = \sum u_i v_i$.

MATRICES AND SOME ARITHMETIC OPERATIONS ON MATRICES

A real matrix is a rectangular array of real numbers. A matrix with m rows and n columns is referred to as an mxn matrix. The element in the ith row and jth column of the matrix is referred to as the ij element of the matrix and is denoted by a_{ij}.

The transpose of a matrix M is denoted by M^T. The element in the ith row and jth column of M^T is the same as the element of M in its jth row and ith column. M^T is obtained from M by interchanging the rows and columns of M. For example, if

$$M = \begin{matrix} 2 & 7 & -3 \\ \\ 4 & 0 & 9 \end{matrix} \text{, then } M^T = \begin{matrix} 2 & 4 \\ 7 & 0 \\ -3 & 9 \end{matrix}$$

If $\mathbf{X}$ is a vector with m components, $x_1, ..., x_m$, then it can be written as a column vector with components listed one below another. It can be written as a row vector, $\mathbf{X} = (x_1, ..., x_m)$. The transpose of a row vector is the column vector with the same components, and the transpose of a column vector is the corresponding row vector.

The addition of matrices is possible if they have the same size, that is, the same number of rows and same number of columns. Then you just add the ij elements of the two matrices to get the ij elements of the sum matrix. For example,

```
3 -4 5     5 2 -3       8 -2  2
        +          =
2  3 7     6 0  4       8  3 11
```

Multiplication is defined for a given pair of matrices, only if a condition on their respective sizes is satisfied. Then too, it is not a commutative operation. This means that if you exchange the matrix on the left with the matrix on the right, the multiplication operation may not be defined, and even if it is, the result may not be the same as before such an exchange.

The condition to be satisfied for multiplying the matrices *A*, *B* as *AB* is, that the number of columns in *A* is equal to the number of rows in *B*. Then to get the *ij* element of the product matrix *AB*, you take the *i*th row of *A* as one vector and the *j*th column of *B* as a second vector and do a dot product of the two. For example, the two matrices given previously to illustrate the addition of two matrices are not compatible for multiplication in whichever order you take them. It is because there are three columns in each, which is different from the number of rows, which is 2 in each. Another example is given as follows.

```
            3 -4 5                  5  6
Let A =               and   B =    2  0
            2  3 7                 -3  4
```

Then *AB* and *BA* are both defined, *AB* is a 2x2 matrix, whereas *BA* is 3x3.

```
            -8  38                 27 -2 67
Also AB =             and BA  =     6 -8 10
            -5  40                 -1 24 13
```

Lyapunov Function

A **Lyapunov** function is a function that decreases with time, taking on non-negative values. It is used to correspond between the state variables of a system and real numbers. The state of the system changes as time changes, and the function decreases. Thus, the **Lyapunov** function decreases with each change of state of the system.

We can construct a simple example of a function with the property of decreasing with each change of state as follows. Suppose a real number, *x*, represents the state of a dynamic system at time t. Also suppose that x is bounded for any t by a positive real number M. That means x is less than M for every value of t.

Then the function,

$$f(x,t) = \exp(-|x|/(M+|x|+t))$$

is non-negative and decreases with increasing t.

LOCAL MINIMUM

A function $f(x)$ is defined to have a local minimum at y, with a value z, if

> $f(y) = z$, and $f(x) \geq z$, for each x, such that there exists a positive real number h such that $y - h \leq x \leq y + h$.

In other words, there is no other value of x in a neighborhood of y, where the value of the function is smaller than z.

There can be more than one local minimum for a function in its domain. A **Step** function (with a graph resembling a staircase) is a simple example of a function with an infinite number of points in its domain with local minima.

GLOBAL MINIMUM

A function $f(x)$ is defined to have a global minimum at y, with a value z, if

> $f(y) = z$, and $f(x) \geq z$, for each x in the domain of the function f.

In other words, there is no other value of x in the domain of the function f, where the value of the function is smaller than z. Clearly, a global minimum is also a local minimum, but a local minimum may not be a global minimum.

There can be more than one global minimum for a function in its domain. The trigonometric function $f(x) = \sin x$ is a simple example of a function with an infinite number of points with global minima. You may recall that $\sin(3\pi/2)$, $\sin(7\pi/2)$, and so on are all -1, the smallest value for the sine function.

KRONECKER DELTA FUNCTION

The Kronecker delta function is a function of two variables. It has a value of 1 if the two arguments are equal, and 0 if they are not. Formally,

$$\delta(x,y) = \begin{cases} 1 \text{ if } x = y \\ 0 \text{ if } x \neq y \end{cases}$$

GAUSSIAN DENSITY DISTRIBUTION

The Gaussian Density distribution, also called the Normal distribution, has a density function of the following form. There is a constant parameter c, which can have any positive value.

$$f(x) = \frac{1}{\sqrt{(2\pi c)}} \exp(-x^2 / c)$$

GLOSSARY

A

Activation　　　　The weighted sum of the inputs to a neuron in a neural network.

Adaline　　　　Adaptive linear element machine.

Adaptive Resonance Theory　　　　Theory developed by Grossberg and Carpenter for categorization of patterns, and to address the stability–plasticity dilemma.

Algorithm　　　　A step-by-step procedure to solve a problem.

Annealing　　　　A process for preventing a network from being drawn into a local minimum.

ART　　　　(Adaptive Resonance Theory) ART1 is the result of the initial development of this theory for binary inputs. Further developments led to ART2 for analog inputs. ART3 is the latest.

Artificial neuron　　　　The primary object in an artificial neural network to mimic the neuron activity of the brain. The artificial neuron is a processing element of a neural network.

Associative memory　　　　Activity of associating one pattern or object with itself or another.

Autoassociative　　　　Making a correspondence of one pattern or object with itself.

B

Backpropagation　　　　A neural network training algorithm for feedforward networks where the errors at the output layer are propagated back to the layer before in learning. If the previous layer is not the input layer, then the errors at this hidden layer are propagated back to the layer before.

BAM　　　　Bidirectional Associative Memory network model.

526

Bias	A value added to the activation of a neuron.
Binary digit	A value of 0 or 1.
Bipolar value	A value of −1 or +1.
Boltzmann machine	A neural network in which the outputs are determined with probability distributions. Trained and operated using simulated annealing.
Brain-State-in-a-Box	Anderson's single-layer, laterally connected neural network model. It can work with inputs that have noise in them or are incomplete.

C

Cauchy machine	Similar to the Boltzmann machine, except that a Cauchy distribution is used for probabilities.
Cognitron	The forerunner to the Neocognitron. A network developed to recognize characters.
Competition	A process in which a winner is selected from a layer of neurons by some criterion. Competition suggests inhibition reflected in some connection weights being assigned a negative value.
Connection	A means of passing inputs from one neuron to another.
Connection weight	A numerical label associated with a connection and used in a weighted sum of inputs.
Constraint	A condition expressed as an equation or inequality, which has to be satisfied by the variables.
Convergence	Termination of a process with a final result.
Crisp	The opposite of fuzzy—usually a specific numerical quantity or value for an entity.

D

Delta rule	A rule for modification of connection weights, using both the output and the error obtained. It is also called the *LMS rule*.

E

Energy function

A function of outputs and weights in a neural network to determine the state of the system, e.g., Lyapunov function.

Excitation

Providing positive weights on connections to enable outputs that cause a neuron to fire.

Exemplar

An example of a pattern or object used in training a neural network.

Expert system

A set of formalized rules that enable a system to perform like an expert.

F

FAM

Fuzzy Associative Memory network. Makes associations between fuzzy sets.

Feedback

The process of relaying information in the opposite direction to the original.

Fit vector

A vector of values of degree of membership of elements of a fuzzy set.

Fully connected network

A neural network in which every neuron has connections to all other neurons.

Fuzzy

As related to a variable, the opposite of crisp. A fuzzy quantity represents a range of value as opposed to a single numeric value, e.g., "hot" vs. $89.4°$.

Fuzziness

Different concepts having an overlap to some extent. For example, descriptions of fair and cool temperatures may have an overlap of a small interval of temperatures.

Fuzzy Associative Memory

A neural network model to make association between fuzzy sets.

Fuzzy equivalence relation

A fuzzy relation (relationship between fuzzy variables) that is reflexive, symmetric, and transitive.

Fuzzy partial order

A fuzzy relation (relationship between fuzzy variables) that is reflexive, antisymmetric, and transitive.

G

Gain

Sometimes a numerical factor to enhance the activation. Sometimes a connection for the same purpose.

Generalized Delta rule

A rule used in training of networks such as backpropagation training where hidden layer weights are modified with backpropagated error.

Global minimum

A point where the value of a function is no greater than the value at any other point in the domain of the function.

H

Hamming distance

The number of places in which two binary vectors differ from each other.

Hebbian learning

A learning algorithm in which Hebb's rule is used. The change in connection weight between two neurons is taken as a constant times the product of their outputs.

Heteroassociative

Making an association between two distinct patterns or objects.

Hidden layer

An array of neurons positioned in between the input and output layers.

Hopfield network

A single layer, fully connected, autoassociative neural network.

I

Inhibition

The attempt by one neuron to diminish the chances of firing by another neuron.

Input layer

An array of neurons to which an external input or signal is presented.

Instar

A neuron that has no connections going from it to other neurons.

L

Lateral connection   A connection between two neurons that belong to the same layer.

Layer   An array of neurons positioned similarly in a network for its operation.

Learning   The process of finding an appropriate set of connection weights to achieve the goal of the network operation.

Linearly separable   Two subsets of a linear set having a linear barrier (hyperplane) between the two of them.

LMS rule   Least mean squared error rule, with the aim of minimizing the average of the squared error. Same as the Delta rule.

Local minimum   A point where the value of the function is no greater than the value at any other point in its neighborhood.

Long-term memory (LTM)   Encoded information that is retained for an extended period.

Lyapunov function   A function that is bounded below and represents the state of a system that decreases with every change in the state of the system.

M

Madaline   A neural network in which the input layer has units that are Adalines. It is a multiple-Adaline.

Mapping   A correspondence between elements of two sets.

N

Neural network   A collection of processing elements arranged in layers, and a collection of connection edges between pairs of neurons. Input is received at one layer, and output is produced at the same or at a different layer.

Noise   Distortion of an input.

Nonlinear optimization   Finding the best solution for a problem that has a nonlinear function in its objective or in a constraint.

O

On center off surround   Assignment of excitatory weights to connections to nearby neurons and inhibitory weights to connections to distant neurons.

Orthogonal vectors   Vectors whose dot product is 0.

Outstar   A neuron that has no incoming connections.

P

Perceptron   A neural network for linear pattern matching.

Plasticity   Ability to be stimulated by new inputs and learn new mappings or modify existing ones.

R

Resonance   The responsiveness of two neurons in different layers in categorizing an input. An equilibrium in two directions.

S

Saturation   A condition of limitation on the frequency with which a neuron can fire.

Self-organization   A process of partitioning the output layer neurons to correspond to individual patterns or categories, also called unsupervised learning or clustering.

Short-term memory (STM)   The storage of information that does not endure long after removal of the corresponding input.

Simulated annealing   An algorithm by which changes are made to decrease energy or temperature or cost.

Stability   Convergence of a network operation to a steady-state solution.

Supervised learning   A learning process in which the exemplar set consists of pairs of inputs and desired outputs.

T

Threshold value   A value used to compare the activation of a neuron to determine if the neuron fires or not. Sometimes a bias value is added to the activation of a neuron to allow the threshold value to be zero in determining the neuron's output.

Training   The process of helping in a neural network to learn either by providing input/output stimuli (supervised training) or by providing input stimuli (unsupervised training), and allowing weight change updates to occur.

U

Unsupervised learning   Learning in the absence of external information on outputs, also called self-organization or clustering.

V

Vigilance parameter   A parameter used in Adaptive Resonance Theory. It is used to selectively prevent the activation of a subsystem in the network.

W

Weight   A number associated with a neuron or with a connection between two neurons, which is used in aggregating the outputs to determine the activation of a neuron.

References

Ahmadian, Mohamad, and Pimmel, Russell, "Recognizing Geometrical Features of Simulated Targets with Neural Networks," Conference Proceedings of the 1992 Artificial Neural Networks in Engineering Conference, V.3, pp. 409–411.

Aleksander, Igor, and Morton, Helen, *An Introduction to Neural Computing*, Chapman and Hall, London, 1990.

Aiken, Milam, "Forecasting T-Bill Rates with a Neural Net," Technical Analysis of Stocks and Commodities, May 1995, Technical Analysis Inc., Seattle.

Anderson, James, and Rosenfeld, Edward, eds., *Neurocomputing: Foundations of Research*, MIT Press, Cambridge, MA, 1988.

Anzai, Yuichiro, *Pattern Recognition and Machine Learning*, Academic Press, Englewood Cliffs, NJ, 1992.

Azoff, E. Michael, *Neural Network Time Series Forecasting of Financial Markets*, John Wiley & Sons, New York, 1994.

Bauer, Richard J., *Genetic Algorithms and Investment Strategies*, John Wiley & Sons, New York, 1994.

Booch, Grady, *Object Oriented Design with Applications*, Benjamin-Cummings, Redwood City, CA, 1991.

Carpenter, Gail A., and Ross, William D., "ART-EMAP: A Neural Network Architecture for Object Recognition by Evidence Accumulation," *IEEE Transactions on Neural Networks*, Vol. 6., No. 4, July 1995, pp. 805–818.

Colby, Robert W., and Meyers, Thomas A., *The Encyclopedia of Technical Market Indicators*, Business One Irwin, Homewood, IL, 1988.

Collard, J. E., "Commodity Trading with a Three Year Old," *Neural Networks in Finance and Investing*, pp. 411–420, Probus Publishing, Chicago, 1993.

Cox, Earl, *The Fuzzy Systems Handbook*, Academic Press, Boston, 1994.

Dagli, Cihan, et al., eds., *Intelligent Engineering Systems through Artificial Neural Networks*, Volume 2, ASME Press, New York, 1992.

Davalo, Eric, and Naim, Patrick, *Neural Networks*, MacMillan, New York, 1991.

Ercal, F., et al., "Diagnosing Malignant Melanoma Using a Neural Network," Conference Proceedings of the 1992 Artificial Neural Networks in Engineering Conference, V.3, pp. 553–555.

Frank, Deborah, and Pletta, J. Bryan, "Neural Network Sensor Fusion for Security Applications," Conference Proceedings of the 1992 Artificial Neural Networks in Engineering Conference, V.3, pp. 745–748.

Freeman, James A., and Skapura, David M., *Neural Networks Algorithms, Applications, and Programming Techniques*, Addison-Wesley, Reading, MA, 1991.

Gader, Paul, et al., "Fuzzy and Crisp Handwritten Character Recognition Using Neural Networks," Conference Proceedings of the 1992 Artificial Neural Networks in Engineering Conference, V.3, pp. 421–424.

Ganesh, C., et al., "A Neural Network-Based Object Identification System," Conference Proceedings of the 1992 Artificial Neural Networks in Engineering Conference, V.3, pp. 471–474.

Glover, Fred, ORSA CSTS Newsletter Vol 15, No 2, Fall 1994.

Goldberg, David E., *Genetic Algorithms in Search, Optimization and Machine Learning*, Addison-Wesley, Reading, MA, 1989.

Grossberg, Stephen, et al., Introduction and Foundations, Lecture Notes, Neural Network Courses and Conference, Boston University, May 1992.

Hammerstrom, Dan, "Neural Networks at Work," *IEEE Spectrum*, New York, June 1993.

Hertz, John, Krogh, Anders, and Palmer, Richard, *Introduction to the Theory of Neural Computation*, Addison-Wesley, Reading, MA, 1991.

Jagota, Arun, "Word Recognition with a Hopfield-Style Net," Conference Proceedings of the 1992 Artificial Neural Networks in Engineering Conference, V.3, pp. 445–448.

Johnson, R. Colin, "Accuracy Moves OCR into the Mainstream," *Electronic Engineering Times*, CMP Publications, Manhasset, NY, January 16, 1995.

Johnson, R. Colin, "Making the Neural-Fuzzy Connection," *Electronic Engineering Times*, CMP Publications, Manhasset, NY, September 27, 1993.

Johnson, R. Colin, "Neural Immune System Nabs Viruses," *Electronic Engineering Times*, CMP Publications, Manhasset, NY, May 8, 1995.

Jurik, Mark, "The Care and Feeding of a Neural Network," *Futures Magazine*, Oster Communications, Cedar Falls, IA, October 1992.

Kimoto, Takashi, et al., "Stock Market Prediction System with Modular Neural Networks," *Neural Networks in Finance and Investing*, pp. 343–356, Probus Publishing, Chicago, 1993.

Kline, J., and Folger, T.A., *Fuzzy Sets, Uncertainty and Information*, Prentice Hall, New York, 1988.

Konstenius, Jeremy G., "Trading the S&P with a Neural Network," Technical Analysis of Stocks and Commodities, October 1994, Technical Analysis Inc., Seattle.

Kosaka, M., et al., "Applications of Fuzzy Logic/Neural Network to Securities Trading Decision Support System," Conference Proceedings of the 1991 IEEE International Conference on Systems, Man and Cybernetics, V.3, pp. 1913–1918.

Kosko, Bart, and Isaka, Satoru, *Fuzzy Logic*, Scientific American, New York, July 1993.

Kosko, Bart, *Neural Networks and Fuzzy Systems: A Dynamical Systems Approach to Machine Intelligence*, Prentice-Hall, New York, 1992.

Laing, Jonathan, "New Brains: How Smart Computers are Beating the Stock Market," *Barron's*, February 27, 1995.

Lederman, Jess, and Klein, Robert, eds., *Virtual Trading*, Probus Publishing, Chicago, 1995.

Lin, C.T. and Lee, C.S.G, "A Multi-Valued Boltzman Machine", IEEE Transactions on Systems, Man, and Cybernetics, Vol. 25, No. 4, April 1995 pp. 660-668.

MacGregor, Ronald J., *Neural and Brain Modeling*, Academic Press, Englewood Cliffs, NJ, 1987.

Mandelman, Avner, "The Computer's Bullish! A Money Manager's Love Affair with Neural Network Programs," *Barron's*, December 14, 1992.

Maren, Alianna, Harston, Craig, and Pap, Robert, *Handbook of Neural Computing Applications*, Academic Press, Englewood Cliffs, NJ, 1990.

Marquez, Leorey, et al., "Neural Network Models as an Alternative to Regression," *Neural Networks in Finance and Investing*, pp. 435–449, Probus Publishing, Chicago, 1993.

Mason, Anthony et al., "Diagnosing Faults in Circuit Boards—A Neural Net Approach," Conference Proceedings of the 1992 Artificial Neural Networks in Engineering Conference, V.3, pp. 839–843.

McNeill, Daniel, and Freiberger, Paul, *Fuzzy Logic*, Simon & Schuster, New York, 1993.

McNeill, F. Martin, and Thro, Ellen, *Fuzzy Logic: A Practical Approach*, Academic Press, Boston, 1994.

Murphy, John J., *Intermarket Technical Analysis*, John Wiley & Sons, New York, 1991.

Murphy, John J., *Technical Analysis of the Futures Markets*, NYIF Corp., New York, 1986.

Nellis, J., and Stonham, T.J., "A Neural Network Character Recognition System that Dynamically Adapts to an Author's Writing Style," Conference Proceedings of the 1992 Artificial Neural Networks in Engineering Conference, V.3, pp. 975–979.

Peters, Edgar E., *Fractal Market Analysis: Applying Chaos Theory to Investment and Economics*, John Wiley & Sons, New York, 1994.

Ressler, James L., and Augusteijn, Marijke F., "Weapon Target Assignment Accessibility Using Neural Networks," Conference Proceedings of the 1992 Artificial Neural Networks in Engineering Conference, V.3, pp. 397–399.

Rao, Satyanaryana S,, and Sethuraman, Sriram, "A Modified Recurrent Learning Algorithm for Nonlinear Time Series Prediction," Conference Proceedings of the 1992 Artificial Neural Networks in Engineering Conference, V.3, pp. 725–730.

Rao, Valluru, and Rao, Hayagriva, *Power Programming Turbo C++*, MIS:Press, New York, 1992.

Ruggiero, Murray, "How to Build an Artificial Trader," *Futures Magazine*, Oster Communications, Cedar Falls, IA, September 1994.

Ruggiero, Murray, "How to Build a System Framework," *Futures Magazine*, Oster Communications, Cedar Falls, IA, November 1994.

Ruggiero, Murray, "Nothing Like Net for Intermarket Analysis," *Futures Magazine*, Oster Communications, Cedar Falls, IA, May 1995.

Ruggiero, Murray, "Putting the Components before the System," *Futures Magazine*, Oster Communications, Cedar Falls, IA, October 1994.

Ruggiero, Murray, "Training Neural Networks for Intermarket Analysis," *Futures Magazine*, Oster Communications, Cedar Falls, IA, August 1994.

Rumbaugh, James, et al, *Object-Oriented Modeling and Design*, Prentice-Hall Inc, Englewood Cliffs, New Jersey, 1991.

Sharda, Ramesh, and Patil, Rajendra, "A Connectionist Approach to Time Series Prediction: An Empirical Test," *Neural Networks in Finance and Investing*, pp. 451–463, Probus Publishing, Chicago, 1993.

Sherry, Clifford, "Are Your Inputs Correlated?" Technical Analysis of Stocks and Commodities, February 1995, Technical Analysis Inc., Seattle.

Simpson, Patrick K., *Artificial Neural Systems: Foundations, Paradigms, Applications and Implementations*, Pergamon Press, London, 1990.

Soucek, Branko, and Soucek, Marina, *Neural and Massively Parallel Computers: The Sixth Generation*, John Wiley & Sons, New York, 1988.

Stein, Jon, "The Trader's Guide to Technical Indicators," *Futures Magazine*, Oster Communications, Cedar Falls, IA, August 1990.

Terano, Toshiro, et al., *Fuzzy Systems Theory and Its Applications*, Academic Press, Boston, 1993.

Trippi, Robert, and Turban, Efraim, eds., *Neural Networks in Finance and Investing*, Probus Publishing, Chicago, 1993.

Wasserman, Gary S., and Sudjianto, Agus, Conference Proceedings of the 1992 Artificial Neural Networks in Engineering Conference, V.3, pp. 901–904.

Wasserman, Philip D., *Advanced Methods in Neural Computing*, Van Nostrand Reinhold, New York, 1993.

Wasserman, Philip D., *Neural Computing*, Van Nostrand Reinhold, New York, 1989.

Wu, Fred, et al., "Neural Networks for EEG Diagnosis," Conference Proceedings of the 1992 Artificial Neural Networks in Engineering Conference, V.3, pp. 559–565.

Yager, R., ed., *Fuzzy Sets and Applications: Selected Papers by L.Z. Zadeh*, Wiley-Interscience, New York, 1987.

Yan, Jun, Ryan, Michael, and Power, James, *Using Fuzzy Logic*, Prentice-Hall, New York, 1994.

Yoon, Y., and Swales, G., "Predicting Stock Price Performance: A Neural Network Approach," *Neural Networks in Finance and Investing*, pp. 329–339, Probus Publishing, Chicago, 1993.

INDEX

D

How to Use Your Disk

The accompanying disk contains all of the source code for this book. Also, you have raw and processed data for the financial forecasting simulation described in Chapter 14. The files are organized according to the chapter in which they are used. There is a directory for each chapter that has source code or data. Each chapter directory is self-contained, and contains all the files you need to compile programs for that chapter.

To compile any program, copy the files from the chapter directory to a directory of your hard disk. Then compile and build the primary .cpp file for that program in your working directory.